Progress in Brain Research
Volume 223

Neuroscience for Addiction Medicine: From Prevention to Rehabilitation - Constructs and Drugs

Serial Editor

Vincent Walsh
Institute of Cognitive Neuroscience
University College London
17 Queen Square
London WC1N 3AR UK

Editorial Board

Progress in Brain Research
Volume 223

Neuroscience for Addiction Medicine: From Prevention to Rehabilitation - Constructs and Drugs

Edited by

Hamed Ekhtiari
Neurocognitive Laboratory, Iranian National Center for Addiction Studies (INCAS); Research Center for Molecular and Cellular Imaging (RCMCI), Tehran University of Medical Sciences, and Translational Neuroscience Program, Institute for Cognitive Science Studies (ICSS), Tehran, Iran

Martin Paulus
Laureate Institute for Brain Research (LIBR), Tulsa, OK, United States

ELSEVIER
AMSTERDAM • BOSTON • HEIDELBERG • LONDON • NEW YORK • OXFORD
PARIS • SAN DIEGO • SAN FRANCISCO • SINGAPORE • SYDNEY • TOKYO

Elsevier
Radarweg 29, PO Box 211, 1000 AE Amsterdam, Netherlands
The Boulevard, Langford Lane, Kidlington, Oxford OX5 1GB, UK
50 Hampshire Street, 5th Floor, Cambridge, MA 02139, USA

First edition 2016

ISBN: 978-0-444-63545-7
ISSN: 0079-6123

For information on all Elsevier publications visit our website at http://store.elsevier.com/

Contributors

Mustafa al'Absi
University of Minnesota School of Medicine, Duluth, MN, USA

Nelly Alia-Klein
Department of Psychiatry, and Department of Neuroscience, Icahn School of Medicine at Mount Sinai, New York, NY, USA

Barbara C. Banz
Department of Psychiatry, Yale University School of Medicine, New Haven, CT, USA

Lucia Bederson
Department of Psychology, New York University, New York, NY, USA

Wade Berrettini
Karl E Rickles Professor of Psychiatry, Center for Neurobiology and Behavior, Perelman School of Medicine, University of Pennsylvania, Philadelphia, PA, USA

Warren K. Bickel
Addiction Recovery Research Center, Virginia Tech Carilion Research Institute, Roanoke, VA, USA

Jean Lud Cadet
Molecular Neuropsychiatry Research Branch, DHHS/NIH/NIDA Intramural Research Program, National Institutes of Health, Baltimore, MD, USA

Bader Chaarani
Department of Psychiatry, Vermont Center on Behavior and Health, University of Vermont, Burlington, VT, USA

Kelly E. Courtney
Department of Psychology, University of California, Los Angeles, CA, USA

W. Miles Cox
Bangor University, Bangor, UK

Anita Cservenka
Departments of Psychiatry, Oregon Health & Science University, Portland, OR, USA

Manoranjan S. D'Souza
Department of Biomedical and Pharmaceutical Sciences, The Raabe College of Pharmacy, Ohio Northern University, Ada, OH, USA

Scott Edwards
Department of Physiology, Alcohol and Drug Abuse Center of Excellence, Neuroscience Center of Excellence, Louisiana State University Health Sciences Center, New Orleans, LA, USA

Hamed Ekhtiari
Research Center for Molecular and Cellular Imaging; Neurocognitive Laboratory, Iranian National Center for Addiction Studies (INCAS); Translational Neuroscience Program, Institute for Cognitive Sciences Studies (ICSS), and Neuroimaging and Analysis Group, Research Center for Molecular and Cellular Imaging (RCMCI), Tehran University of Medical Sciences, Tehran, Iran

Javad Salehi Fadardi
Ferdowsi University of Mashhad; Bangor University, Bangor, UK, and Addiction Research Centre, Mashhad University of Medical Sciences, Mashhad, Iran

Shelly B. Flagel
Department of Psychiatry, and Molecular and Behavioral Neuroscience Institute, University of Michigan, Ann Arbor, MI, USA

John J. Foxe
Department of Pediatrics, and Department of Neuroscience, Albert Einstein College of Medicine, Bronx, NY, USA

Hugh Garavan
Department of Psychiatry, Vermont Center on Behavior and Health, and Department of Psychological Science, University of Vermont, Burlington, VT, USA

Ashley N. Gearhardt
Department of Psychology, University of Michigan, Ann Arbor, MI, USA

Rita Z. Goldstein
Department of Psychiatry, and Department of Neuroscience, Icahn School of Medicine at Mount Sinai, New York, NY, USA

Colleen A. Hanlon
Medical University of South Carolina, Charleston, SC, USA

Kelsey E. Hudson
Department of Psychological Science, University of Vermont, Burlington, VT, USA

Andrine Lemieux
University of Minnesota School of Medicine, Duluth, MN, USA

Francesco Leri
Department of Psychology, University of Guelph, Guelph, ON, Canada

Scott J. Moeller
Department of Psychiatry, and Department of Neuroscience, Icahn School of Medicine at Mount Sinai, New York, NY, USA

Seyed Mohammad Ahmadi Soleimani
Neurocognitive Laboratory, Iranian National Center for Addiction Studies (INCAS), Tehran University of Medical Sciences, and Department of Physiology, Faculty of Medical Sciences, Tarbiat Modares University, Tehran, Iran

Azarkhsh Mokri
Clinical Department, Iranian National Center for Addiction Studies (INCAS), Tehran University of Medical Sciences, Tehran, Iran

John Monterosso
Neuroscience Graduate Program; Department of Psychology, and Brain and Creativity Institute, University of Southern California, Los Angeles, CA, USA

Jonathan D. Morrow
Department of Psychiatry, University of Michigan, Ann Arbor, MI, USA

Bonnie J. Nagel
Departments of Psychiatry, and Behavioral Neuroscience, Oregon Health & Science University, Portland, OR, USA

Padideh Nasseri
Neurocognitive Laboratory, Iranian National Center for Addiction Studies (INCAS), Tehran University of Medical Sciences, and Translational Neuroscience Program, Institute for Cognitive Science Studies (ICSS), Tehran, Iran

Marc N. Potenza
Department of Psychiatry; Department of Neurobiology, Child Study Center, and CASAColumbia, and Connecticut Mental Health Center, Yale University School of Medicine, New Haven, CT, USA

Alexandra Potter
Department of Psychiatry, Vermont Center on Behavior and Health, and Department of Psychological Science, University of Vermont, Burlington, VT, USA

Amanda J. Quisenberry
Addiction Recovery Research Center, Virginia Tech Carilion Research Institute, Roanoke, VA, USA

Arash Rahmani
Iranian National Center for Addiction Studies, Tehran University of Medical Sciences, Tehran, Iran

Lara A. Ray
Department of Psychology, University of California, Los Angeles, CA, USA

Erica M. Schulte
Department of Psychology, University of Michigan, Ann Arbor, MI, USA

Sarah E. Snider
Addiction Recovery Research Center, Virginia Tech Carilion Research Institute, Roanoke, VA, USA

Philip A. Spechler
Department of Psychiatry, Vermont Center on Behavior and Health, and Department of Psychological Science, University of Vermont, Burlington, VT, USA

Jeffrey S. Stein
Addiction Recovery Research Center, Virginia Tech Carilion Research Institute, Roanoke, VA, USA

Jane R. Taylor
Department of Psychiatry, Yale University, New Haven, CT, USA

Mary M. Torregrossa
Department of Psychiatry, University of Pittsburgh, Pittsburgh, PA, USA

Yvonne H.C. Yau
Department of Neurology and Neurosurgery, Montreal Neurological Institute, McGill University, and Montreal Neurological Institute, 3801 Rue University, Montréal, QC, Canada

Fatemeh Yavari
Neurocognitive Laboratory, Iranian National Center for Addiction Studies (INCAS), Tehran University of Medical Sciences, Tehran, Iran

Sarah W. Yip
Department of Psychiatry, Yale University School of Medicine, New Haven, CT, USA

Sonja Yokum
Oregon Research Institute, Eugene, OR, USA

Yan Zhou
The Laboratory of the Biology of Addictive Diseases, The Rockefeller University, New York, NY, USA

Contents

Preface: Neuroscience for Addiction Medicine: From Prevention to Rehabilitation

It is estimated that a total of 246 million people, i.e., over 5% of the world's adult population, have used an illicit drug during the last year. Meanwhile, more than 10% of these drug users are suffering from drug use disorders and the number of drug-related deaths is estimated to be over 187,000 annually (UN Office of Drugs and Crime, 2015). Adding disorders related to the nonpharmacologic or behavioral addictions such as pathological gambling, Internet and gaming addictions, overeating and obesity, and compulsive sexual behaviors to the drug addictions comprises a group of brain disorders that contribute as one of the major challenges for humankind in the current millennium.

Addiction medicine has been regarded as a stand-alone specialty among other medical professions in several countries; however, there are still serious concerns regarding the availability and effectiveness of interventions in a wide range from prevention to rehabilitation in addiction medicine. Accumulating pathophysiological evidences for "Addiction as a Brain Disorder" during last 20 years is extending expectations from neuroscience to contribute more seriously in the routine clinical practices during prevention, assessment, treatment, and rehabilitation of addictive disorders. Neuroscience has made tremendous progress toward understanding basic neural processes; however, there is still a lot of progress needed to be made in utilizing neuroscience approaches in clinical medicine in general and addiction medicine in particular.

The basic idea of a book to provide the current status of the field of neuroscience of addiction with particular emphasis on potential applications in a clinical setting was jumped out during meetings in the 2nd Basic and Clinical Neuroscience Congress in October 2013 in Tehran with Professor Vincent Walsh, the *Progress in Brain Research*, PBR, Editor in Chief. We, Martin and Hamed, started to work together for a proposal to the PBR advisory board to compile a volume of reviews in June 2014 in the Laureate Institute for Brain Research, Tulsa, OK. After receiving the green lights from the PBR office, the invitations went out to the senior scholars in the field from October 2014. We received overwhelming positive feedbacks from over 120 contributors from 90 institutes in 14 countries that ended up with 36 chapters in two volumes in October 2015. During this 1 year of intensive efforts, all the chapters were peer reviewed and revised accordingly to meet high-quality standards of the PBR and our vision for the whole concept of the volumes. The first volume, PBR Vol. 223, is mainly focused on the basic neurocognitive constructs contributing to pathophysiological basis of pharmacological and behavioral addictions, and the second volume, PBR Vol. 224, depicts the contribution of neuroscience methods and interventions in the future of clinical practices in addiction medicine.

The goal of these two volumes is to provide readers with insights into current gaps and possible directions of research that would address impactful questions. The fundamental question that is addressed in these volumes is "how can neuroscience be used to make a real difference in addiction medicine"? To that end, we asked the contributors to:

(1) review the recent literature with a time horizon of approximately 5–10 years,
(2) identify current gaps in our knowledge that contribute to the limited impact of the area of research to clinical practice, and
(3) provide a perspective where the field is heading and how impactful questions can be addressed to change the practice of addiction medicine.

We envision that both neuroscientists and clinical investigators will be the primary audience of these two volumes. Moreover, the common interest of these individuals will be the application of neuroscience approaches in studies to assess or treat individuals with addictive disorders. We think that these PBR volumes will provide the audiences with most recent evidences from different disciplines in brain studies on the wide range of addictive disorders in an integrative way toward "Neuroscience for Addiction Medicine: From Prevention to Rehabilitation." The hope is that the information provided in the series of chapters in these two volumes will trigger new researches that will help to connect basic neuroscience to clinical addiction medicine.

The Editors
Hamed Ekhtiari, MD,
Iranian National Center for Addiction Studies
Martin Paulus, MD,
Laureate Institute for Brain Research

REFERENCE

UN Office of Drugs and Crime, 2015. World Drug Report 2015. United Nation Publication, Vienna.

PART

1

Neuroscience Constructs for Addiction Medicine

CHAPTER

1

Neuroscience of resilience and vulnerability for addiction medicine: From genes to behavior

Jonathan D. Morrow[*,1], **Shelly B. Flagel**[*,†]

*Department of Psychiatry, University of Michigan, Ann Arbor, MI, USA

†Molecular and Behavioral Neuroscience Institute, University of Michigan, Ann Arbor, MI, USA

[1]Corresponding author: Tel.: +1-734-764-0231; Fax: +1-734-232-0244, e-mail address: jonmorro@umich.edu

Abstract

Addiction is a complex behavioral disorder arising from roughly equal contributions of genetic and environmental factors. Behavioral traits such as novelty-seeking, impulsivity, and cue-reactivity have been associated with vulnerability to addiction. These traits, at least in part, arise from individual variation in functional neural systems, such as increased striatal dopaminergic activity and decreased prefrontal cortical control over subcortical emotional and motivational responses. With a few exceptions, genetic studies have largely failed to consistently identify specific alleles that affect addiction liability. This may be due to the multifactorial nature of addiction, with different genes becoming more significant in certain environments or in certain subsets of the population. Epigenetic mechanisms may also be an important source of risk. Adolescence is a particularly critical time period in the development of addiction, and environmental factors at this stage of life can have a large influence on whether inherited risk factors are actually translated into addictive behaviors. Knowledge of how individual differences affect addiction liability at the level of genes, neural systems, behavioral traits, and sociodevelopmental trajectories can help to inform and improve clinical practice.

Keywords

Addiction, Individual differences, Cue-reactivity, Impulsivity, Dopamine, Neural circuits, Genetics

There is considerable variability in the likelihood of developing addiction upon exposure to drugs of abuse. This is evidenced by the fact that over 90% of Americans have used alcohol, but only 8–12% ever meet criteria for alcohol dependence (Anthony et al., 1994). Determining what factors render certain individuals more

Progress in Brain Research, Volume 223, ISSN 0079-6123, http://dx.doi.org/10.1016/bs.pbr.2015.09.004

susceptible to addiction has proven difficult to discern because of the array of variables involved. Over the past few decades, we have learned that there is a complex interplay of genes and environment that govern the neurobiological and behavioral processes relevant to addiction. However, there are, unquestionably, multiple algorithms by which these factors may be combined to alter addiction liability. Below we will briefly review findings from both human and animal studies that highlight some of the behavioral, neural, and genetic variables believed to contribute to addiction liability.

1 BEHAVIORAL TRAITS

Despite the oft-repeated adage that "there is no addictive personality," there is a clear association between addiction and certain personality traits. For example, clinical studies have found that the trait known as neuroticism or negative emotionality is associated with substance use disorders as well as depressive and anxiety disorders (Kotov et al., 2010; Terracciano et al., 2008). The mechanisms underlying this association are not well-characterized, but are thought to include increased stress sensitivity (Ersche et al., 2012). Another personality trait associated with addiction is the "externalizing" phenotype, characterized by novelty- and sensation-seeking behavior, hypersensitivity to rewards, and insensitivity to punishment (Dick et al., 2013; Hicks et al., 2013; Pingault et al., 2013). Evidence from animal models suggests that the sensation-seeking trait may specifically increase the propensity to initiate and continue drug use, as opposed to predisposing toward compulsive use that would meet criteria for substance dependence (Belin et al., 2008; Deroche-Gamonet et al., 2004; Piazza et al., 1989), and some human studies have substantiated this finding (Ersche et al., 2013). Trait impulsivity, otherwise known as disinhibition or lack of constraint, has perhaps the strongest evidence for an association with addiction. In the animal literature, the transition to compulsive drug use can be predicted by measures of impulsivity (Belin et al., 2008; Dalley et al., 2007); specifically the inability to withhold a prepotent response (e.g., 5-choice serial reaction time task). Similar tasks have been used with human subjects in the laboratory to assess disinhibition or lack of constraint—and, in agreement with the rodent studies, these studies have largely shown evidence for an association between trait impulsivity and addiction (for review, see Verdejo-Garcia et al., 2008). Another addiction-related trait is "cue-reactivity"; perhaps not surprisingly, as relapse is most often triggered by cues (e.g., people, places, paraphernalia) in the environment that have been previously associated with the drug-taking experience. Indeed, both human studies and animal models suggest that individuals for whom the cue attains incentive motivational value or incentive salience are the individuals most likely to exhibit relapse (e.g., see Carter and Tiffany, 1999; Janes et al., 2010; Rohsenow et al., 1990; Saunders and Robinson, 2010, 2011). These different personality traits have not only been associated with different phases of addiction but also with different types of drugs of abuse. For example, cocaine addicts tend to be more impulsive than heroin addicts; whereas heroin addicts are more anxious than cocaine addicts (Bornovalova et al., 2005; Lejuez et al., 2005, 2006). These data beg the question of whether certain personality traits predispose an individual to a particular phase (e.g., initiation vs. relapse) of addiction or type of drug (e.g., psychostimulants vs. opioids), or if it is the drugs themselves—via alteration of brain function—that cause the behavioral traits.

2 NEUROBIOLOGICAL FACTORS

Although it has been difficult to parse cause from consequence when it comes to elucidating the neurobiological mechanisms underlying addiction, there is general agreement as to what neurotransmitter systems and brain regions are involved. All drugs of abuse share the ability to elevate dopamine transmission, either directly or indirectly (Hyman et al., 2006). It is therefore not surprising that dopamine and the mesocorticolimbic "reward" circuitry have been a primary focus of neuroscience research related to addiction. The most consistent findings to emerge from imaging studies of addicted patients are decreased dopamine type 2/3 (D2/3) receptor binding capacity, particularly in the striatum, and decreased activity in prefrontal cortical (PFC) areas that normally provide "top-down" executive control over striatal activity (Volkow et al., 1993; Wang et al., 2012a). Decreased striatal D2/3 receptor binding has also been reliably associated with novelty-seeking and impulsivity in both human and animal studies (Buckholtz et al., 2010; Dalley et al., 2011; Leyton et al., 2002; Zald et al., 2008), as has increased dopaminergic activity in the striatum at baseline and in response to various stimuli in rats (Hooks et al., 1991; Piazza et al., 1991). Further, human studies have shown that, in addition to lower levels of functional activity in PFC areas, impulsive individuals exhibit decreased functional connectivity between the PFC and subcortical structures, including the amygdala and ventral striatum (Davis et al., 2013; Schmaal et al., 2012). Fewer studies have investigated the neurobiological basis of "cue-reactivity," though existing evidence from both humans and animals suggests increased mesolimbic dopaminergic activity in cue-reactive individuals (Flagel et al., 2011; Jasinska et al., 2014). Thus, a simplified picture has emerged that individuals predisposed toward addiction are characterized neurobiologically by relatively high dopaminergic activity, coupled with decreased "top-down" cortical control.

3 GENETICS

Twin studies have yielded heritability estimates of 30–70% for addiction (Agrawal and Lynskey, 2008). Most of the genetic influences on substance use appear to be shared across different classes of substances (Kendler et al., 2008; Tsuang et al., 1998). However, the most robust findings from candidate gene and from genome-wide association studies (GWAS) have been specific to certain classes of drugs.

For example, polymorphisms affecting the function of the alcohol dehydrogenase and aldehyde dehydrogenase are some of the oldest and most potent known genetic risk/resilience factors for any psychiatric disorder, but these are genes that specifically affect alcohol metabolism and are therefore specifically related to alcohol use disorders (Hurley and Edenberg, 2012). To our knowledge, the only other association reliably and convincingly detected by both GWAS and candidate gene studies is that of nicotine dependence with variants of nicotinic acetylcholine receptor (nAChR) subunit genes (Bierut et al., 2008). Although genes affecting several other proteins have been associated with addiction, including gamma-amino butyric acid (GABA) receptors, opioid receptors, and cannabinoid receptors, these findings have been inconsistent across studies and generally specific to one or a few substances (Hall et al., 2013; Wang et al., 2012b). Even studies of genes involved in dopamine transmission have yielded mixed results, despite the fact that augmentation of dopamine transmission in the ventral striatum is a mechanistic pathway common to all drugs of abuse (Hyman et al., 2006). Difficulties in the replication of candidate gene findings do not necessarily mean that the associations are invalid; instead, it may indicate that individual genetic effects are limited to specific populations and endophenotypes. Indeed, transgenic animal studies of candidate genes generally show much more consistent and robust effects on drug-taking behaviors than human association studies would otherwise suggest. Thus, like most psychiatric disorders, addiction appears to be highly heritable, but the multifactorial and polygenic nature of the disorder makes specific gene associations very difficult to detect.

4 EPIGENETICS

Intriguingly, emerging evidence from the animal literature is implicating transgenerational epigenetic mechanisms as possible contributors to the heritability of addictive disorders (Vassoler and Sadri-Vakili, 2014; Yohn et al., 2015). Epigenetic changes are experience-dependent chemical alterations to chromosomes that affect gene expression. The most widely studied epigenetic markers are DNA methylation and histone methylation and acetylation. Although there have been a number of studies demonstrating epigenetic modifications in response to drugs of abuse (for review, see Renthal and Nestler, 2008), few, to our knowledge, have identified epigenetic mechanisms that contribute to addiction vulnerability. Thus, for the purpose of this chapter, we will focus on transgenerational epigenetic mechanisms, that is, those that are retained throughout embryonic development, and thereby passed on from parent to offspring. For example, exposure to alcohol causes several epigenetic changes to be passed on to offspring and successive generations of rodents, including demethylation of the imprinted gene *H19* (Ouko et al., 2009), demethylation of the promoter region of exon IV of the brain-derived neurotrophic factor (*Bdnf*) gene (Finegersh and Homanics, 2014), increased methylation of the dopamine transporter (*Dat*) promoter (Kim et al., 2014), and methylation of the pro-opioid melanocortin (*Pomc*) promoter in the arcuate nucleus (Govorko et al., 2012). Remarkably, there are a number of common associations of these epigenetic changes, including increased *Bdnf* expression in the ventral tegmental area (VTA), decreased DAT in the cortex and striatum, decreased hypothalamic *Pomc* (Govorko et al., 2012), decreased fear behaviors, increased aggression and impulsivity (Meek et al., 2007), and attention deficits (Kim et al., 2014).

There is also evidence of transgenerational epigenetic changes induced by other substances. For example, rats exposed to opioids have progeny that exhibit altered responses to dopaminergic agents (Byrnes et al., 2013; Vyssotski, 2011). Offspring of dams exposed to nicotine are hyperactive and inattentive, and have increased methylation of the *Bdnf* promoter and decreased BDNF levels in the frontal cortex (Toledo-Rodriguez et al., 2010; Yochum et al., 2014; Zhu et al., 2014). In contrast to changes induced by other substances, the transgenerational effects of cocaine exposure may actually be protective, as the progeny of cocaine-exposed rodents have increased acetylated histone 3 associated with *Bdnf* exon IV, increased BDNF expression in the medial prefrontal cortex, and reduced cocaine self-administration (Vassoler et al., 2013). Though many mechanistic details for these effects remain to be discovered, and all of the epigenetic findings mentioned here await further confirmation from other groups, transgenerational epigenetic inheritance of risk may prove to be an important component of individual differences in vulnerability to addiction.

5 DEVELOPMENTAL FACTORS

Environmental factors and life experiences also play a large role in determining an individual's risk for developing an addictive disorder. Several studies have shown that the younger a person is upon first exposure to drugs or alcohol, the higher their risk of addiction, even after controlling for other variables (e.g., Chen et al., 2009; Dawson et al., 2008; King and Chassin, 2007). Similarly, animal studies have shown that exposure to stress, particularly in the prenatal or early childhood period, increases the risk of addiction (Deminiere et al., 1992; Henry et al., 1995; Kippin et al., 2008). Human imaging studies show that the adolescent brain is also particularly responsive to stressful stimuli (Gunnar et al., 2009; Stroud et al., 2009). Human and animal studies have shown that stress very early in life will sensitize the hypothalamic-pituitary-adrenal axis, such that later stress responses become exaggerated (Higley et al., 1991; Liu et al., 1997; Tarullo and Gunnar, 2006). In addition, dopaminergic activity increases in the striatum and decreases in cortical regions after early life stress in both humans and animals (Blanc et al., 1980; Brake et al., 2004; Pruessner et al., 2004). Importantly, animal studies indicate that many of these changes can be mitigated by increased maternal care or environmental enrichment (Barbazanges et al., 1996; Plotsky and Meaney, 1993; Solinas et al., 2010). Genetic studies in humans have shown that childhood experiences moderate the effects of several genes on addiction, including polymorphisms in the serotonin transporter, dopamine type 2 receptor, monoamine oxidase, and corticotrophin releasing hormone receptor 1 (Bau et al., 2000; Bjork et al., 2010; Blomeyer et al., 2008). Thus, many genetic risk factors may only become relevant in the setting of known environmental stressors such as parental divorce, migration, and comorbid psychiatric illness; conversely, genetic influences may be reduced by protective environmental factors such as marriage, religiosity, and parental involvement (Dick et al., 2007a,b; Heath et al., 1989; Koopmans et al., 1999).

The contributions of genetic and environmental risk factors vary over the course of development, and multiple lines of evidence from the human and animal literature implicate adolescence as a critical period in the development of addictive disorders (Adriani and Laviola, 2004; Belsky et al., 2013; Vrieze et al., 2012). As with most psychiatric disorders, the onset of addictive disorders peaks in adolescence (SAMSHA, 2014). Brain maturation takes place unevenly throughout the brain, with basic motivational regions such as the striatum developing well before more cognitive PFC regions that are involved in exerting control over appetitive urges (Dahl, 2008; Gogtay et al., 2004; Sowell et al., 2003). Dopaminergic activity throughout the limbic system is increased during adolescence (McCutcheon et al., 2012; Rosenberg and Lewis, 1994). In addition, glutamatergic connections between the prefrontal cortex and subcortical structures, including the ventral striatum and amygdala, are reduced in adolescents (Brenhouse et al., 2008; Cunningham et al., 2002). Hence, the adolescent brain is sometimes described as a high-performance sports car with faulty brakes. As might be expected based on these neurobiological characteristics, adolescents are more impulsive and sensation-seeking than adults (Adriani and Laviola, 2003; Adriani et al., 1998; Romer et al., 2009). They are also more likely to engage in risky behaviors, including taking drugs more often and in larger quantities, than adults (Merrick et al., 2004; SAMSHA, 2014; Steinberg, 2008).

It is interesting to note that risk-taking behavior may also serve important, adaptive functions for adolescents. The transition to independence requires stepping outside of one's comfort zone in order to achieve a sense of competence in adult situations. Risky activities such as substance use may contribute to social development, as teens who experiment with drugs are more socially competent and accepted by their peers than abstainers (Spear, 2000). Social aspects of the environment are more emotionally salient for adolescents, and this sensitivity is reflected by increased limbic activity in response to social cues (Choudhury et al., 2006; Monk et al., 2003; Yang et al., 2003). Perhaps unsurprisingly, then, substance use and antisocial behavior among peers is a strong risk factor for the development of addiction in adolescence (Dick et al., 2007a,b; Harden et al., 2008). Hormonal influences are also likely to play a role in addiction during this time period, as testosterone contributes to synaptic pruning during adolescence (Nguyen et al., 2013). Women, though less likely overall to develop addictive disorders, generally have a more severe and treatment-resistant course of illness, more stress-related comorbidities, and faster transitions to compulsive drug use than men, again highlighting the influence of hormones on drug-taking behavior (Kuhn, 2015; Nguyen et al., 2013). These findings, taken together, illustrate that adolescence is an extraordinarily sensitive time window with regard to the development of addiction.

6 CONCLUSION AND FUTURE DIRECTIONS

The information garnered from research into addiction vulnerability has the potential to inform and improve treatment of addictive disorders in several ways. For instance, there is considerable interest in using biomarkers to identify individuals who are at high risk of developing addiction. Theoretically, information about a person's dopaminergic activity, functional connectivity patterns, or even BDNF expression patterns in the brain could be used to estimate risk, but currently none of these indicators are sensitive or specific enough to serve as true biomarkers. Genetic information has the potential to be very informative, as heredity can account for upward of 70% of an individual's risk for addiction. However, other than a handful of substance-specific genes, genetic studies have so far not been very successful at consistently finding particular genotypes that contribute to addiction liability. Because of the multifactorial nature of addiction, future genetic studies may need to focus on particular subpopulations, endophenotypes, or subtypes of addiction, in addition to better accounting for environmental modifiers of genetic risk, in order to identify clinically relevant risk alleles. Emerging evidence from the animal literature suggests that epigenomic association studies may also be useful for accounting for the heritable portion of addiction vulnerability.

However, despite gaps in our knowledge of the specific genes and neural circuitry involved in addiction liability, existing information is often enough to produce clinically relevant estimates of an individual's risk of developing an addictive disorder. For example, we already know that an impulsive, sensation-seeking individual, whose parents and grandparents suffered from addiction, who undergoes neglect or other trauma at an early age, and who is surrounded by peers engaging in high-risk substance use, is very likely to develop an addictive disorder. We can even predict with considerable confidence that the disorder will emerge sometime between the ages of 12 and 25. The question then becomes, how do we use this information to improve clinical outcomes? First, do no harm. In 2013, the leading cause of accidental death in the United States was drug overdose, and over 50% of the drugs involved were prescription opioids and benzodiazepines (CDC, 2014, 2015). Prescribing physicians should make a concerted effort to limit access to drugs with addictive potential for individuals *and relatives* of individuals at high risk of developing addictive disorders, because the vast majority of abused prescription drugs are prescribed either to the user themselves or to a relative of the user (SAMSHA, 2014). Patients should be educated about their own risk profile and that of their family members, so that they can make informed decisions about the way they use potentially addictive substances. Formal prevention programs aimed at adolescents have largely failed to influence substance use rates, but parental behaviors often have a profound effect on teenage substance use (SAMSHA, 2014). Thus, parents of adolescents who are at high risk of developing addiction should be encouraged to take steps that are known to reduce the risk of addiction, such as explicitly discouraging drug use, monitoring the child's peers and activities, actively involving themselves in the child's

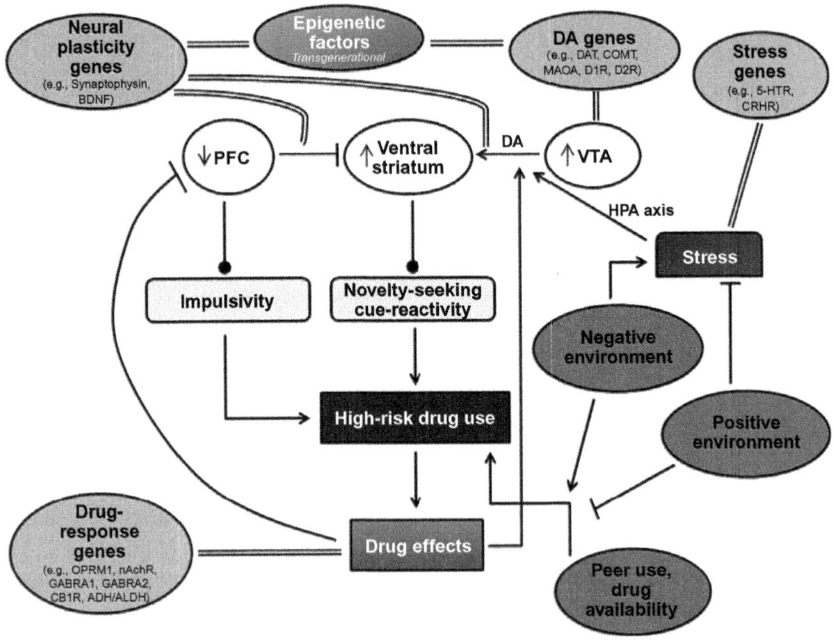

FIGURE 1

Addiction vulnerability at multiple, interacting levels. High-risk drug use (red; black in the print version) is potentiated by personality traits (green; light gray in the print version) including impulsivity, novelty-seeking, and cue-reactivity. These personality traits, in turn, reflect neurobiological traits (yellow; white in the print version) including increased dopaminergic activity and decreased prefrontal cortical control over ventral striatal impulses. Addictive drugs (purple; dark gray in the print version) directly affect this neural circuitry, which is one driver of the cycle of addiction. Stress (black), acting through the hypothalamic pituitary adrenal (HPA) axis, predisposes toward addictive behavior by enhancing dopaminergic activity. Environmental factors (gray) affect vulnerability either through their effects on stress, or via a more direct effect on the probability of drug use. Genetic polymorphisms (blue; light gray in the print version) affect this system in a variety of ways. "Drug–response genes" modulate the pharmacologic effects of drug use, while other genes modulate dopaminergic activity, stress reactivity, or corticolimbic connectivity patterns. Transgenerational epigenetic influences (orange; dark gray in the print version) may be mediated by these same gene families, with most of the evidence so far implicating dopaminergic genes and synaptic plasticity genes. Definitions of connectors: arrows indicate one variable potentiating the other; lines terminating with a hash bar indicate an inhibitory relationship; lines terminating with a circle indicate a positive association; double-hashed lines indicate a relationship that can be either positive or negative, depending on the allele. Abbreviations: 5-HTR, serotonin receptor; ADH, alcohol dehydrogenase; ALDH, aldehyde dehydrogenase; BDNF, brain-derived neurotrophic factor; CB1R, cannabinoid type 1 receptor; COMT, catechol-*O*-methyl transferase; CRHR, corticotrophin-releasing hormone receptor; D1R, dopamine type 1 receptor; D2R, dopamine type 2 receptor; DAT, dopamine transporter; GABRA1, gamma-aminobutyric acid (GABA) receptor subunit alpha-1; GABRA2, GABA receptor subunit alpha-2; HPA, hypothalamic-pituitary-adrenal; MAOA, monoamine oxidase A; nAChR, nicotinic acetylcholine receptor; OPRM1, opioid receptor mu 1; PFC, prefrontal cortex; VTA, ventral tegmental area.

homework and other activities, providing a stable family life, and involving the child in religious activities.

Treatment of patients who already have addiction may also benefit from knowledge of specific vulnerability factors. For example, personality traits associated with addiction can, in some cases, be targeted by specific clinical interventions. To date, few studies have taken this approach, but one indication of its potential utility is the finding that, for individuals with addiction and comorbid attention deficit hyperactivity disorder, treatment of their impulsivity with potentially addictive psychostimulants paradoxically reduces their risk of relapse (Levin et al., 2007). Selective serotonin reuptake inhibitors (SSRIs) have largely been disappointing as a treatment for addiction (Nunes and Levin, 2004) but because they actually reduce the neuroticism trait (Tang et al., 2009), SSRIs might be useful in treating a subset of patients for whom neuroticism is a primary driver of their addiction. Information about personality traits and other neurobiological factors might also be used to tailor specific treatment interventions; for example, emphasizing stress reduction in individuals with high neuroticism, or focusing more on identifying and avoiding cues for individuals with markers of excessive cue-reactivity. Sophisticated methods (e.g., optogenetics, designer receptors exclusively activated by designer drugs—DREADDs) are being developed in rodents to directly manipulate the neural circuitry responsible for individual differences in cue-reactivity and other behavioral traits, but because many of these approaches involve genetic modification of neurons, they are many years away from being available for clinical trials.

As research progresses, the multifactorial nature of addiction becomes even more apparent. Yet, remarkably, as outlined above, there are a number of vulnerability factors that repeatedly appear in the literature, common to both human and animal studies, and linked at multiple levels of analysis (e.g., genetic and neurobiological; see Fig. 1 for a simplified visual summary). Moving forward, the advent and accessibility of new technology (e.g., Saunders et al., 2015) will allow increasingly precise analysis of the neurobiological factors contributing to addiction liability. For example, chemogenetic approaches could be used to manipulate "top-down" cortical circuits in order to "switch" the behavioral phenotype of an animal from one that is addiction-prone, to one that is addiction-resilient. A continuing challenge for the field will be integrating this new knowledge with the other layers of genetic, epigenetic, developmental, and environmental factors that interact in multiple ways with this neural circuitry in order to determine an individual's risk for addiction.

REFERENCES

Adriani, W., Laviola, G., 2003. Elevated levels of impulsivity and reduced place conditioning with d-amphetamine: two behavioral features of adolescence in mice. Behav. Neurosci. 117 (4), 695–703.

Adriani, W., Laviola, G., 2004. Windows of vulnerability to psychopathology and therapeutic strategy in the adolescent rodent model. Behav. Pharmacol. 15 (5-6), 341–352.

Adriani, W., Chiarotti, F., Laviola, G., 1998. Elevated novelty seeking and peculiar d-amphetamine sensitization in periadolescent mice compared with adult mice. Behav. Neurosci. 112 (5), 1152–1166.

Agrawal, A., Lynskey, M.T., 2008. Are there genetic influences on addiction: evidence from family, adoption and twin studies. Addiction 103 (7), 1069–1081.

Anthony, J.C., Warner, S.A., Kessler, R.C., 1994. Comparative epidemiology of dependence on tobacco, alcohol, controlled substances, and inhalants: basic findings from the National Comorbidity Survey. Exp. Clin. Psychopharmacol. 2 (3), 244–268.

Barbazanges, A., Vallee, M., Mayo, W., Day, J., Simon, H., Le Moal, M., Maccari, S., 1996. Early and later adoptions have different long-term effects on male rat offspring. J. Neurosci. 16 (23), 7783–7790.

Bau, C.H., Almeida, S., Hutz, M.H., 2000. The TaqI A1 allele of the dopamine D2 receptor gene and alcoholism in Brazil: association and interaction with stress and harm avoidance on severity prediction. Am. J. Med. Genet. 96 (3), 302–306.

Belin, D., Mar, A.C., Dalley, J.W., Robbins, T.W., Everitt, B.J., 2008. High impulsivity predicts the switch to compulsive cocaine-taking. Science 320 (5881), 1352–1355.

Belsky, D.W., Moffitt, T.E., Baker, T.B., Biddle, A.K., Evans, J.P., Harrington, H., et al., 2013. Polygenic risk and the developmental progression to heavy, persistent smoking and nicotine dependence: evidence from a 4-decade longitudinal study. JAMA Psychiatry 70 (5), 534–542.

Bierut, L.J., Stitzel, J.A., Wang, J.C., Hinrichs, A.L., Grucza, R.A., Xuei, X., et al., 2008. Variants in nicotinic receptors and risk for nicotine dependence. Am. J. Psychiatry 165 (9), 1163–1171.

Bjork, K., Hansson, A.C., Sommer, W.H., 2010. Genetic variation and brain gene expression in rodent models of alcoholism implications for medication development. Int. Rev. Neurobiol. 91, 129–171.

Blanc, G., Herve, D., Simon, H., Lisoprawski, A., Glowinski, J., Tassin, J.P., 1980. Response to stress of mesocortico-frontal dopaminergic neurones in rats after long-term isolation. Nature 284 (5753), 265–267.

Blomeyer, D., Treutlein, J., Esser, G., Schmidt, M.H., Schumann, G., Laucht, M., 2008. Interaction between CRHR1 gene and stressful life events predicts adolescent heavy alcohol use. Biol. Psychiatry 63 (2), 146–151.

Bornovalova, M.A., Daughters, S.B., Hernandez, G.D., Richards, J.B., Lejuez, C.W., 2005. Differences in impulsivity and risk-taking propensity between primary users of crack cocaine and primary users of heroin in a residential substance-use program. Exp. Clin. Psychopharmacol. 13 (4), 311–318.

Brake, W.G., Zhang, T.Y., Diorio, J., Meaney, M.J., Gratton, A., 2004. Influence of early postnatal rearing conditions on mesocorticolimbic dopamine and behavioural responses to psychostimulants and stressors in adult rats. Eur. J. Neurosci. 19 (7), 1863–1874.

Brenhouse, H.C., Sonntag, K.C., Andersen, S.L., 2008. Transient D1 dopamine receptor expression on prefrontal cortex projection neurons: relationship to enhanced motivational salience of drug cues in adolescence. J. Neurosci. 28 (10), 2375–2382.

Buckholtz, J.W., Treadway, M.T., Cowan, R.L., Woodward, N.D., Li, R., Ansari, M.S., et al., 2010. Dopaminergic network differences in human impulsivity. Science 329 (5991), 532.

Byrnes, J.J., Johnson, N.L., Carini, L.M., Byrnes, E.M., 2013. Multigenerational effects of adolescent morphine exposure on dopamine D2 receptor function. Psychopharmacology (Berl) 227 (2), 263–272.

Carter, B.L., Tiffany, S.T., 1999. Cue-reactivity and the future of addiction research. Addiction 94 (3), 349–351.

CDC, 2014. Centers for Disease Control and Prevention. Web-based Injury Statistics Query and Reporting System (WISQARS). http://www.cdc.gov/injury/wisqars/fatal.html.

CDC, 2015. Centers for Disease Control and Prevention. National Vital Statistics System Mortality Data. http://www.cdc.gov/nchs/deaths.htm.

Chen, C.Y., Storr, C.L., Anthony, J.C., 2009. Early-onset drug use and risk for drug dependence problems. Addict. Behav. 34 (3), 319–322.

Choudhury, S., Blakemore, S.J., Charman, T., 2006. Social cognitive development during adolescence. Soc. Cogn. Affect. Neurosci. 1 (3), 165–174.

Cunningham, M.G., Bhattacharyya, S., Benes, F.M., 2002. Amygdalo-cortical sprouting continues into early adulthood: implications for the development of normal and abnormal function during adolescence. J. Comp. Neurol. 453 (2), 116–130.

Dahl, R.E., 2008. Biological, developmental, and neurobehavioral factors relevant to adolescent driving risks. Am. J. Prev. Med. 35 (3 Suppl.), S278–S284.

Dalley, J.W., Fryer, T.D., Brichard, L., Robinson, E.S., Theobald, D.E., Laane, K., et al., 2007. Nucleus accumbens D2/3 receptors predict trait impulsivity and cocaine reinforcement. Science 315 (5816), 1267–1270.

Dalley, J.W., Everitt, B.J., Robbins, T.W., 2011. Impulsivity, compulsivity, and top-down cognitive control. Neuron 69 (4), 680–694.

Davis, F.C., Knodt, A.R., Sporns, O., Lahey, B.B., Zald, D.H., Brigidi, B.D., Hariri, A.R., 2013. Impulsivity and the modular organization of resting-state neural networks. Cereb. Cortex 23 (6), 1444–1452.

Dawson, D.A., Goldstein, R.B., Chou, S.P., Ruan, W.J., Grant, B.F., 2008. Age at first drink and the first incidence of adult-onset DSM-IV alcohol use disorders. Alcohol. Clin. Exp. Res. 32 (12), 2149–2160.

Deminiere, J.M., Piazza, P.V., Guegan, G., Abrous, N., Maccari, S., Le Moal, M., Simon, H., 1992. Increased locomotor response to novelty and propensity to intravenous amphetamine self-administration in adult offspring of stressed mothers. Brain Res. 586 (1), 135–139.

Deroche-Gamonet, V., Belin, D., Piazza, P.V., 2004. Evidence for addiction-like behavior in the rat. Science 305 (5686), 1014–1017.

Dick, D.M., Pagan, J.L., Holliday, C., Viken, R., Pulkkinen, L., Kaprio, J., Rose, R.J., 2007a. Gender differences in friends' influences on adolescent drinking: a genetic epidemiological study. Alcohol. Clin. Exp. Res. 31 (12), 2012–2019.

Dick, D.M., Viken, R., Purcell, S., Kaprio, J., Pulkkinen, L., Rose, R.J., 2007b. Parental monitoring moderates the importance of genetic and environmental influences on adolescent smoking. J. Abnorm. Psychol. 116 (1), 213–218.

Dick, D.M., Aliev, F., Latendresse, S.J., Hickman, M., Heron, J., Macleod, J., et al., 2013. Adolescent alcohol use is predicted by childhood temperament factors before age 5, with mediation through personality and peers. Alcohol. Clin. Exp. Res. 37 (12), 2108–2117.

Ersche, K.D., Turton, A.J., Chamberlain, S.R., Muller, U., Bullmore, E.T., Robbins, T.W., 2012. Cognitive dysfunction and anxious-impulsive personality traits are endophenotypes for drug dependence. Am. J. Psychiatry 169 (9), 926–936.

Ersche, K.D., Jones, P.S., Williams, G.B., Smith, D.G., Bullmore, E.T., Robbins, T.W., 2013. Distinctive personality traits and neural correlates associated with stimulant drug use versus familial risk of stimulant dependence. Biol. Psychiatry 74 (2), 137–144.

Finegersh, A., Homanics, G.E., 2014. Paternal alcohol exposure reduces alcohol drinking and increases behavioral sensitivity to alcohol selectively in male offspring. PLoS One 9 (6), e99078.

Flagel, S.B., Clark, J.J., Robinson, T.E., Mayo, L., Czuj, A., Willuhn, I., et al., 2011. A selective role for dopamine in stimulus-reward learning. Nature 469 (7328), 53–57.

Gogtay, N., Giedd, J.N., Lusk, L., Hayashi, K.M., Greenstein, D., Vaituzis, A.C., et al., 2004. Dynamic mapping of human cortical development during childhood through early adulthood. Proc. Natl. Acad. Sci. U. S. A. 101 (21), 8174–8179.

Govorko, D., Bekdash, R.A., Zhang, C., Sarkar, D.K., 2012. Male germline transmits fetal alcohol adverse effect on hypothalamic proopiomelanocortin gene across generations. Biol. Psychiatry 72 (5), 378–388.

Gunnar, M.R., Wewerka, S., Frenn, K., Long, J.D., Griggs, C., 2009. Developmental changes in hypothalamus-pituitary-adrenal activity over the transition to adolescence: normative changes and associations with puberty. Dev. Psychopathol. 21 (1), 69–85.

Hall, F.S., Drgonova, J., Jain, S., Uhl, G.R., 2013. Implications of genome wide association studies for addiction: are our a priori assumptions all wrong? Pharmacol. Ther. 140 (3), 267–279.

Harden, K.P., Hill, J.E., Turkheimer, E., Emery, R.E., 2008. Gene-environment correlation and interaction in peer effects on adolescent alcohol and tobacco use. Behav. Genet. 38 (4), 339–347.

Heath, A.C., Jardine, R., Martin, N.G., 1989. Interactive effects of genotype and social environment on alcohol consumption in female twins. J. Stud. Alcohol 50 (1), 38–48.

Henry, C., Guegant, G., Cador, M., Arnauld, E., Arsaut, J., Le Moal, M., Demotes-Mainard, J., 1995. Prenatal stress in rats facilitates amphetamine-induced sensitization and induces long-lasting changes in dopamine receptors in the nucleus accumbens. Brain Res. 685 (1–2), 179–186.

Hicks, B.M., Foster, K.T., Iacono, W.G., McGue, M., 2013. Genetic and environmental influences on the familial transmission of externalizing disorders in adoptive and twin offspring. JAMA Psychiatry 70 (10), 1076–1083.

Higley, J.D., Hasert, M.F., Suomi, S.J., Linnoila, M., 1991. Nonhuman primate model of alcohol abuse: effects of early experience, personality, and stress on alcohol consumption. Proc. Natl. Acad. Sci. U. S. A. 88 (16), 7261–7265.

Hooks, M.S., Jones, G.H., Smith, A.D., Neill, D.B., Justice Jr., J.B., 1991. Response to novelty predicts the locomotor and nucleus accumbens dopamine response to cocaine. Synapse 9 (2), 121–128.

Hurley, T.D., Edenberg, H.J., 2012. Genes encoding enzymes involved in ethanol metabolism. Alcohol Res. 34 (3), 339–344.

Hyman, S.E., Malenka, R.C., Nestler, E.J., 2006. Neural mechanisms of addiction: the role of reward-related learning and memory. Annu. Rev. Neurosci. 29, 565–598.

Janes, A.C., Pizzagalli, D.A., Richardt, S., deB Frederick, B., Chuzi, S., Pachas, G., et al., 2010. Brain reactivity to smoking cues prior to smoking cessation predicts ability to maintain tobacco abstinence. Biol. Psychiatry 67 (8), 722–729.

Jasinska, A.J., Stein, E.A., Kaiser, J., Naumer, M.J., Yalachkov, Y., 2014. Factors modulating neural reactivity to drug cues in addiction: a survey of human neuroimaging studies. Neurosci. Biobehav. Rev. 38, 1–16.

Kendler, K.S., Schmitt, E., Aggen, S.H., Prescott, C.A., 2008. Genetic and environmental influences on alcohol, caffeine, cannabis, and nicotine use from early adolescence to middle adulthood. Arch. Gen. Psychiatry 65 (6), 674–682.

Kim, P., Choi, C.S., Park, J.H., Joo, S.H., Kim, S.Y., Ko, H.M., et al., 2014. Chronic exposure to ethanol of male mice before mating produces attention deficit hyperactivity disorder-like phenotype along with epigenetic dysregulation of dopamine transporter expression in mouse offspring. J. Neurosci. Res. 92 (5), 658–670.

King, K.M., Chassin, L., 2007. A prospective study of the effects of age of initiation of alcohol and drug use on young adult substance dependence. J. Stud. Alcohol Drugs 68 (2), 256–265.

Kippin, T.E., Szumlinski, K.K., Kapasova, Z., Rezner, B., See, R.E., 2008. Prenatal stress enhances responsiveness to cocaine. Neuropsychopharmacology 33 (4), 769–782.

Koopmans, J.R., Slutske, W.S., van Baal, G.C., Boomsma, D.I., 1999. The influence of religion on alcohol use initiation: evidence for genotype X environment interaction. Behav. Genet. 29 (6), 445–453.

Kotov, R., Gamez, W., Schmidt, F., Watson, D., 2010. Linking "big" personality traits to anxiety, depressive, and substance use disorders: a meta-analysis. Psychol. Bull. 136 (5), 768–821.

Kuhn, C., 2015. Emergence of sex differences in the development of substance use and abuse during adolescence. Pharmacol. Ther. 153, 55–78.

Lejuez, C.W., Bornovalova, M.A., Daughters, S.B., Curtin, J.J., 2005. Differences in impulsivity and sexual risk behavior among inner-city crack/cocaine users and heroin users. Drug Alcohol Depend. 77 (2), 169–175.

Lejuez, C.W., Paulson, A., Daughters, S.B., Bornovalova, M.A., Zvolensky, M.J., 2006. The association between heroin use and anxiety sensitivity among inner-city individuals in residential drug use treatment. Behav. Res. Ther. 44 (5), 667–677.

Levin, F.R., Evans, S.M., Brooks, D.J., Garawi, F., 2007. Treatment of cocaine dependent treatment seekers with adult ADHD: double-blind comparison of methylphenidate and placebo. Drug Alcohol Depend. 87 (1), 20–29.

Leyton, M., Boileau, I., Benkelfat, C., Diksic, M., Baker, G., Dagher, A., 2002. Amphetamine-induced increases in extracellular dopamine, drug wanting, and novelty seeking: a PET/[11C]raclopride study in healthy men. Neuropsychopharmacology 27 (6), 1027–1035.

Liu, D., Diorio, J., Tannenbaum, B., Caldji, C., Francis, D., Freedman, A., et al., 1997. Maternal care, hippocampal glucocorticoid receptors, and hypothalamic-pituitary-adrenal responses to stress. Science 277 (5332), 1659–1662.

McCutcheon, J.E., Conrad, K.L., Carr, S.B., Ford, K.A., McGehee, D.S., Marinelli, M., 2012. Dopamine neurons in the ventral tegmental area fire faster in adolescent rats than in adults. J. Neurophysiol. 108 (6), 1620–1630.

Meek, L.R., Myren, K., Sturm, J., Burau, D., 2007. Acute paternal alcohol use affects offspring development and adult behavior. Physiol. Behav. 91 (1), 154–160.

Merrick, J., Kandel, I., Birnbaum, L., Hyam, E., Press, J., Morad, M., 2004. Adolescent injury risk behavior. Int. J. Adolesc. Med. Health 16 (3), 207–213.

Monk, C.S., McClure, E.B., Nelson, E.E., Zarahn, E., Bilder, R.M., Leibenluft, E., et al., 2003. Adolescent immaturity in attention-related brain engagement to emotional facial expressions. Neuroimage 20 (1), 420–428.

Nguyen, T.V., McCracken, J., Ducharme, S., Botteron, K.N., Mahabir, M., Johnson, W., et al., 2013. Testosterone-related cortical maturation across childhood and adolescence. Cereb. Cortex 23 (6), 1424–1432.

Nunes, E.V., Levin, F.R., 2004. Treatment of depression in patients with alcohol or other drug dependence: a meta-analysis. JAMA 291 (15), 1887–1896.

Ouko, L.A., Shantikumar, K., Knezovich, J., Haycock, P., Schnugh, D.J., Ramsay, M., 2009. Effect of alcohol consumption on CpG methylation in the differentially methylated regions of H19 and IG-DMR in male gametes: implications for fetal alcohol spectrum disorders. Alcohol. Clin. Exp. Res. 33 (9), 1615–1627.

Piazza, P.V., Deminiere, J.M., Le Moal, M., Simon, H., 1989. Factors that predict individual vulnerability to amphetamine self-administration. Science 245 (4925), 1511–1513.

Piazza, P.V., Rouge-Pont, F., Deminiere, J.M., Kharoubi, M., Le Moal, M., Simon, H., 1991. Dopaminergic activity is reduced in the prefrontal cortex and increased in the nucleus accumbens of rats predisposed to develop amphetamine self-administration. Brain Res. 567 (1), 169–174.

Pingault, J.B., Cote, S.M., Galera, C., Genolini, C., Falissard, B., Vitaro, F., Tremblay, R.E., 2013. Childhood trajectories of inattention, hyperactivity and oppositional behaviors and prediction of substance abuse/dependence: a 15-year longitudinal population-based study. Mol. Psychiatry 18 (7), 806–812.

Plotsky, P.M., Meaney, M.J., 1993. Early, postnatal experience alters hypothalamic corticotropin-releasing factor (CRF) mRNA, median eminence CRF content and stress-induced release in adult rats. Brain Res. Mol. Brain Res. 18 (3), 195–200.

Pruessner, J.C., Champagne, F., Meaney, M.J., Dagher, A., 2004. Dopamine release in response to a psychological stress in humans and its relationship to early life maternal care: a positron emission tomography study using [11C]raclopride. J. Neurosci. 24 (11), 2825–2831.

Renthal, W., Nestler, E.J., 2008. Epigenetic mechanisms in drug addiction. Trends Mol. Med. 14 (8), 341–350.

Rohsenow, D.J., Niaura, R.S., Childress, A.R., Abrams, D.B., Monti, P.M., 1990. Cue reactivity in addictive behaviors: theoretical and treatment implications. Int. J. Addict. 25 (7A–8A), 957–993.

Romer, D., Betancourt, L., Giannetta, J.M., Brodsky, N.L., Farah, M., Hurt, H., 2009. Executive cognitive functions and impulsivity as correlates of risk taking and problem behavior in preadolescents. Neuropsychologia 47 (13), 2916–2926.

Rosenberg, D.R., Lewis, D.A., 1994. Changes in the dopaminergic innervation of monkey prefrontal cortex during late postnatal development: a tyrosine hydroxylase immunohistochemical study. Biol. Psychiatry 36 (4), 272–277.

SAMSHA, 2014. Results from the 2013 National Survey on Drug Use and Health: Summary of National Findings. Substance Abuse and Mental Health Services Administration, Rockville, MD.

Saunders, B.T., Robinson, T.E., 2010. A cocaine cue acts as an incentive stimulus in some but not others: implications for addiction. Biol. Psychiatry 67 (8), 730–736.

Saunders, B.T., Robinson, T.E., 2011. Individual variation in the motivational properties of cocaine. Neuropsychopharmacology 36 (8), 1668–1676.

Saunders, B.T., Richard, J.M., Janak, P.H., 2015. Contemporary approaches to neural circuit manipulation and mapping: focus on reward and addiction. Philos. Trans. R. Soc. Lond. B Biol. Sci. 370 (1677).

Schmaal, L., Goudriaan, A.E., van der Meer, J., van den Brink, W., Veltman, D.J., 2012. The association between cingulate cortex glutamate concentration and delay discounting is mediated by resting state functional connectivity. Brain Behav. 2 (5), 553–562.

Solinas, M., Thiriet, N., Chauvet, C., Jaber, M., 2010. Prevention and treatment of drug addiction by environmental enrichment. Prog. Neurobiol. 92 (4), 572–592.

Sowell, E.R., Peterson, B.S., Thompson, P.M., Welcome, S.E., Henkenius, A.L., Toga, A.W., 2003. Mapping cortical change across the human life span. Nat. Neurosci. 6 (3), 309–315.

Spear, L.P., 2000. The adolescent brain and age-related behavioral manifestations. Neurosci. Biobehav. Rev. 24 (4), 417–463.

Steinberg, L., 2008. A social neuroscience perspective on adolescent risk-taking. Dev. Rev. 28 (1), 78–106.

Stroud, L.R., Foster, E., Papandonatos, G.D., Handwerger, K., Granger, D.A., Kivlighan, K.T., Niaura, R., 2009. Stress response and the adolescent transition: performance versus peer rejection stressors. Dev. Psychopathol. 21 (1), 47–68.

Tang, T.Z., DeRubeis, R.J., Hollon, S.D., Amsterdam, J., Shelton, R., Schalet, B., 2009. Personality change during depression treatment: a placebo-controlled trial. Arch. Gen. Psychiatry 66 (12), 1322–1330.

Tarullo, A.R., Gunnar, M.R., 2006. Child maltreatment and the developing HPA axis. Horm. Behav. 50 (4), 632–639.

Terracciano, A., Lockenhoff, C.E., Crum, R.M., Bienvenu, O.J., Costa Jr., P.T., 2008. Five-Factor Model personality profiles of drug users. BMC Psychiatry 8, 22.

Toledo-Rodriguez, M., Lotfipour, S., Leonard, G., Perron, M., Richer, L., Veillette, S., et al., 2010. Maternal smoking during pregnancy is associated with epigenetic modifications of the brain-derived neurotrophic factor-6 exon in adolescent offspring. Am. J. Med. Genet. B Neuropsychiatr. Genet. 153B (7), 1350–1354.

Tsuang, M.T., Lyons, M.J., Meyer, J.M., Doyle, T., Eisen, S.A., Goldberg, J., et al., 1998. Co-occurrence of abuse of different drugs in men: the role of drug-specific and shared vulnerabilities. Arch. Gen. Psychiatry 55 (11), 967–972.

Vassoler, F.M., Sadri-Vakili, G., 2014. Mechanisms of transgenerational inheritance of addictive-like behaviors. Neuroscience 264, 198–206.

Vassoler, F.M., White, S.L., Schmidt, H.D., Sadri-Vakili, G., Pierce, R.C., 2013. Epigenetic inheritance of a cocaine-resistance phenotype. Nat. Neurosci. 16 (1), 42–47.

Verdejo-Garcia, A., Lawrence, A.J., Clark, L., 2008. Impulsivity as a vulnerability marker for substance-use disorders: review of findings from high-risk research, problem gamblers and genetic association studies. Neurosci. Biobehav. Rev. 32 (4), 777–810.

Volkow, N.D., Fowler, J.S., Wang, G.J., Hitzemann, R., Logan, J., Schlyer, D.J., et al., 1993. Decreased dopamine D2 receptor availability is associated with reduced frontal metabolism in cocaine abusers. Synapse 14 (2), 169–177.

Vrieze, S.I., McGue, M., Iacono, W.G., 2012. The interplay of genes and adolescent development in substance use disorders: leveraging findings from GWAS meta-analyses to test developmental hypotheses about nicotine consumption. Hum. Genet. 131 (6), 791–801.

Vyssotski, D.L., 2011. Transgenerational epigenetic compensation. Evolocus 1, 1–6.

Wang, G.J., Smith, L., Volkow, N.D., Telang, F., Logan, J., Tomasi, D., et al., 2012a. Decreased dopamine activity predicts relapse in methamphetamine abusers. Mol. Psychiatry 17 (9), 918–925.

Wang, J.C., Kapoor, M., Goate, A.M., 2012b. The genetics of substance dependence. Annu. Rev. Genomics Hum. Genet. 13, 241–261.

Yang, T.T., Menon, V., Reid, A.J., Gotlib, I.H., Reiss, A.L., 2003. Amygdalar activation associated with happy facial expressions in adolescents: a 3-T functional MRI study. J. Am. Acad. Child Adolesc. Psychiatry 42 (8), 979–985.

Yochum, C., Doherty-Lyon, S., Hoffman, C., Hossain, M.M., Zelikoff, J.T., Richardson, J.R., 2014. Prenatal cigarette smoke exposure causes hyperactivity and aggressive behavior: role of altered catecholamines and BDNF. Exp. Neurol. 254, 145–152.

Yohn, N.L., Bartolomei, M.S., Blendy, J.A., 2015. Multigenerational and transgenerational inheritance of drug exposure: The effects of alcohol, opiates, cocaine, marijuana, and nicotine. Prog. Biophys. Mol. Biol. 118 (1-2), 21–33.

Zald, D.H., Cowan, R.L., Riccardi, P., Baldwin, R.M., Ansari, M.S., Li, R., et al., 2008. Midbrain dopamine receptor availability is inversely associated with novelty-seeking traits in humans. J. Neurosci. 28 (53), 14372–14378.

Zhu, J., Lee, K.P., Spencer, T.J., Biederman, J., Bhide, P.G., 2014. Transgenerational transmission of hyperactivity in a mouse model of ADHD. J. Neurosci. 34 (8), 2768–2773.

CHAPTER

2

Drug-induced neurotoxicity in addiction medicine: From prevention to harm reduction

S. Mohammad Ahmadi Soleimani*,†, Hamed Ekhtiari*,‡,§, Jean Lud Cadet¶,1

**Neurocognitive Laboratory, Iranian National Center for Addiction Studies (INCAS), Tehran University of Medical Sciences, Tehran, Iran*

†Department of Physiology, Faculty of Medical Sciences, Tarbiat Modares University, Tehran, Iran

‡Translational Neuroscience Program, Institute for Cognitive Science Studies (ICSS), Tehran, Iran

§Research Center for Molecular and Cellular Imaging (RCMCI), Tehran University of Medical Sciences, Tehran, Iran

¶Molecular Neuropsychiatry Research Branch, DHHS/NIH/NIDA Intramural Research Program, National Institutes of Health, Baltimore, MD, USA

[1]Corresponing author: e-mail address: jcadet@intra.nida.nih.gov

Abstract

Neurotoxicity is considered as a major cause of neurodegenerative disorders. Most drugs of abuse have nonnegligible neurotoxic effects many of which are primarily mediated by several dopaminergic and glutamatergic neurotransmitter systems. Although many researchers have investigated the medical and cognitive consequences of drug abuse, the neurotoxicity induced by these drugs still requires comprehensive attention. The science of neurotoxicity promises to improve preventive and therapeutic strategies for brain disorders such as Alzheimer disease and Parkinson's disease. However, its clinical applications for addiction medicine remain to be defined adequately. This chapter reviews the most commonly discussed mechanisms underlying neurotoxicity induced by common drugs of abuse including amphetamines, cocaine, opiates, and alcohol. In addition, the known factors that trigger and/or predispose to drug-induced neurotoxicity are discussed. These factors include drug-related, individual-related, and environmental insults. Moreover, we introduce some of the potential pharmacological antineurotoxic interventions deduced from experimental animal studies. These interventions involve various targets such as dopaminergic system, mitochondria, cell death signaling, and NMDA receptors, among others. We conclude the chapter with a discussion of addicted patients who might benefit from such interventions.

Keywords

Neurotoxicity, Drugs of abuse, Neuroprotection, Addiction medicine

Progress in Brain Research, Volume 223, ISSN 0079-6123, http://dx.doi.org/10.1016/bs.pbr.2015.07.004

1 INTRODUCTION

Neurotoxicity is defined as any adverse effect on the structure or function of the central and peripheral nervous systems at the result of a diversity of biological, chemical, or physical agents. Based on the location or the severity of neurotoxic damages, these can be accompanied by neurocognitive deficits that impact various aspects of daily life activities. Although neurotoxicity is a well-known contributory factor in some neurodegenerative including Alzheimer's disease and Parkinson's disease, the impact of drug-induced damage in addicted patients remains to be fully evaluated. Nevertheless, the accumulated evidence suggests that drug-induced neurotoxicity is mediated by activation of several neurotransmitter systems including dopamine and glutamate that work in concert to damage the brain (Cadet et al., 2014). These drug-induced damages can negatively impact various neurocognitive spheres that include memory, attention, decision-making, and executive functions (Cadet and Bisagno, 2014). There is, at present, a burgeoning literature on the influence of these cognitive deficits on social and psychological functioning. Nevertheless, much remains to be done to provide a detailed hypothesis that might provide a link between these drug-induced brain changes and treatment responses. Here, we provide a brief review of the existing literature on the mechanisms of neurotoxicity consequent to commonly abused drugs (amphetamines, cocaine, opiates, and alcohol). The chapter also discusses identified predisposing factors, potentials for preventive, and therapeutic approaches, as well as future challenges to apply basic science knowledge of drug-induced neurotoxic damage to clinical practice.

2 DRUG-INDUCED NEUROTOXICITY: MECHANISMS AND PATHWAYS

During the past three decades, the efforts of several groups of investigators have led to the identification of several cellular and molecular mechanisms of drug neurotoxicity. This chapter presents the bases of toxicity produced by amphetamine, amphetamine derivatives, cocaine, and opiates.

2.1 OXIDATIVE STRESS

The increase in extracellular monoamines caused by drugs of abuse is thought to be responsible for their addictive properties. Importantly, however, the increased dopamine (DA) in the synaptic cleft might also be responsible for the neurotoxic damage caused by several of these agents (Cadet and Brannock, 1998; Cadet et al., 2007). In fact, this might provide a partial explanation for the original report of methamphetamine-induced toxicity in brain regions with high monoaminergic content (Gibb and Kogan, 1979). Dopamine by itself is neurotoxic both *in vitro* and *in vivo* (Graham et al., 1978). It is easily oxidized via enzymatic and nonenzymatic mechanisms and then induces oxidative stress (Cadet and Brannock, 1998). Amphetamine,

amphetamine derivatives, cocaine, 3,4-methylenedioxy-methamphetamine (MDMA), and opiates have all been reported to produce oxidative stress within the nervous system (Yamamoto and Bankson, 2005). Active metabolites of dopamine and/or related substances might cause oxidative stress by forming free radicals via the formation of quinones and the generation of quinone cascades secondary to MDMA metabolism (Lyles and Cadet, 2003). Cocaine exposure causes oxidative stress by increasing H_2O_2 concentration and decreasing catalase activity in rat prefrontal cortex and striatum (Dietrich et al., 2005; Macêdo et al., 2005). Cocaine also causes decreased levels of antioxidants such as glutathione (GSH) or vitamin E (Lipton et al., 2003; Poon et al., 2007). In contrast to the situation for the psychostimulants, much less is known about opiate-induced oxidative stress. However, heroin has been reported to decrease the activities of superoxide dismutase (SOD), catalase, and glutathione peroxidase (GPx) in the mouse brain. Heroin exposure is reported to increase oxidative DNA damage, protein oxidation, and lipid peroxidation (Qiusheng et al., 2005; Xu et al., 2006). Finally, morphine was shown to reduce fatty acid contents in spinal cord and brain by causing oxidative stress (Ozmen et al., 2007).

2.2 APOPTOTIC PROCESSES

There is convincing evidence that some drugs of abuse can cause neuronal apoptotic cell death. Cells undergoing apoptosis are characterized by morphological and biochemical hallmarks that include cell shrinkage, chromatin condensation, and fragmentation into membrane-bound apoptotic bodies. Cell death is triggered by intrinsic and extrinsic molecular pathways that include increased permeability of mitochondrial membrane and activation of death receptors (Jayanthi et al., 2005). Death pathways also involve activation of cysteine aspartic proteases (caspases) and caspase-independent pathways (Kroemer and Martin, 2005). Experiments published in the Cadet laboratory were among the first ones to show that amphetamine and amphetamine derivatives could induce apoptosis *in vitro* and *in vivo* models (Cadet et al., 2007). These observations have been extensively replicated (Cunha-Oliveira et al., 2006; Dietrich et al., 2005; Oliveira et al., 2002). Amphetamine exposure leads to caspase activation in various brain regions (Cunha-Oliveira et al., 2006; Krasnova et al., 2005; Waren et al., 2007). Amphetamine exposure stimulates mitochondrial pathways that lead to caspase activation. Mitochondria-dependent death pathways involve the release of cytochrome *c*, decrease in mitochondrial potential, and increased Bax/Bcl2 ratios (Imam et al., 2005; Krasnova et al., 2005; Oliveira et al., 2003). Other studies have identified p53 as an important regulator of D-amphetamine-induced cell death (Krasnova et al., 2005). MDMA can also induce apoptosis in rat cortical neurons by activation of 5-HT2A receptors (Capela et al., 2006). Exposure to cocaine can also activates biochemical mechanisms involved in apoptosis without leading to morphological apoptotic characteristics (Cunha-Oliveira et al., 2006; Dey et al., 2007; Imam et al., 2005; Mitchell and Snyder-Keller, 2003; Oliveira et al., 2003). Interestingly, cocaine produces apoptosis in human neuronal progenitor cells by generating oxidative stress (Poon et al., 2007). Opiates may also cause apoptosis in humans and in animal models (Cunha-Oliveira et al., 2007; Hu et al., 2002; Mao et al., 2002; Tramullas et al., 2008). Heroin and morphine cause caspase activation and cytochrome *c* release from mitochondria (Cunha-Oliveira et al., 2007; Oliveira et al., 2003) as well as increased Bax/Bcl2 ratios (Cunha-Oliveira et al., 2007; Mao et al., 2002). Chronic heroin exposure upregulates proapoptotic proteins (Fas, FasL, and Bad) in the cortex and hippocampus of mice (Tramullas et al., 2008). Morphological hallmarks of apoptosis have also been observed *in vitro* following exposure to heroin (Cunha-Oliveira et al., 2007; Oliveira et al., 2002).

2.3 EXCITOTOXICITY

Excitotoxicity refers to cell death due to the toxic effects of excitatory amino acids. This happens at the result of massive Ca^{2+} influx secondary to the overactivation of *N*-methyl-D-aspartate (NMDA) glutamate receptors. Methamphetamine induces excitotoxicity by glutamate release and activation of glutamate receptors (Yamamoto and Bankson, 2005). Administration of glutamate receptor antagonists including MK-801 or dizocilpine reduces methamphetamine-induced neurodegeneration in different parts of the brain (Battaglia et al., 2002; Bowyer et al., 2001; Chipana et al., 2008; Fuller et al., 1992; Gołembiowska et al., 2003; Ohmori et al., 1993; Sonsalla et al., 1989; Weihmuller et al., 1992). The neurotoxic effects of opiates may also be mediated by activation of NMDA receptors (Mao et al., 2002). Crack abuse may also lead to excitotoxic damage (Oliveira et al., 2011). Amphetamine (Reid et al., 1997; Wolf et al., 2000) and cocaine (Williams and Steketee, 2004) both increase extracellular glutamate concentrations in the nucleus accumbens, ventral tegmental area (VTA), striatum, and prefrontal cortex. In addition, long-term cocaine exposure also influences glutamate functions in the VTA and nucleus accumbens. These alterations include changes in synaptic plasticity (i.e., increasing the number of dendritic spines), changes in glutamate homeostasis, and activation of postsynaptic glutamatergic signaling (Uys and Reissner, 2011). In addition, cocaine increases intracellular Ca^{2+} concentration in rat cortical neurons (Cunha-Oliveira et al., 2010). This leads to the activation of several Ca^{2+}-dependent enzymes that cause degradation of proteins, phospholipids, and nucleic acids (Rego and Oliveira, 2003). The adverse effects of alcohol may also involve hyperexcitability during the process of alcohol withdrawal. This increase in glutamatergic transmission may result from a combination of changes including increased NMDA receptor activation, decreased GABA receptor activation, and enhanced function of voltage-activated calcium channels (Dolin et al., 1987; Koppi et al., 1987; Little et al., 1986; Lovinger, 1993; Skattebol and Rabin, 1987). Another important aspect of alcohol withdrawal is thiamine deficiency (Martin et al., 1991). This vitamin acts as a cofactor in several enzymatic reactions. In animal models, severe thiamine deficiency causes neurological symptoms such as convulsions. There is also evidence supporting the link between excitotoxicity and thiamine deficiency (Langlais and Mair, 1990). Specifically, thiamine deficiency-induced neuronal loss and convulsions are diminished by the administration of the NMDA receptor antagonist, MK-801, in experimental animals.

2.4 INVOLVEMENT OF OTHER BIOCHEMICAL MECHANISMS

In addition to the mentioned mechanisms, other biochemical pathways may also serve as triggers of drug-induced neurotoxicity. For example, activation of microglia can lead to the release of proinflammatory mediators that may compromise neuronal viability (Domercq and Matute, 2004). In the case of drug toxicity, methamphetamine exposure leads to microglial activation that appears in conjunction with dopaminergic toxicity in the dorsal striatum (Bowyer et al., 1994; Escubedo et al., 1998; Guilarte et al., 2003; Thomas and Kuhn, 2005; Thomas et al., 2004a,b). Importantly, the time course of methamphetamine-induced microglial activation appears to coincide or to precede methamphetamine toxicity, supporting the notion of the involvement of microglial cells in methamphetamine toxicity (Thomas et al., 2004b). Of clinical relevance is the fact human methamphetamine addicts show widespread microglial activation in their brains (Sekine et al., 2008).

Hyperthermia is another proposed mechanism for methamphetamine neurotoxicity both in humans (Kalant and Kalant, 1975) and rodents (Sandoval et al., 2000). Hyperthermia may potentiate drug-induced dopamine and tyrosine hydroxylase depletion by increasing oxidative stress (Lin et al., 1991; Omar et al., 1987). In general, biochemical reactions are sensitive to temperature changes including those occurring in the brain. An additional organelle that is involved in methamphetamine toxicity is the endoplasmic reticulum (ER) (Jayanthi et al., 2004, 2009). Methamphetamine-induced ER stress is thought to be the earliest factor leading to apoptosis in the mouse brain after drug administration. Specifically, methamphetamine has been shown to cause neuronal apoptosis through cross talks between ER and mitochondria-mediated death cascades. This cross talk triggers both caspase-dependent and -independent death pathways (Jayanthi et al., 2004) and appears to depend on activation of DA D1 receptors (Jayanthi et al., 2009).

In the case of cocaine, administration of the drug produces increased synaptic serotonin levels and changes in serotonin transporters (Cunningham et al., 1992; Levy et al., 1994). Increased brain concentrations of serotonin can disrupt the blood–brain barrier (BBB) (Sharma et al., 1990) and can cause hyperthermia (Capela et al., 2009; Sharma, 2007). These findings are consistent with reports of psychostimulant-induced hyperthermia (Hawkins and Davis, 2005; Hawkins et al., 2004; Kousik et al., 2011; Lin et al., 1992; Monks et al., 2004; Sharma and Ali, 2008). The issue of hyperthermia as adverse consequences of drug abuse is of clinical relevance because they can impact the clinical course of patients who present with drug intoxication after either suicidal attempts or accidental overdoses.

3 DRUG-INDUCED NEUROTOXICITY: TRIGGERING AND SUSCEPTIBILITY FACTORS

The neurotoxicity induced by drugs of abuse is primarily mediated by alterations in several neurotransmitter systems. However, the severity of these neurotoxic effects may be significantly affected by a variety of other factors as described below (Fig. 1).

3.1 DRUG-RELATED FACTORS

3.1.1 Active metabolites and adulterants

Neurotoxicity induced by drugs of abuse is influenced by the production of metabolites that can cross the BBB. For example, the metabolism of cocaine results in the production of neurotoxic compounds including benzoylecgonine, norcocaine, and cocaethylene that have their own toxicity profiles (Milhazes et al., 2006; Nassogne et al., 1998). Metabolism of the amphetamines can produce other active metabolites that are known to impact neurotransmitter release or reuptake (Smoluch et al., 2014). Heroin is metabolized to 6-monoacetylmorphine and morphine with potential neurotoxic consequences (Hu et al., 2002; Mao et al., 2002). Adulterants also play a role in drug-induced neurotoxicity including the toxicity of heroin that produces more toxicity in PC12 cells depending on the level of the purity of drugs available to drug addicts (Oliveira et al., 2002). Highly purified heroin produces less caspase activation than less pure heroin (Cunha-Oliveira et al., 2007).

3.1.2 Polydrug abuse

The use of multiple drugs by addicts can also impact their clinical presentation and the adverse consequences of the drugs used by these patients. In addition to their drugs of choice (primary drug), addicts may use other substance to potentiate or attenuate the behavioral of the primary drug. There are various patterns of polydrug

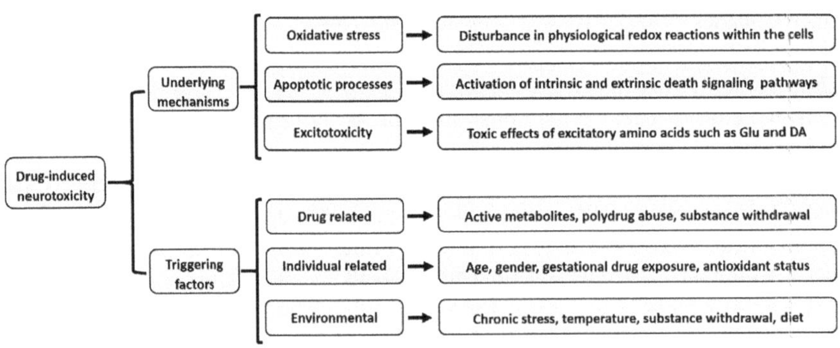

FIGURE 1

Drug-induced neurotoxicity at a glance: underlying mechanisms and triggering factors. Glu and DA represent glutamate and dopamine, respectively.

abuse (Connor et al., 2014) and the resulting effects of these combinations depend on the biochemical cascades that are impacted by each one. For example, benzodiazepines are known to influence the pharmacokinetics of opioids (Jones et al., 2012). Diazepam acts as both a noncompetitive inhibitor of methadone metabolism (Jones et al., 2012) and a competitive inhibitor of hepatic *N*-demethylation of methadone (Jones et al., 2012). These interactions increase methadone concentration in brain tissues and may therefore increase the neurotoxic profile of methadone. Mephedrone, a methamphetamine analog, does not seem to cause neurotoxicity by itself but increases the neurotoxicity of other drugs of abuse including methamphetamine, amphetamine, and MDMA.

3.1.3 Substance withdrawal

Long-term METH withdrawal sensitizes NMDA receptors to agonist exposure (Smith et al., 2008). Mechanistically, METH withdrawal decreases Mg^{2+} blockade of NMDA receptors and results in increased excitatory postsynaptic potentials (Moriguchi et al., 2002), a phenomenon that may potentiate NMDA-induced neurotoxicity. Ethanol withdrawal also leads to neuronal hyperexcitability that manifests as seizures during various intervals of alcohol withdrawal (Hoffman and Tabakoff, 1994; Lovinger, 1993).

3.2 ENVIRONMENTAL FACTORS

3.2.1 Chronic stress

Chronic stress can alter the neurochemical responses to drugs of abuse. For example, drug-induced dopamine release increases in several brain regions following preexposure to stress (Hamamura and Fibiger, 1993; Kalivas and Duffy, 1989; Rouge-Pont et al., 1995). Stressful events enhance rats tendency to self-administer drugs of abuse (Covington and Miczek, 2001; Piazza and Le Moal, 1998), thereby increasing potential risks of drug-induced neurotoxicity. Chronic stress also increases the hyperthermic response to methamphetamine (Tata and Yamamoto, 2008). These data, taken together, suggest long-term stress can potentiate the vulnerability of brain cells to the neurotoxic effects of psychostimulants (Matuszewich and Yamamoto, 2004).

3.2.2 Ambient temperature

The neurotoxicity induced by several drugs of abuse including amphetamine, methamphetamine, and 3,4-MDMA are affected by environmental temperature (Bowyer and Holson, 1995; Bowyer et al., 2001; Miller and O'Callaghan, 1994). Even relatively small variations in ambient temperature can significantly impact neurotoxicity caused by the amphetamines. Specifically, it has been suggested that increasing the environmental temperature following D-methamphetamine abuse is equivalent to increasing the dose of the drug (Miller and O'Callaghan, 2003). Similar effects of environmental temperature have been reported for MDMA. In contrast, lowering environmental temperature can provide substantial degree of protection (Johnson et al., 2000). This phenomenon might play a significant role in clinical emergencies reported in places where young drug abusers meet to dance and take amphetamine-like compounds (Chadwick et al., 1991; Randall, 1992).

3.2.3 Diet and nutritional supplies

Nutritional deficiencies may also impact drug toxicity. For example, selenium deficiency potentiates methamphetamine-induced depletion of tyrosine hydroxylase immunoreactivity, DA, and its metabolites (Kim et al., 2000). Vitamin E deficiency also enhances susceptibility to the neurotoxicity induced by D-MDMA in mice (Johnson et al., 2002). Thiamine deficiency produces mitochondrial dysfunction, glutamate excitotoxicity, and oxidative stress in different parts of the brain (Todd and Butterworth, 1999, 2001). This is important because chronic alcoholic patients commonly suffer from thiamine deficiency (Kopelman et al., 2009; Victor et al., 1989) due, in part, to the fact that alcohol interferes with the intestinal absorption of dietary nutrients.

3.3 INDIVIDUAL-RELATED FACTORS

3.3.1 The role of age

Drug-induced neurotoxicity varies in severity according to age (Teuchert-Noodt and Dawirs, 1991). For example, brain amphetamine levels in old rats are twice as high as the levels in young ones (Truex and Schmidt, 1980). In fact, older rats experience greater methamphetamine neurotoxicity than younger animals (Krasnova and Cadet, 2009). Older mice experience methamphetamine toxicity even after low doses of the drug, whereas younger rodents show very little or no toxicity even at higher doses (Miller et al., 2000).

3.3.2 The role of gender

A number of animal studies have reported that methamphetamine induces greater neurotoxicity in males than females (Miller et al., 1998). MDMA also causes greater lethality in male mice (Miller and O'Callaghan, 1995). In contrast, women are more susceptible than men to the possible complications of alcohol abuse (Alfonso-Loeches et al., 2013).

3.3.3 Gestational drug exposure

Prenatal exposure to methamphetamine increases the risk of neurotoxicity in offspring (Heller et al., 2001). The mechanism by which prenatal methamphetamine exposure could potentiate drug-induced neurotoxicity is not well understood. Because methamphetamine toxicity is dependent on the functions of DAT and VMAT-2 (Fumagalli et al., 1999), prenatal exposure to methamphetamine may decrease the ability of DAT VMAT-2 to maintain DA homeostasis in DA axon terminals (Heller et al., 2001). Much remains to be done on this subject.

3.3.4 Antioxidant status

Cells contain various biochemical agents that serve to protect them against the toxic effects of oxygen and it metabolites (Cadet and Brannock, 1998). These antioxidants include GPx, catalase, and SOD that protect against the toxicity of hydrogen peroxide and superoxide radicals (Cadet and Brannock, 1998). Of relevance to this discussion, mice deficient in GPx are more susceptible to the adverse effects of neurotoxins (Zhang et al., 2000). Interestingly, neurons that survive in neurodegenerative diseases express high concentration of SOD (Browne et al., 1999). The importance of the balance between toxic prooxidants and innate antioxidant defense mechanisms has been tested by genetic elimination and augmentation of these pathways. Downregulation of Cu/Zn-SOD increases neuronal death both *in vivo* and *in vitro* (Kondo et al., 1997; Troy et al., 1996). Also, deficiency in α-tocopherol (vitamin E) transport protein produces neurodegeneration (Yokota et al., 2001). Increasing the expression of SOD and GPx as well as using SOD mimetics has been reported to be neuroprotective (Pineda et al., 2001; Pong et al., 2000). Importantly, mice with high levels of CuZn-SOD are protected against the toxicity of methamphetamine and MDMA (Cadet et al., 1994a,b).

4 DRUG-INDUCED NEUROTOXICITY: POTENTIAL PREVENTIVE STRATEGIES

Although mechanisms underlying drug-induced neurotoxic effects are not perfectly understood, pharmacologic approaches have been proposed for their prevention Table 1.

Table 1 Potential Pharmacologic Interventions to Prevent Drug-Induced Neurotoxicity

Antineurotoxic Interventions	Pharmacologic Agents	Drug Name
Modulating dopamine system	Dopamine receptor agonists	Pramipexole
	Dopamine receptor antagonists	Eticlopride
Addressing oxidative challenge	Artificial antioxidants	*N*-acetyl-L-cysteine (NAC)
	Natural antioxidants	Ascorbic acid (vitamin C), vitamin E
NMDA receptor blockade	NMDA receptor antagonists	Memantine, ketamine
Antiapoptotic approach	Agents with antiapoptotic property	Calpastatin, minocycline
Drug rotation approach	Opioids	Methadone, morphine, hydromorphone
Anti-inflammatory approach	COX inhibitors	Ketoprofen, indomethacin
Thermoregulatory interventions	Barbiturates	Phenobarbital
	Benzodiazepines	Diazepam

4.1 MODULATING BRAIN DOPAMINE LEVELS

As mentioned above, dopamine plays a pivotal role in mediating methamphetamine-induced neurotoxicity by causing production of dopamine-related reactive oxygen species (ROS) and oxidative stress. Agents decreasing brain dopamine levels such as tyrosine hydroxylase inhibitor and α-methyl-*p*-tyrosine have indeed shown to exert protective effects against the neurotoxicity induced by methamphetamine in striatal dopaminergic axons (Axt et al., 1990; Gibb and Kogan, 1979; Hotchkiss and Gibb, 1980; Schmidt and Gibb, 1985; Thomas et al., 2008). Pramipexole, a dopamine D_2/D_3 receptor agonist, may cause neuroprotection against methamphetamine-induced toxicity by reducing dopamine turnover by stimulation of presynaptic dopamine receptors or by increasing antioxidant and trophic properties of the brain (Hall et al., 1996).

4.2 ADDRESSING OXIDATIVE CHALLENGE

Pretreatment with antioxidants such as *N*-acetyl-L-cysteine, ascorbic acid, and vitamin E can protect against psychostimulant-induced neurotoxicity (De Vito and Wagner, 1989; Fukami et al., 2004; Hashimoto et al., 2004; Wagner et al., 1985). Increasing mitochondrial energy metabolism through pre- and posttreatment of mice with L-carnitine that coordinates beta-oxidation in mitochondria C significantly attenuates methamphetamine-induced production of the neurotoxin, 3-nitropropionic acid in the striatum (Virmani et al., 2002). Formation of peroxynitrite production can be inhibited by pretreatment with some selective antioxidants (selenium and melatonin), several peroxynitrite decomposition catalysts, and selective neuronal nitric oxide synthase inhibitors (Imam et al., 2001). Vitamin D has also been shown to exert protection against methamphetamine toxicity (Cass et al., 2006). This is thought to be mediated by upregulation of glial cell line-derived neurotrophic factor (Cass et al., 2006). Vitamin D also enhances glutathione levels and suppresses the production of inducible nitric oxide synthase (Cass et al., 2006).

4.3 ANTIAPOPTOTIC APPROACH

Preventing the activation of apoptotic processes may also be an effective approach to protect against drug-induced neurotoxicity. For example, administration of the dopamine type 1 receptor antagonist (SCH23390) attenuates the activation of Fas-mediated cell death (Jayanthi et al., 2005). Melatonin, working as a direct free radical scavenger, was shown to protect against methamphetamine-induced cell death (Wisessmith et al., 2009). Melatonin reverses the methamphetamine-induced decrease in mitochondrial function and phosphorylation of tyrosine hydroxylase in dopaminergic-cultured cells (Suwanjang et al., 2010). It reduces induction of Bax, caspase, and cell death in these neurons (Suwanjang et al., 2010). Desipramine which is a monoamine uptake inhibitor that prevents methamphetamine toxicity (Wisessmith et al., 2009). Calpastatin, an endogenous protease inhibitor, was also shown to reverse methamphetamine-induced activation of death pathways in dopaminergic cell lines (Chetsawang et al., 2012; Suwanjang et al., 2012).

4.4 NMDA RECEPTOR ANTAGONISM

Antagonism of NMDA receptors with ketamine or modulation of glutamate transporter activity in spinal cord was shown to prevent opiate-mediated neurotoxicity (Bruera and Kim, 2003). Memantine is another NMDA receptor antagonist with well-known neuroprotective properties (Turski et al., 1991). Memantine is thought to prevent the cellular damage following activation of NMDA receptors by glutamate. This drug has also been approved in Europe as a therapeutic agent for moderately severe to severe Alzheimer's disease (Doraiswamy, 2002). In addition, antagonism of metabotropic glutamate receptor 5 (mGluR5) has been shown to prevent the degeneration of dopaminergic neurons induced by methamphetamine in rats (Gołembiowska et al., 2003).

4.5 ROTATION IN DRUGS

Opioid rotation refers to a shift from one opioid to another with the aim of improving therapeutic effectiveness or reducing adverse effects (Quigley, 2004; Thomsen et al., 1999). It is a well-accepted clinical method for decreasing drug-induced neurotoxic effects. Using equipotent therapeutic and nontoxic doses of other opioids can reduce signs and symptoms of opioid toxicity. Previous studies propose that a variety of two or three opioids are essential to reach satisfactory long-term effectiveness. Best results are obtained using morphine, hydromorphone, and methadone in majority of cases (de Stoutz et al., 1995). For example, in a patient suffering from morphine-induced neurotoxicity, it was observed that rotation to methadone, which is an opioid with NMDA antagonistic properties, significantly reduces the morphine neurotoxic effects (Tarumi et al., 2002). The mechanistic rationale for this approach is that opioid metabolites are involved in the development of opioid-induced neurotoxicity, and opioid rotation may allow for clearance of toxic metabolites while the analgesic effect is maintained (de Stoutz et al., 1995).

4.6 ANTI-INFLAMMATORY APPROACH

Neuroinflammatory processes have been reported to be involved in neurotoxicity induced by methamphetamine treatment. Cyclooxygenase (COX) is one of the main inflammatory mediators that act as the rate-limiting enzyme in prostaglandin biosynthesis. In recent years, there has been an increased interest in use of COX inhibitors as a therapeutic approach to protect against neurodegeneration (Etminan et al., 2003; Gasparini et al., 2004; Hoffmann, 2000; Mhatre et al., 2004). In this regard, several anti-inflammatory agents including ketoprofen and indomethacin protect against methamphetamine-induced microgliosis and neurotoxicity (Asanuma et al., 2003, 2004). In contrast, a recent study has suggested that COX-2-containing cells appear to undergo damage during the early stages of methamphetamine-induced neurotoxicity and that the selective inhibition of this enzyme may actually be detrimental rather than protective after exposure to toxic doses of methamphetamine (Zhang et al., 2007). One possible reason for these discrepancies may be related to differences in methamphetamine metabolism among different animal species since the latter study was conducted in mice. Thus, it is important to conduct toxicity studies in rodents that have more similar metabolic pathways to those observed in humans (Caldwell et al., 1972; Yanagisawa et al., 1997).

4.7 THERMOREGULATORY INTERVENTIONS

Hyperthermia is considered to be an influencing factor in mediating methamphetamine neurotoxicity by facilitating ROS production and increasing dopamine oxidation. Increase in body temperature can be attenuated by administration of dopamine receptor antagonists (Albers and Sonsalla, 1995; Broening et al., 2005; He et al., 2004). L-Lobeline, a nicotinic receptor ligand, has both temperature-dependent and -independent neuroprotective effects against methamphetamine neurotoxicity. These protective effects may be secondary to the fact that lobeline attenuates methamphetamine-induced changes in dopamine release, hyperthermia, and the long-term depletion of striatal dopamine and 5-HT content (Eyerman and Yamamoto, 2005). There is also an evidence indicating that keeping animals in low environmental temperatures or pretreatment with pharmacologic agents that produce hypothermia such as MK-801, diazepam, and phenobarbital reduces METH neurotoxicity (Ali et al., 1994).

5 DRUG-INDUCED NEUROTOXICITY IN ADDICTION MEDICINE

Different populations with substance use disorders could be potential targets for neuroprotective interventions.

5.1 TREATMENT SEEKERS

Complete abstinence is not a reasonable and achievable goal for the first few weeks of treatment (Shoptaw et al., 1994). High levels of compliance for medications and dietary supplements may facilitate therapeutic plans for neuroprotection against lapse-induced toxicity.

5.2 NONTREATMENT SEEKERS

There is a long-term interval between initiation of drug use, progression to substance use disorder, and seeking treatment (Power et al., 1992). This interval could be a critical period to reduce harm of neurotoxicity by different educational and pharmacologic interventions.

5.3 RELAPSE-PRONE PATIENTS

Patients who receive residential abstinence-based treatment programs for a period of time are vulnerable to lapses after their discharges from residential centers (Arbour et al., 2011). A period of abstinence reduces the natural barriers against neurotoxicity and reexposure to illicit drugs could have serious negative effects.

5.4 INTOXICATED PATIENTS DURING OVERDOSE OR BINGE EPISODES

Large doses of illicit drugs during overdose or binge drug use could activate multiple pathways of neurotoxicity. Considering pharmacologic and nonpharmacologic (such as reducing core body temperature) neuroprotective interventions within drug overdose management protocols could reduce harm and long-lasting brain sequelae (Rolland et al., 2011).

5.5 CLIENTS SUFFERING FROM SEVERE DRUG WITHDRAWAL SYMPTOMS

Approaches to treat potential neurological damage during alcohol withdrawal syndrome are often considered during the treatment of these patients (Adinoff, 1994). However, neuroprotective approaches to other addicted patient populations are often neglected. This needs to be remedied by development of medications that address not only drug self-administration but also withdrawal-associated damage to the brain.

6 CONCLUSION AND FUTURE DIRECTIONS

This review has focused on the published literature dealing with the toxic effects of various licit and illicit drugs. This chapter suggests that the development of agents that only address self-administration aspects of drug addiction may not be sufficient to reduce neuropathological complications of these drugs. In reality, the clinical course of addicted patients is intimately linked to the neurological functioning of these patients and depends on the abused drugs in question. Therefore, the addition of neuroprotective agents in conjunction with antiaddictive therapies is warranted in cases such as methamphetamine addiction that is accompanied by the development of Parkinsonism in some older patients. Nevertheless, because drugs of abuse appear to exert their neurotoxic effects through distinct molecular pathways, understanding the precise biochemical substrates for each agent is of paramount importance. Future studies are needed to develop strategies that might improve the recovery of brain systems affected by repeated exposure to substances of abuse.

ACKNOWLEDGMENTS

This chapter was supported, in part, by funds of the NIDA Intramural Research Program (J.L.C.). H.E. has received funds from Tehran University of Medical Sciences for his contribution in this chapter.

REFERENCES

Adinoff, B., 1994. The alcohol withdrawal syndrome. Neurobiology of treatment and toxicity. Am. J. Addict. 3, 277–288.

Albers, D.S., Sonsalla, P.K., 1995. Methamphetamine-induced hyperthermia and dopaminergic neurotoxicity in mice: pharmacological profile of protective and nonprotective agents. J. Pharmacol. Exp. Ther. 275 (3), 1104–1114.

Alfonso-Loeches, S., Pascual, M., Guerri, C., 2013. Gender differences in alcohol-induced neurotoxicity and brain damage. Toxicology 311 (1–2), 27–34.

Ali, S.F., et al., 1994. Low environmental temperatures or pharmacologic agents that produce hypothermia decrease methamphetamine neurotoxicity in mice. Brain Res. 658 (1–2), 33–38.

Arbour, S., Hambley, J., Ho, V., 2011. Predictors and outcome of aftercare participation of alcohol and drug users completing residential treatment. Subst. Use Misuse 46 (10), 1275–1287.

Asanuma, M., et al., 2003. Methamphetamine-induced neurotoxicity in mouse brain is attenuated by ketoprofen, a non-steroidal anti-inflammatory drug. Neurosci. Lett. 352 (1), 13–16.

Asanuma, M., et al., 2004. Specific gene expression and possible involvement of inflammation in methamphetamine-induced neurotoxicity. Ann. N. Y. Acad. Sci. 1025, 69–75.

Axt, K.J., et al., 1990. alpha-Methyl-p-tyrosine pretreatment partially prevents methamphetamine-induced endogenous neurotoxin formation. Brain Res. 515 (1–2), 269–276.

Battaglia, G., Fornai, F., Busceti, C., 2002. Selective blockade of mGlu5 metabotropic glutamate receptors is protective against methamphetamine neurotoxicity. J. Neurosci. 22 (6), 2135–2141. Available at: http://www.jneurosci.org/content/22/6/2135.short. accessed March 27, 2015.

Bowyer, J.F., Holson, R.R., 1995. Methamphetamine and amphetamine neurotoxicity. In: Chang, L.W., Dyer, R.S. (Eds.), Handbook of Neurotoxicology. Marcel Dekker, New York.

Bowyer, J.F., et al., 1994. Further studies of the role of hyperthermia in methamphetamine neurotoxicity. J. Pharmacol. Exp. Ther. 268 (3), 1571–1580.

Bowyer, J.F., et al., 2001. Phenobarbital and dizocilpine can block methamphetamine-induced neurotoxicity in mice by mechanisms that are independent of thermoregulation. Brain Res. 919 (1), 179–183.

Broening, H.W., Morford, L.L., Vorhees, C.V., 2005. Interactions of dopamine D1 and D2 receptor antagonists with D-methamphetamine-induced hyperthermia and striatal dopamine and serotonin reductions. Synapse 56 (2), 84–93.

Browne, S.E., Ferrante, R.J., Beal, M.F., 1999. Oxidative stress in Huntington's disease. Brain Pathol. (Zurich, Switzerland) 9 (1), 147–163.

Bruera, E., Kim, H.N., 2003. Cancer pain. JAMA 290 (18), 2476–2479.

Cadet, J.L., Bisagno, V., 2014. Glial-neuronal ensembles: partners in drug addiction-associated synaptic plasticity. Front. Pharmacol. 5, 204.

Cadet, J.L., Brannock, C., 1998. Free radicals and the pathobiology of brain dopamine systems. Neurochem. Int. 32 (2), 117–131.

Cadet, J.L., Sheng, P., et al., 1994a. Attenuation of methamphetamine-induced neurotoxicity in copper/zinc superoxide dismutase transgenic mice. J. Neurochem. 62 (1), 380–383.

Cadet, J.L., Ali, S., Epstein, C., 1994b. Involvement of oxygen-based radicals in methamphetamine-induced neurotoxicity: evidence from the use of CuZnSOD transgenic mice. Ann. N. Y. Acad. Sci. 738, 388–391.

Cadet, J.L., Krasnova, I.N., Jayanthi, S., Lyles, J., 2007. Neurotoxicity of substituted amphetamines: molecular and cellular mechanisms. Neurotox. Res. 11 (3–4), 183–202.

Cadet, J.L., Bisagno, V., Milroy, C.M., 2014. Neuropathology of substance use disorders. Acta Neuropathol. 127 (1), 91–107.

Caldwell, J., Dring, L.G., Williams, R.T., 1972. Metabolism of (14 C)methamphetamine in man, the guinea pig and the rat. Biochem. J. 129 (1), 11–22.

Capela, J.P., Ruscher, K., Lautenschlager, M., Freyer, D., Dirnagl, U., Gaio, A.R., Bastos, M.L., Meisel, A., Carvalho, F., 2006. Ecstasy-induced cell death in cortical neuronal cultures is serotonin 2A-receptor-dependent and potentiated under hyperthermia. Neuroscience 139 (3), 1069–1081.

Capela, J.P., et al., 2009. Molecular and cellular mechanisms of ecstasy-induced neurotoxicity: an overview. Mol. Neurobiol. 39 (3), 210–271.

Cass, W.A., Smith, M.P., Peters, L.E., 2006. Calcitriol protects against the dopamine- and serotonin-depleting effects of neurotoxic doses of methamphetamine. Ann. N. Y. Acad. Sci. 1074, 261–271.

Chadwick, I.S., et al., 1991. Ecstasy, 3-4 methylenedioxymethamphetamine (MDMA), a fatality associated with coagulopathy and hyperthermia. J. R. Soc. Med. 84 (6), 371.

Chetsawang, J., et al., 2012. Calpastatin reduces methamphetamine-induced induction in c-Jun phosphorylation, Bax and cell death in neuroblastoma SH-SY5Y cells. Neurosci. Lett. 506 (1), 7–11.

Chipana, C., Torres, I., Camarasa, J., Pubill, D., Escubedo, E., 2008. Memantine protects against amphetamine derivatives-induced neurotoxic damage in rodents. Neuropharmacology 54 (8), 1254–1263.

Connor, J.P., et al., 2014. Polysubstance use: diagnostic challenges, patterns of use and health. Curr. Opin. Psychiatry 27 (4), 269–275.

Covington 3rd., H.E., Miczek, K.A., 2001. Repeated social-defeat stress, cocaine or morphine. Effects on behavioral sensitization and intravenous cocaine self-administration "binges". Psychopharmacology 158 (4), 388–398.

Cunha-Oliveira, T., Rego, A.C., Cardoso, S.M., Borges, F., Swerdlow, R.H., Macedo, T., de Oliveira, C.R., 2006. Mitochondrial dysfunction and caspase activation in rat cortical neurons treated with cocaine or amphetamine. Brain Res. 1089 (1), 44–54.

Cunha-Oliveira, T., et al., 2007. Street heroin induces mitochondrial dysfunction and apoptosis in rat cortical neurons. J. Neurochem. 101 (2), 543–554. Available at: http://www.ncbi.nlm.nih.gov/pubmed/17250679.

Cunha-Oliveira, T., et al., 2010. Neurotoxicity of heroin-cocaine combinations in rat cortical neurons. Toxicology 276 (1), 11–17. Available at: http://www.ncbi.nlm.nih.gov/pubmed/20600547. accessed March 27, 2015.

Cunningham, K.A., Paris, J.M., Goeders, N.E., 1992. Chronic cocaine enhances serotonin autoregulation and serotonin uptake binding. Synapse 11 (2), 112–123.

De Stoutz, N.D., Bruera, E., Suarez-Almazor, M., 1995. Opioid rotation for toxicity reduction in terminal cancer patients. J. Pain Symptom Manag. 10 (5), 378–384.

De Vito, M.J., Wagner, G.C., 1989. Methamphetamine-induced neuronal damage: a possible role for free radicals. Neuropharmacology 28 (10), 1145–1150.

Dey, S., et al., 2007. Cocaine exposure *in vitro* induces apoptosis in fetal locus coeruleus neurons by altering the Bax/Bcl-2 ratio and through caspase-3 apoptotic signaling. Neuroscience 144 (2), 509–521. Available at: http://www.pubmedcentral.nih.gov/articlerender.fcgi?artid=2562674&tool=pmcentrez&rendertype=abstract.

Dietrich, J.-B., et al., 2005. Acute or repeated cocaine administration generates reactive oxygen species and induces antioxidant enzyme activity in dopaminergic rat brain structures. Neuropharmacology 48 (7), 965–974. Available at: http://www.ncbi.nlm.nih.gov/pubmed/15857623.

Dolin, S., et al., 1987. Increased dihydropyridine-sensitive calcium channels in rat brain may underlie ethanol physical dependence. Neuropharmacology 26 (2–3), 275–279.

Domercq, M., Matute, C., 2004. Neuroprotection by tetracyclines. Trends Pharmacol. Sci. 25 (12), 609–612. Available at: http://www.ncbi.nlm.nih.gov/pubmed/15530637.

Doraiswamy, P.M., 2002. Non-cholinergic strategies for treating and preventing Alzheimer's disease. CNS Drugs 16 (12), 811–824.

Escubedo, E., et al., 1998. Microgliosis and down-regulation of adenosine transporter induced by methamphetamine in rats. Brain Res. 814 (1–2), 120–126.

Etminan, M., Gill, S., Samii, A., 2003. Effect of non-steroidal anti-inflammatory drugs on risk of Alzheimer's disease: systematic review and meta-analysis of observational studies. BMJ 327 (7407), 128.

Eyerman, D.J., Yamamoto, B.K., 2005. Lobeline attenuates methamphetamine-induced changes in vesicular monoamine transporter 2 immunoreactivity and monoamine depletions in the striatum. J. Pharmacol. Exp. Ther. 312 (1), 160–169.

Fukami, G., et al., 2004. Effect of antioxidant N-acetyl-L-cysteine on behavioral changes and neurotoxicity in rats after administration of methamphetamine. Brain Res. 1016 (1), 90–95.

Fuller, R.W., Hemrick-Luecke, S.K., Ornstein, P.L., 1992. Protection against amphetamine-induced neurotoxicity toward striatal dopamine neurons in rodents by LY274614, an excitatory amino acid antagonist. Neuropharmacology 31 (10), 1027–1032.

Fumagalli, F., et al., 1999. Increased methamphetamine neurotoxicity in heterozygous vesicular monoamine transporter 2 knock-out mice. J. Neurosci. Off. J. Soc. Neurosci. 19 (7), 2424–2431.

Gasparini, L., et al., 2004. Modulation of beta-amyloid metabolism by non-steroidal anti-inflammatory drugs in neuronal cell cultures. J. Neurochem. 88 (2), 337–348.

Gibb, J.W., Kogan, F.J., 1979. Influence of dopamine synthesis on methamphetamine-induced changes in striatal and adrenal tyrosine hydroxylase activity. Naunyn Schmiedeberg's Arch. Pharmacol. 310 (2), 185–187. Available at: http://link.springer.com/10.1007/BF00500283.

Gołembiowska, K., et al., 2003. Neuroprotective action of MPEP, a selective mGluR5 antagonist, in methamphetamine-induced dopaminergic neurotoxicity is associated with a decrease in dopamine outflow and inhibition of hyperthermia in rats. Neuropharmacology

45 (4), 484–492. Available at: http://www.sciencedirect.com/science/article/pii/S0028390803002090. accessed March 27, 2015.

Graham, D.G., Tiffany, S.M., Bell Jr., W.R., Gutknecht, W.F., 1978. Autoxidation versus covalent binding of quinones as the mechanism of toxicity of dopamine, 6-hydroxydopamine, and related compounds toward C1300. Mol. Pharmacol. 14 (4), 644–653. Available at: http://molpharm.aspetjournals.org/content/14/4/644.short. accessed March 27, 2015.

Guilarte, T.R., et al., 2003. Methamphetamine-induced deficits of brain monoaminergic neuronal markers: distal axotomy or neuronal plasticity. Neuroscience 122 (2), 499–513.

Hall, E.D., et al., 1996. Neuroprotective effects of the dopamine D2/D3 agonist pramipexole against postischemic or methamphetamine-induced degeneration of nigrostriatal neurons. Brain Res. 742 (1–2), 80–88.

Hamamura, T., Fibiger, H.C., 1993. Enhanced stress-induced dopamine release in the prefrontal cortex of amphetamine-sensitized rats. Eur. J. Pharmacol. 237 (1), 65–71.

Hashimoto, K., et al., 2004. Protective effects of N-acetyl-L-cysteine on the reduction of dopamine transporters in the striatum of monkeys treated with methamphetamine. Neuropsychopharmacology 29 (11), 2018–2023.

Hawkins, B.T., Davis, T.P., 2005. The blood–brain barrier/neurovascular unit in health and disease. Pharmacol. Rev. 57 (2), 173–185.

Hawkins, B.T., et al., 2004. Nicotine increases *in vivo* blood–brain barrier permeability and alters cerebral microvascular tight junction protein distribution. Brain Res. 1027 (1–2), 48–58.

He, J., et al., 2004. Neuroprotective effects of olanzapine on methamphetamine-induced neurotoxicity are associated with an inhibition of hyperthermia and prevention of Bcl-2 decrease in rats. Brain Res. 1018 (2), 186–192.

Heller, A., et al., 2001. Gender-dependent enhanced adult neurotoxic response to methamphetamine following fetal exposure to the drug. J. Pharmacol. Exp. Ther. 298 (2), 769–779.

Hoffman, P.L., Tabakoff, B., 1994. The role of the NMDA receptor in ethanol withdrawal. EXS 71, 61–70.

Hoffmann, C., 2000. COX-2 in brain and spinal cord implications for therapeutic use. Curr. Med. Chem. 7 (11), 1113–1120.

Hotchkiss, A.J., Gibb, J.W., 1980. Long-term effects of multiple doses of methamphetamine on tryptophan hydroxylase and tyrosine hydroxylase activity in rat brain. J. Pharmacol. Exp. Ther. 214 (2), 257–262.

Hu, S., et al., 2002. Morphine induces apoptosis of human microglia and neurons. Neuropharmacology 42 (6), 829–836. Available at: http://linkinghub.elsevier.com/retrieve/pii/S0028390802000308.

Imam, S.Z., et al., 2001. Methamphetamine-induced dopaminergic neurotoxicity: role of peroxynitrite and neuroprotective role of antioxidants and peroxynitrite decomposition catalysts. Ann. N. Y. Acad. Sci. 939, 366–380.

Imam, S.Z., Duhart, H.M., Skinner, J.T., Ali, S.F., 2005. Cocaine induces a differential dose-dependent alteration in the expression profile of immediate early genes, transcription fact ors, and caspases in PC12 cells: a possible mechanism of neurotoxic damage in cocaine addiction. Ann. N. Y. Acad. Sci. 1053, 482–490.

Jayanthi, S., et al., 2004. Methamphetamine induces neuronal apoptosis via cross-talks between endoplasmic reticulum and mitochondria-dependent death cascades. FASEB J. 18 (2), 238–251.

Jayanthi, S., et al., 2005. Calcineurin/NFAT-induced up-regulation of the Fas ligand/Fas death pathway is involved in methamphetamine-induced neuronal apoptosis. Proc. Natl. Acad. Sci. USA 102 (3), 868–873.
Jayanthi, S., et al., 2009. Methamphetamine induces dopamine D1 receptor-dependent endoplasmic reticulum stress-related molecular events in the rat striatum. PLoS One 4 (6), e6092.
Johnson, E.A., Sharp, D.S., Miller, D.B., 2000. Restraint as a stressor in mice: against the dopaminergic neurotoxicity of D-MDMA, low body weight mitigates restraint-induced hypothermia and consequent neuroprotection. Brain Res. 875 (1–2), 107–118.
Johnson, E.A., et al., 2002. d-MDMA during vitamin E deficiency: effects on dopaminergic neurotoxicity and hepatotoxicity. Brain Res. 933 (2), 150–163.
Jones, J.D., Mogali, S., Comer, S.D., 2012. Polydrug abuse: a review of opioid and benzodiazepine combination use. Drug Alcohol Depend. 125 (1–2), 8–18.
Kalant, H., Kalant, O.J., 1975. Death in amphetamine users: causes and rates. Can. Med. Assoc. J. 112 (3), 299–304.
Kalivas, P.W., Duffy, P., 1989. Similar effects of daily cocaine and stress on mesocorticolimbic dopamine neurotransmission in the rat. Biol. Psychiatry 25 (7), 913–928.
Kim, H., et al., 2000. Selenium deficiency potentiates methamphetamine-induced nigral neuronal loss; comparison with MPTP model. Brain Res. 862 (1–2), 247–252.
Kondo, T., et al., 1997. Reduction of CuZn-superoxide dismutase activity exacerbates neuronal cell injury and edema formation after transient focal cerebral ischemia. J. Neurosci. Off. J. Soc. Neurosci. 17 (11), 4180–4189.
Kopelman, M.D., et al., 2009. The Korsakoff syndrome: clinical aspects, psychology and treatment. Alcohol Alcohol. 44 (2), 148–154.
Koppi, S., et al., 1987. Calcium-channel-blocking agent in the treatment of acute alcohol withdrawal—caroverine versus meprobamate in a randomized double-blind study. Neuropsychobiology 17 (1–2), 49–52.
Kousik, S.M., et al., 2011. Methamphetamine-induced vascular changes lead to striatal hypoxia and dopamine reduction. Neuroreport 22 (17), 923–928.
Krasnova, I.N., Cadet, J.L., 2009. Methamphetamine toxicity and messengers of death. Brain Res. Rev. 60 (2), 379–407.
Krasnova, I.N., Ladenheim, B., Cadet, J.L., 2005. Amphetamine induces apoptosis of medium spiny striatal project ion neurons via the mitochondria-dependent pathway. FASEB J. 19 (7), 851–853.
Kroemer, G., Martin, S.J., 2005. Caspase-independent cell death. Nat. Med. 11 (7), 725–730. Available at: http://www.ncbi.nlm.nih.gov/pubmed/16015365.
Langlais, P., Mair, R., 1990. Protective effects of the glutamate antagonist MK-801 on pyrithiamine-induced lesions and amino acid changes in rat brain. J. Neurosci. 10 (5), 1664–1674. Available at: http://www.jneurosci.org/content/10/5/1664.short. accessed March 27, 2015.
Levy, A.D., Baumann, M.H., Van de Kar, L.D., 1994. Monoaminergic regulation of neuroendocrine function and its modification by cocaine. Front. Neuroendocrinol. 15 (2), 85–156.
Lin, P.S., et al., 1991. Hyperthermia enhances the cytotoxic effects of reactive oxygen species to Chinese hamster cells and bovine endothelial cells *in vitro*. Radiat. Res. 126 (1), 43–51.
Lin, S.J., et al., 1992. Long-term nicotine exposure increases aortic endothelial cell death and enhances transendothelial macromolecular transport in rats. Arterioscler. Thromb. 12 (11), 1305–1312.

Lipton, J.W., et al., 2003. Prenatal cocaine administration increases glutathione and alpha-tocopherol oxidation in fetal rat brain. Dev. Brain Res. 147 (1–2), 77–84. Available at: http://linkinghub.elsevier.com/retrieve/pii/S0165380603002475. accessed March 27, 2015.

Little, H.J., Dolin, S.J., Halsey, M.J., 1986. Calcium channel antagonists decrease the ethanol withdrawal syndrome. Life Sci. 39 (22), 2059–2065.

Lovinger, D.M., 1993. Excitotoxicity and alcohol-related brain damage. Alcohol. Clin. Exp. Res. 17 (1), 19–27.

Lyles, J., Cadet, J.L., 2003. Methylenedioxymethamphetamine (MDMA, Ecstasy) neurotoxicity: cellular and molecular mechanisms. Brain Res. Brain Res. Rev. 42 (2), 155–168.

Macêdo, D.S., et al., 2005. Cocaine alters catalase activity in prefrontal cortex and striatum of mice. Neurosci. Lett. 387 (1), 53–56. Available at: http://www.ncbi.nlm.nih.gov/pubmed/16085363.

Mao, J., et al., 2002. Neuronal apoptosis associated with morphine tolerance: evidence for an opioid-induced neurotoxic mechanism. J. Neurosci. 22 (17), 7650–7661. Available at: http://www.jneurosci.org/content/22/17/7650.short. accessed March 27, 2015.

Martin, D., Cohen, S., Morrisett, R.A., 1991. Ethanol effects upon rat hippocampal epileptiform activity. Alcohol. Clin. Exp. Res. 15, 324.

Matuszewich, L., Yamamoto, B.K., 2004. Chronic stress augments the long-term and acute effects of methamphetamine. Neuroscience 124 (3), 637–646.

Mhatre, M., Floyd, R.A., Hensley, K., 2004. Oxidative stress and neuroinflammation in Alzheimer's disease and amyotrophic lateral sclerosis: common links and potential therapeutic targets. J. Alzheimer's Dis. 6 (2), 147–157.

Milhazes, N., et al., 2006. Synthesis and cytotoxic profile of 3,4-methylenedioxymethamphetamine ("ecstasy") and its metabolites on undifferentiated PC12 cells: a putative structure-toxicity relationship. Chem. Res. Toxicol. 19 (10), 1294–1304.

Miller, D.B., O'Callaghan, J.P., 1994. Environment-, drug- and stress-induced alterations in body temperature affect the neurotoxicity of substituted amphetamines in the C57BL/6J mouse. J. Pharmacol. Exp. Ther. 270 (2), 752–760.

Miller, D.B., O'Callaghan, J.P., 1995. The role of temperature, stress, and other factors in the neurotoxicity of the substituted amphetamines 3,4-methylenedioxymethamphetamine and fenfluramine. Mol. Neurobiol. 11 (1–3), 177–192.

Miller, D.B., O'Callaghan, J.P., 2003. Elevated environmental temperature and methamphetamine neurotoxicity. Environ. Res. 92 (1), 48–53.

Miller, D.B., et al., 1998. The impact of gender and estrogen on striatal dopaminergic neurotoxicity. Ann. N. Y. Acad. Sci. 844, 153–165.

Miller, D.B., O'Callaghan, J.P., Ali, S.F., 2000. Age as a susceptibility factor in the striatal dopaminergic neurotoxicity observed in the mouse following substituted amphetamine exposure. Ann. N. Y. Acad. Sci. 914, 194–207.

Mitchell, E.S., Snyder-Keller, A., 2003. c-fos and cleaved caspase-3 expression after perinatal exposure to ethanol, cocaine, or the combination of both drugs. Dev. Brain Res. 147 (1–2), 107–117. Available at: http://linkinghub.elsevier.com/retrieve/pii/S0165380603002700.

Monks, T.J., et al., 2004. The role of metabolism in 3,4-(+)-methylenedioxyamphetamine and 3,4-(+)-methylenedioxymethamphetamine (ecstasy) toxicity. Ther. Drug Monit. 26 (2), 132–136.

Moriguchi, S., et al., 2002. Enhancement of N-methyl-D-aspartate receptor-mediated excitatory postsynaptic potentials in the neostriatum after methamphetamine sensitization. An *in vitro* slice study. Exp. Brain Res. 144 (2), 238–246.

Nassogne, M.C., Evrard, P., Courtoy, P.J., 1998. Selective direct toxicity of cocaine on fetal mouse neurons. Teratogenic implications of neurite and apoptotic neuronal loss. Ann. N. Y. Acad. Sci. 846, 51–68.

Ohmori, T., Koyama, T., Muraki, A., Yamashita, I., 1993. Competitive and noncompetitive N-methyl-D-aspartate antagonists protect dopaminergic and serotonergic neurotoxicity produced by methamphetamine in various brain regions. J. Neural Transm. Gen. Sect. 92 (2–3), 970106.

Oliveira, M.T., et al., 2002. Toxic effects of opioid and stimulant drugs on undifferentiated PC12 cells. Ann. N. Y. Acad. Sci. 965, 487–496. Available at: http://doi.wiley.com/10.1111/j.1749-6632.2002.tb04190.x.

Oliveira, M.T., et al., 2003. Drugs of abuse induce apoptotic features in PC12 cells. Ann. N. Y. Acad. Sci. 1010 (1), 667–670. Available at: http://doi.wiley.com/10.1196/annals.1299.121.

Oliveira, M.F., Norremose, K.A., Oliveira, O.P., Guimarães, W.V., Sekeff, F.A., 2011. Cartas aos editores. Glutamatergic excitotoxicity secondary to status epilepticus after crack abuse: a case report Excitotoxicidade glutamatérgica secundária ao uso de crack: relato de caso. Rev. Bras. Psiquiatr. 33 (2), 209–210.

Omar, R.A., Yano, S., Kikkawa, Y., 1987. Antioxidant enzymes and survival of normal and simian virus 40-transformed mouse embryo cells after hyperthermia. Cancer Res. 47 (13), 3473–3476.

Ozmen, I., Naziroğlu, M., Alici, H.A., Sahin, F., Cengiz, M., Eren, I., 2007. Spinal morphine administration reduces the fatty acid contents in spinal cord and brain by increasing oxidative stress. Neurochem. Res. 32 (1), 19–25.

Piazza, P.V., Le Moal, M., 1998. The role of stress in drug self-administration. Trends Pharmacol. Sci. 19 (2), 67–74.

Pineda, J.A., et al., 2001. Extracellular superoxide dismutase overexpression improves behavioral outcome from closed head injury in the mouse. J. Neurotrauma 18 (6), 625–634.

Pong, K., Doctrow, S.R., Baudry, M., 2000. Prevention of 1-methyl-4-phenylpyridinium- and 6-hydroxydopamine-induced nitration of tyrosine hydroxylase and neurotoxicity by EUK-134, a superoxide dismutase and catalase mimetic, in cultured dopaminergic neurons. Brain Res. 881 (2), 182–189.

Poon, H.F., et al., 2007. Cocaine-induced oxidative stress precedes cell death in human neuronal progenitor cells. Neurochem. Int. 50 (1), 69–73. Available at: http://linkinghub.elsevier.com/retrieve/pii/S0197018606002403.

Power, R., Hartnoll, R., Chalmers, C., 1992. Help-seeking among illicit drug users: some differences between a treatment and nontreatment sample. Int. J. Addict. 27 (8), 887–904.

Qiusheng, Z., Yuntao, Z., Rongliang, Z., Dean, G., Changling, L., 2005. Effects of verbascoside and luteolin on oxidative damage in brain of heroin treated mice. Pharmazie 60 (7), 539–543.

Quigley, C., 2004. Opioid switching to improve pain relief and drug tolerability. Cochrane Database Syst. Rev. 2004 (3), CD004847.

Randall, T., 1992. Ecstasy-fueled "rave" parties become dances of death for English youths. JAMA 268 (12), 1505–1506.

Rego, A., Oliveira, C., 2003. Mitochondrial dysfunction and reactive oxygen species in excitotoxicity and apoptosis: implications for the pathogenesis of neurodegenerative diseases. Neurochem. Res. 28 (10), 1563–1574. Available at: http://link.springer.com/article/10.1023/A:1025682611389. accessed March 27, 2015.

Reid, M., Hsu, K., Berger, S., 1997. Cocaine and amphetamine preferentially stimulate glutamate release in the limbic system: studies on the involvement of dopamine. Synapse 27 (2), 95–105. Available at: http://onlinelibrary.wiley.com/doi/10.1002/(SICI)1098-2396(199710)27:2%3C95::AID-SYN1%3E3.0.CO;2-6/abstract. accessed March 27, 2015.

Rolland, B., et al., 2011. Pharmaceutical approaches of binge drinking. Curr. Pharm. Des. 17 (14), 1333–1342.

Rouge-Pont, F., et al., 1995. Stress-induced sensitization and glucocorticoids. II. Sensitization of the increase in extracellular dopamine induced by cocaine depends on stress-induced corticosterone secretion. J. Neurosci. Off. J. Soc. Neurosci. 15 (11), 7189–7195.

Sandoval, V., Hanson, G.R., Fleckenstein, A.E., 2000. Methamphetamine decreases mouse striatal dopamine transporter activity: roles of hyperthermia and dopamine. Eur. J. Pharmacol. 409 (3), 265–271.

Schmidt, C.J., Gibb, J.W., 1985. Role of the dopamine uptake carrier in the neurochemical response to methamphetamine: effects of amfonelic acid. Eur. J. Pharmacol. 109 (1), 73–80.

Sekine, Y., et al., 2008. Methamphetamine causes microglial activation in the brains of human abusers. J. Neurosci. Off. J. Soc. Neurosci. 28 (22), 5756–5761.

Sharma, H.S., 2007. Methods to produce hyperthermia-induced brain dysfunction. Prog. Brain Res. 162, 173–199.

Sharma, H.S., Ali, S.F., 2008. Acute administration of 3,4-methylenedioxymethamphetamine induces profound hyperthermia, blood–brain barrier disruption, brain edema formation, and cell injury. Ann. N. Y. Acad. Sci. 1139, 242–258.

Sharma, H.S., Olsson, Y., Dey, P.K., 1990. Changes in blood–brain barrier and cerebral blood flow following elevation of circulating serotonin level in anesthetized rats. Brain Res. 517 (1–2), 215–223.

Shoptaw, S., et al., 1994. The Matrix model of outpatient stimulant abuse treatment: evidence of efficacy. J. Addict. Dis. 13 (4), 129–141.

Skattebol, A., Rabin, R.A., 1987. Effects of ethanol on 45Ca2+ uptake in synaptosomes and in PC12 cells. Biochem. Pharmacol. 36 (13), 2227–2229.

Smith, K.J., et al., 2008. Potentiation of N-methyl-D-aspartate receptor-mediated neuronal injury during methamphetamine withdrawal *in vitro* requires co-activation of IP3 receptors. Brain Res. 1187, 67–73.

Smoluch, M., et al., 2014. Determination of psychostimulants and their metabolites by electrochemistry linked on-line to flowing atmospheric pressure afterglow mass spectrometry. Analyst 139 (17), 4350–4355.

Sonsalla, P., Nicklas, W., Heikkila, R., 1989. Role for excitatory amino acids in methamphetamine-induced nigrostriatal dopaminergic toxicity. Science 243 (4889), 398–400. Available at: http://www.sciencemag.org/cgi/doi/10.1126/science.2563176.

Suwanjang, W., et al., 2010. The protective effect of melatonin on methamphetamine-induced calpain-dependent death pathway in human neuroblastoma SH-SY5Y cultured cells. J. Pineal Res. 48 (2), 94–101.

Suwanjang, W., et al., 2012. Calpastatin reduces calpain and caspase activation in methamphetamine-induced toxicity in human neuroblastoma SH-SY5Y cultured cells. Neurosci. Lett. 526 (1), 49–53.

Tarumi, Y., Pereira, J., Watanabe, S., 2002. Methadone and fluconazole: respiratory depression by drug interaction. J. Pain Symptom Manag. 23 (2), 148–153.

Tata, D.A., Yamamoto, B.K., 2008. Chronic stress enhances methamphetamine-induced extracellular glutamate and excitotoxicity in the rat striatum. Synapse 62 (5), 325–336.

Teuchert-Noodt, G., Dawirs, R.R., 1991. Age-related toxicity in prefrontal cortex and caudate-putamen complex of gerbils (Meriones unguiculatus) after a single dose of methamphetamine. Neuropharmacology 30 (7), 733–743.

Thomas, D.M., Kuhn, D.M., 2005. Attenuated microglial activation mediates tolerance to the neurotoxic effects of methamphetamine. J. Neurochem. 92 (4), 790–797.

Thomas, D.M., Francescutti-Verbeem, D.M., et al., 2004a. Identification of differentially regulated transcripts in mouse striatum following methamphetamine treatment—an oligonucleotide microarray approach. J. Neurochem. 88 (2), 380–393.

Thomas, D.M., Walker, P.D., et al., 2004b. Methamphetamine neurotoxicity in dopamine nerve endings of the striatum is associated with microglial activation. J. Pharmacol. Exp. Ther. 311 (1), 1–7.

Thomas, D.M., Francescutti-Verbeem, D.M., Kuhn, D.M., 2008. The newly synthesized pool of dopamine determines the severity of methamphetamine-induced neurotoxicity. J. Neurochem. 105 (3), 605–616.

Thomsen, A.B., Becker, N., Eriksen, J., 1999. Opioid rotation in chronic non-malignant pain patients. A retrospective study. Acta Anaesthesiol. Scand. 43 (9), 918–923.

Todd, K., Butterworth, R.F., 1999. Mechanisms of selective neuronal cell death due to thiamine deficiency. Ann. N. Y. Acad. Sci. 893, 404–411.

Todd, K.G., Butterworth, R.F., 2001. *In vivo* microdialysis in an animal model of neurological disease: thiamine deficiency (Wernicke) encephalopathy. Methods (San Diego, Calif.) 23 (1), 55–61.

Tramullas, M., Martínez-Cué, C., Hurlé, M.A., 2008. Chronic administration of heroin to mice produces up-regulation of brain apoptosis-related proteins and impairs spatial learning and memory. Neuropharmacology 54 (4), 640–652. Available at: http://www.ncbi.nlm.nih.gov/pubmed/18201731. accessed March 27, 2015.

Troy, C.M., et al., 1996. Downregulation of Cu/Zn superoxide dismutase leads to cell death via the nitric oxide-peroxynitrite pathway. J. Neurosci. Off. J. Soc. Neurosci. 16 (1), 253–261.

Truex, L.L., Schmidt, M.J., 1980. 3H-amphetamine concentrations in the brains of young and aged rats: implications for assessment of drug effects in aged animals. Neurobiol. Aging 1 (1), 93–95.

Turski, L., et al., 1991. Protection of substantia nigra from MPP+ neurotoxicity by N-methyl-D-aspartate antagonists. Nature 349 (6308), 414–418.

Uys, J.D., Reissner, K.J., 2011. Glutamatergic neuroplasticity in cocaine addiction. Prog. Mol. Biol. Transl. Sci. 98, 367–400. Available at: http://www.ncbi.nlm.nih.gov/pubmed/21199777.

Victor, M., Adams, R.D., Collins, H., 1989. Wernicke-Korsakof Syndrome and Related Neurologic Disorders due to Alcoholism and Malnutrition. FA Davies Company, Philadelphia, PA, USA.

Virmani, A., et al., 2002. The protective role of L-carnitine against neurotoxicity evoked by drug of abuse, methamphetamine, could be related to mitochondrial dysfunction. Ann. N. Y. Acad. Sci. 965, 225–232.

Wagner, G.C., Carelli, R.M., Jarvis, M.F., 1985. Pretreatment with ascorbic acid attenuates the neurotoxic effects of methamphetamine in rats. Res. Commun. Chem. Pathol. Pharmacol. 47 (2), 221–228.

Waren, M.W., et al., 2007. Calpain and caspase proteolytic markers co-localize with rat cortical neurons after exposure to methamphetamine and MDMA. Acta Neuropathol. 114 (3), 277086.

Weihmuller, F.B., O'Dell, S.J., Marshall, J.F., 1992. MK-801 protection against methamphetamine-induced striatal dopamine terminal injury is associated with attenuated dopamine overflow. Synapse 11 (2), 155–163.

Williams, J.M., Steketee, J.D., 2004. Cocaine increases medial prefrontal cortical glutamate overflow in cocaine-sensitized rats: a time course study. Eur. J. Neurosci. 20 (6), 1639–1646. Available at: http://www.ncbi.nlm.nih.gov/pubmed/15355331.

Wisessmith, W., et al., 2009. Melatonin reduces induction of Bax, caspase and cell death in methamphetamine-treated human neuroblastoma SH-SY5Y cultured cells. J. Pineal Res. 46 (4), 433–440.

Wolf, M.E., et al., 2000. Amphetamine increases glutamate efflux in the rat ventral tegmental area by a mechanism involving glutamate transporters and reactive oxygen species. J. Neurochem. 75 (4), 1634–1644. Available at: http://doi.wiley.com/10.1046/j.1471-4159.2000.0751634.x.

Xu, B., Wang, Z., Li, G., Li, B., Lin, H., Zheng, R., Zheng, Q., 2006. Heroin-administered mice involved in oxidative stress and exogenous antioxidant-alleviated withdrawal syndrome. Basic Clin. Pharmacol. Toxicol. 99 (2), 153–161.

Yamamoto, B.K., Bankson, M.G., 2005. Amphetamine neurotoxicity: cause and consequence of oxidative stress. Crit. Rev. Neurobiol. 17 (2), 87–117. Available at: http://www.dl.begellhouse.com/journals/7b004699754c9fe6,04c698ea6529efba,51741a1b2a4b5a60.html.

Yanagisawa, Y., Nakazato, K., Nagai, T., 1997. Binding of methamphetamine to serum albumin in various species *in vitro*. Pharmacol. Res. 35 (2), 99–102.

Yokota, T., et al., 2001. Delayed-onset ataxia in mice lacking alpha-tocopherol transfer protein: model for neuronal degeneration caused by chronic oxidative stress. Proc. Natl. Acad. Sci. USA 98 (26), 15185–15190.

Zhang, J., et al., 2000. Enhanced N-methyl-4-phenyl-1,2,3,6-tetrahydropyridine toxicity in mice deficient in CuZn-superoxide dismutase or glutathione peroxidase. J. Neuropathol. Exp. Neurol. 59 (1), 53–61.

Zhang, X., et al., 2007. Selective inhibition of cyclooxygenase-2 exacerbates methamphetamine-induced dopamine depletion in the striatum in rats. Neuroscience 150 (4), 950–958.

CHAPTER

3

Stress psychobiology in the context of addiction medicine: from drugs of abuse to behavioral addictions

Andrine Lemieux, Mustafa al'Absi[1]

University of Minnesota School of Medicine, Duluth, MN, USA

[1]*Corresponding author: Tel.: +1-218-726-7122; Fax: +1-218-726-7559*

e-mail address: malabsi@umn.edu

Abstract

In this chapter, we briefly review the basic biology of psychological stress and the stress response. We propose that psychological stress and the neurobiology of the stress response play in substance use initiation, maintenance, and relapse. The proposed mechanisms for this include, on the one hand, the complex interactions between biological mediators of the stress response and the dopaminergic reward system and, on the other hand, mediators of the stress response and other systems crucial in moderating key addiction-related behaviors such as endogenous opioids, the sympathetic-adrenal-medullary system, and endocannabinoids. Exciting new avenues of study including genomics, sex as a moderator of the stress response, and behavioral addictions (gambling, hypersexuality, dysfunctional internet use, and food as an addictive substance) are also briefly presented within the context of stress as a moderator of the addictive process.

Keywords

Stress, Stress response pathways, Relapse, Emotions, Addictive behaviors, Hypothalamic-pituitary-adrenocortical axis, Sympathetic-adrenal-medullary response, Addictive behaviors, Cortisol

1 INTRODUCTION

The stress–addiction connection has been well established in the literature. The solid foundation of research upon which the connection rests and multiple advances in the fields of endocrinology, neuroimaging, computer science, and clinical practice have

Progress in Brain Research, Volume 223, ISSN 0079-6123, http://dx.doi.org/10.1016/bs.pbr.2015.08.001

resulted in several new avenues of study relevant to stress and addiction. In this chapter, we will review the basic psychology and biology of the stress response. From there, we will review how stress may be a vulnerability factor for substance use initiation, maintenance of continued substance use and, finally, relapse following an attempt to quit chronic substance abuse. In addition to stress as a vulnerability factor, we will also review other moderators of the addiction cycle. We briefly review some of the newer forms of appetitive or "addictive" behaviors including gambling, sex, and excessive internet use. This review will conclude with a model of stress and addiction.

1.1 DEFINITIONS

Stress is an elusive, often overused, concept. Although "stress" has become the common cold of the twenty-first century evoked to explain an ever increasing array of modern maladies, in this review, we will refer to "stress" as the lived, psychological, and phenomenological experience. In contrast, the "stress response" will be used to describe the physiological and neurophysiological changes that accompany the stress experience. A "stressor" refers to the circumstance, encounter, environment, or situation that evokes both "stress" and a "stress response" in an individual or organism.

As with the concept of stress, "addiction" is a highly complex behavioral phenomenon involving multiple components including behavioral, emotional, cognitive, and physiological systems. Of particular importance are the symptoms of withdrawal, negative affect, and craving that accompany abstinence (both acute and prolonged), and the high produced by the consumption of the addictive substance. While a variety of stressors external to the addiction process may facilitate initial experimentation, chronic use, or relapse, the withdrawal and negative affect that accompanies withdrawal are in themselves an aversive and stressful experience (Kassel et al., 2007). Use in response to negative affect will result in negative reinforcement (removal of an aversive stimulus) which, in turn, increases the probability of subsequent use and, if repeated, chronic use. Positive reinforcement stems from the high or the pleasure experienced in response to use of an addictive substance. The Diagnostic and Statistical Manual-IV (American Psychiatric Association, 2000) defined addiction or substance dependence as a "maladaptive pattern of substance use leading to clinically significant impairment or distress" (American Psychiatric Association, 2000). The current DSM 5 (American Psychiatric Association, 2014) has redefined the definition of addiction to include both substance abuse and dependence that occur along a natural continuum from mild to severe. Qualifying for a diagnosis under this new system the words "addiction" and "dependence" are avoided in favor of the more broad substance use disorders.

If indeed stress is a critical factor at all stages of the addiction cycle from experimentation, to maintenance of chronic use, and to relapse upon attempted cessation, then we should see evidence of a stress effect at all levels of analysis from the macro (behavior) to the micro (cellular, genetic, and molecular). We would expect there to be a convergence of evidence from an analysis of such domains as behavior, cognition, emotion, and physiology, both peripheral and central. Given the wide range of analyses and the highly complex nature of each domain, support for this supposition will necessarily be multidisciplinary, varied in methodologies, and incremental.

2 STRESS RESPONSE SYSTEMS

Early in the twentieth century stress was viewed as a nonspecific physiological response to demands placed on a body. This nonspecific response was studied in animals, and later humans, with particular emphasis on changes in either the adrenal cortex (Hans Selye) or the medulla (Walter B. Cannon (Hinkle, 1987)). Both Selye and Cannon were interested in not only the multisystemic effects of stress, but they were also among the first to coin terms and concepts still in use in the study of stress.

2.1 SYMPATHETIC-ADRENAL-MEDULLARY AXIS

Walter B. Cannon was the first to characterize the acute stress response as the "flight or fight" response, or the automatic behavioral reaction induced, or so it was believed, by the production of adrenalin (later termed epinephrine) by the adrenal medulla of an organism when faced with a threatening situation or foe. While Cannon's research was limited to sympathectomized animal preparations, we now know the sympathetic-adrenal-medullary (SAM) axis stimulates the production of neurotransmitters norepinephrine (NE) and epinephrine (EPI). The locus coeruleus is dense with catecholamine neurons, and activation of these neurons increases sympathetic nervous system (SNS) activity which is propagated via sympathetic preganglion neurons. Stimulation of these neurons in turn stimulates the adrenal medulla to release NE and EPI into circulation. These neurotransmitters activate receptors such as smooth and cardiac muscles, other endocrine glands, the immune system, and adipose tissue.

The study of SAM functions and emotional states got its start with Cannon's animal studies in 1911, but it was not until technological advancements in the 1950s that allowed for the detailed study of negative emotions and catecholamines in human urine. Since that time the physiological consequences of emotion-induced SNS arousal have been extensively studied, in particular, the effects of negative emotions on cardiovascular functions (Goldstein and Kopin, 2008; Martens et al., 2008). This early work led to the supposition that different emotional states have different catecholamine profiles (Goldstein and Kopin, 2008). Some of these early conceptualizations, such as the differential catecholamine profile for anger directed inward as opposed to outward, continue to play a role in the study of stress and addiction (al'Absi and Bongard, 2006; al'Absi et al., 2007).

2.2 HYPOTHALAMIC-PITUITARY-ADRENAL AXIS

In addition to the SAM system, others have focused on the hypothalamic-pituitary-adrenal (HPA) axis. Hans Selye, often referred to as the father of stress studies, outlined a series of stages through which the challenged individual (animal or human) goes through when faced with prolonged stress. He called this the general adaptation syndrome (GAS) which consists of alarm, resistance, and exhaustion. Selye identified the adrenal cortex with its production of mineralocorticoids and glucocorticoids as the key mediators of the stress response. In the alarm phase, the cascade of events begins with the detection of a threat and the production of corticotropin-releasing factor (CRF) and vasopressin from neurons in the paraventricular nucleus (PVN). This CRF is transported from the median eminence of the hypothalamus to the anterior pituitary where it stimulates the production of proopiomelanocortin (POMC). POMC, a relatively large precursor protein, is enzymatically cleaved into beta-endorphin (β-endorphin) and adrenocorticotropic hormone (ACTH). The ACTH enters into peripheral circulation and stimulates the cortex of the adrenal gland to produce glucocorticoids (cortisol in humans, corticosterone in rodents). Selye's resistance phase occurs when an organism seeks to reduce or eradicate the source of the threat or stress that initiated the GAS. In both humans and animals, a variety of physically and psychologically demanding behaviors are elicited during this phase. Many of these attempts to cope with or eliminate the stressor during the resistance phase are either unsuccessful, or in the case of substance use in humans, harmful. When coping and resistance fail, the glucocorticoid production persists due to failure of the negative feedback regulation of the system until physiological and emotional exhaustion occurs. Selye was among the first to study extensively the wide range of negative physiological consequences of stress during the exhaustion phase, including thymic atrophy, gastric and duodenal ulcers, and even death. Contemporary researchers have documented important negative impacts of CRF and cortisol on neuronal morphology and functions all along the developmental trajectory (Buss et al., 2012; Lupien et al., 2009). The timing of glucocorticoid exposure during development is an important predictor of CNS morphology impact and it has been proposed that early exposure to glucocorticoids increases the risk of later emotional and behavioral problems (Lupien et al., 2009). Glucocorticoid production in response to threat is not, however, universal nor is it necessarily detrimental. Cortisol plays a critical role in multiple adaptive body functions and the absence, as well as the overproduction, of cortisol is a clinical condition requiring intervention and treatment (Allolio, 2015). Selye was also the first to distinguish individual differences in stress reactivity which included "distress" or negative reactions and "eustress" or positive reactions.

2.3 THE INTEGRATION OF THE STRESS RESPONSE

The HPA and SAM systems do not act in isolation. Instead, there is a highly complex and dynamic CNS system of integration. This system can be simplified into several basic components, one of which includes arousal. All stressful situations begin with heightened arousal which moderates the behavioral response to the stress. Neurophysiologically, this requires the coordinated efforts of hypothalamic CRF and locus coeruleus-derived NE systems. For example, neurons project bidirectionally between CRF-producing hypothalamic periventricular nucleus neurons to the brain stem areas rich in NE including the locus coeruleus. Neurochemical and pharmacological studies using agonists and antagonists for both CRF and NE further support the tight interconnection between these two regions (Dunn et al., 2004).

This dynamic relationship between the PVN and locus coeruleus is, in turn, in dynamic relationship with other brain regions implicated in the process of addiction. For example, the hypothalamic PVN receives input directly or indirectly from multiple regions including the prefrontal cortex, lateral septum, amygdala, hippocampus, and the bed nucleus of the stria terminalis (BNST; Herman et al., 2002). Thus, upstream brain regions implicated in emotions (limbic: septum, amygdala), decision making (prefrontal cortex), memory (hippocampus), and stress (BNST) all relay signals capable of modifying activity within the PVN. In addition to CRF, these circuits have their own mediators which include other neuromodulators and neurotransmitters such as glutamate, GABA, substance P, NPY, dopamine, endogenous opioids, endocannabinoids (eCBs), and serotonin (Fride et al., 2009; Herman et al., 2004; Korte et al., 1991). A complete overview of cortical neuromodulators and neurotransmitters is beyond the scope of this review, but the curious reader may find the reviews by Kavalali or Wang and Lupica helpful (Kavalali, 2015; Wang and Lupica, 2014).

Likewise, the locus coeruleus receives afferent projects from peripheral inputs (e.g., vagal and somatosensory afferents), the brainstem and higher cortical regions such as the amygdala, hypothalamus, and prefrontal cortex. In turn, the locus coeruleus sends projects widely throughout the cortex, limbic system, thalamus, and hypothalamus and the brainstem and it has been shown to modulate prefrontal cortex independent of hypothalamic inputs (Chandler et al., 2014). Its functional targets include regions mediating sleep–wake cycles, neuroendocrine functioning (including the HPA), and autonomic, motor, and sensory functions. As an integrator of the stress response, the locus coeruleus and noradrenergic inputs regulate many broad behaviors as sleep, arousal, neuropathology, cognition, and pain modulation (Van Bockstaele and Valentino, 2013).

In addition to complex dynamics of CRF and NE systems, other complex, distributed networks in the brain help to regulate the stress response. These include serotonergic, endogenous opioid, and eCB systems. There is some evidence that the LC–NE system response to stress is moderated by both CRF and endogenous opioids. Social stress in rats results in persistent decrease in LC activation which is moderated by endogenous μ-opioid receptor functioning (Chaijale et al., 2013; Curtis et al., 2012). Serotonin, like CRF, is increased by a variety of stress paradigms, CRF signaling, or electrical stimulation of the PVN and stimulation of serotonin, in turn, heightens both the HPA axis and the SAM system (Kondo et al., 2015). Finally, eCBs have recently been shown to moderate the stress response. For example, the eCB system (Di Marzo et al., 2005) consists of two endogenous ligands, AEA

(*N*-arachidonylethanolamine anandamide) and 2-AG (2-arachidonoylglycerol). Acute stress in humans leads to increased production and release of eCBs (Dlugos et al., 2012). Likewise, eCBs may also act to regulate the stress-related HPA axis activation (Hill and McEwen, 2010). The system is highly dynamic as evident by the fact that CNS eCB signaling has been shown to regulate both the activation and termination of stress through decreases in brain AEA and increases in AG-1 (Steiner and Wotjak, 2008).

This overview suggests that the HPA and LC–NE systems are dynamically interdependent and both systems, in turn, are affected by other stress-related systems including serotonergic, endogenous opioid, and eCB systems. As we will see in subsequent sections, each of these systems, in turn, moderate a variety of behaviors that have been implicated or observed as a component of the addiction process.

2.4 COMMON PATHWAYS BETWEEN STRESS AND ADDICTION

There is evidence that, in many ways, the dynamic neurophysiology of the stress response mirrors that of the neurophysiology evident in humans and animals who have been chronically exposed to drugs of abuse. For example, both stress and addiction share similar changes in behavior, similar neurophysiological changes in the HPA, LC–NE, autonomic, and eCB systems, and similar risk profiles (sex, psychopathology, etc.). Chronic social stress in animals and humans leads to increases in anxiety, negative affect, and changes in sleep and eating (Adam and Epel, 2007; Akerstedt, 2006; Chida and Hamer, 2008), all of which are common in persistent substance abuse. The same is true for disruptions in attention, concentration, memory, and decision making (Het et al., 2005). From a neurophysiological perspective, there are also many common pathways. As indicated above, both chronic social stress and chronic exposure to drugs of abuse such as morphine leads to changes in LC–NE functioning that appears to be dependent upon endogenous opioid functions (Chaijale et al., 2013; Curtis et al., 2012). In general, though there is some variability based on the chemistry of the abused substance, acute drug use also leads to increased HPA and SNS functioning in much the same way as stress (al'Absi et al., 2008; Fox et al., 2006; Hamidovic et al., 2010; Mick et al., 2013).

One common pathway that has received the most intense research attention is the role of dopaminergic reward pathways in the brain. As stated above, drugs of abuse heighten the activity of the HPA, SNS, and endogenous opioid systems in much the same way as chronic stress. The behavioral effects are, in turn, moderated by multiple neurobiological systems including the catecholamines: dopamine, NE, and serotonin (Salamone and Correa, 2013). The HPA and dopaminergic systems are interdependent (Boyson et al., 2014). Dopamine in particular has been linked to the reward properties of drug use. For example, pharmacological studies have shown that stress increases dopamine production via glucocorticoid receptor activation (Boyson et al., 2014). In particular, increased central CRF activity potentiates *N*-methyl-D-aspartate receptor activity which, in turn, results in increased dopaminergic transmission (Marinelli, 2007). Support for the role of HPA activation in increasing dopaminergic activity has been evident in studies using a variety of methodologies (Barrot et al., 2000; Graf et al., 2013). Through this research, critical CNS reward pathways and structures have identified including the ventral tegmental area, nucleus accumbens, and prefrontal cortex (Baik, 2013; Kringelbach et al., 2012; Lawrence and Brooks, 2014).

In addition to the dopaminergic reward pathway, both stress and drugs of abuse negatively impact the serotonergic pathway consisting of the raphe nucleus, striatum, nucleus accumbens, and the entire neocortex. The effects of altered serotonergic functioning are expressed as changes in mood, memory, sleep, and cognition; all of which are evident in chronic stress states and drug abuse (Kirby et al., 2011; Meerlo et al., 2008; Meneses, 2013).

3 STRESS AS A VULNERABILITY FACTOR FOR INITIATION AND MAINTAINING SUBSTANCE ABUSE

3.1 DEVELOPMENTAL STUDIES IN ANIMALS

Evidence of developmental stress as a vulnerability factor has been provided for nearly all substances of abuse, though this differs based on the timing of the stressor (prenatal versus postnatal) and the developmental stage in which substance exposure is tested. For example, maternal stress during gestation can be induced by a variety of experimental paradigms including restraint stress, noise stress, and social stressors using both rodents and nonhuman primates, each of which may produce different outcomes (Nylander and Roman, 2013). This prenatal maternal stress leads to disruption in both basal and stress responsivity along multiple systems including the HPA, dopamine, NE, acetylcholinergic, opioid, and eCB systems (Carboni et al., 2010; Fride et al., 2009; Hausknecht et al., 2013). Neuroanatomical studies have further verified that the ventral tegmental area, nucleus accumbens, and prefrontal cortex show cellular and genetic changes consistent with the type of changes seen in chronic substance use (Carboni et al., 2010; Hausknecht et al., 2013; Kippin et al., 2008). Behaviorally, prenatally stressed offspring demonstrate an increased likelihood of dysphoria and anxiety. When tested later in life for substance use adoption, such prenatally stressed offspring more rapidly adopt a variety of abusive substances including, among others, nicotine, cocaine, marijuana, alcohol, benzodiazepines, and amphetamines (Lakehayli et al., 2015; Reynaert et al., 2015; Rokyta et al., 2008; Silvagni et al., 2008). Further, animals including rats and nonhuman primates that are separated from their mother early in life demonstrate HPA dysregulations, behavioral changes indicative of distress and persistent mood disorder, and increased proclivity toward heightened use of addictive substances (Brenhouse et al., 2013; Coplan et al., 2011). Some have suggested that prenatal stress does not directly impact alcohol preference but instead alters the response to alcohol consumption (Van Waes et al., 2011), though others do report that prenatal stress affects preference for alcohol (Campbell et al., 2009).

3.2 DEVELOPMENTAL STUDIES IN HUMANS

Although limited due to ethical restrictions, there is cross-sectional and quasi-experimental support for a similar effect of early life stress and addiction in humans. For example, chronic stress, in particular stress during early development, increases the risk for psychopathology during adulthood, and, in turn, both chronic stress and psychopathology are risk factors for the initiation, maintenance, and relapse of drug abuse (Brenhouse et al., 2013; Enoch, 2011; Lovallo, 2013). Adverse childhood events (abuse, neglect, separation, and loss during childhood) are associated with early onset of substance use and increased risk of addiction (Banducci et al., 2014; Heron et al., 2013; Hyman et al., 2006; Mason and Spoth, 2012). This has been shown via large epidemiological studies (Heron et al., 2013), studies of drug-dependent patients, psychiatry patients or dual diagnosis patients (Banducci et al., 2014; Brensilver et al., 2013; Evren et al., 2013; Hyman et al., 2006; Kim et al., 2013; Walsh and Cawthon, 2014), and subsequent case follow-up studies of children with officially documented abuse (Macleod et al., 2013).

The likely mechanisms of early stress and addiction potential, similar to the animal research, include dysregulations of the HPA axis and endogenous opioid functions (Dai et al., 2007; Gianoulakis et al., 2005). The offspring of substance using parents demonstrate hyporesponsiveness of cortisol and β-endorphine in response to stress and exaggerated β-endorphine in response to alcohol use (Mick et al., 2013). Significant progress has also been made in identifying genetic vulnerability factors including receptors for GABA A, dopamine r, acetylcholine, and serotonin (for review, see Schuckit, 2014). Although as yet unclear, work with the dopamine and nicotinic acetylcholine gene variants suggests an involvement of central reward systems while the GABA and NPY gene variants suggest a direct or indirect involvement of affect dysregulation as well. A full understanding of the complexity of multiple interconnections between drug effect, reward, and affect is lacking.

4 STRESS AS A RISK FACTOR FOR RELAPSE

Once use becomes chronic and problematic, cessation or abstinence is often desired but exceedingly difficult to achieve. For example, with tobacco smoking the success rate following an attempted cessation is quite low with the majority ($\geq$80%) relapsing within the first month of abstinence. One common explanation for smoking relapse is stress (Cohen and Lichtenstein, 1990). Smoking relapse due to psychosocial stress moves faster than relapse due to other explanations such as alcohol, eating, or drinking (Shiffman et al., 1996). Various physiological markers of stress predict relapse. For example, smokers who demonstrate an attenuated sympathetic and HPA stress response during the first 24 h of relapse have an increased risk of relapse status at 4 weeks postquit (al'Absi, 2006; al'Absi et al., 2004, 2005; Ceballos and al'Absi, 2006) as does heightened negative affect (al'Absi et al., 2004). Smokers report significant craving, distress, anger, and physical symptoms during withdrawal (Hughes and Hatsukami, 1986; Tiffany and Drobes, 1991). Likewise, craving or distress-induced relapse is a powerful negative reinforcement for persistent smoking (Ahmed and Koob, 2005). Although this experience is psychologically stressful, the seemingly counterintuitive attenuated response of those who relapse appears to be related, at least in part, to CRF (Erb, 2007). It is important to recall that acute exposure to nicotine induces a rise in HPA activity which, when chronic, leads to elevation of HPA function. In time this may trigger the HPA's negative feedback loop, of which glucocorticoids and the activation of the glucocorticoid receptors are the central driver. In addition to the HPA axis, catecholamines and glutamate within the nucleus accumbens and prefrontal cortex are induced by drug cues, drug use, and/or stress. These factors are also related to negative affect and, in turn, addiction, impulsivity, or poor decision making (Naqvi and Bechara, 2010). Here, we see the convergence of the experience of stress, either exogenous or specific to the withdrawal experience, negative affect, and executive cognitive skills working in concert to enhance an addict's risk for relapse during a cessation attempt (see Fig. 1).

Although we began this section using nicotine addiction as an example, it is important to point out that stress and abnormal cortisol responses have also been linked to relapse with other substances cessation attempts including cocaine, opiates, alcohol, amphetamines, and marijuana (Fox et al., 2013; Hamidovic et al., 2010; Higley et al., 2011; Sinha, 2011). With some abuse substances such as heroin, however, the cortisol response is elevated rather than attenuated, especially in response to drug paraphernalia cues (Fatseas et al., 2011). Regardless of the direction of the HPA changes, there is abundant evidence that drug use is related to dysregulation of the HPA axis, perhaps in concert with dysregulation in the emotional regulation, central reward, and executive function systems.

5 STRESS AND BEHAVIORAL ADDICTIONS

For the first time, gambling disorder was added to the DSM 5 in 2013 due to its clear addictive behavior patterns (Hasin et al., 2013). If, as we have argued above, psychological stress is a key factor in the initiation, maintenance, and relapse for all addictions, then it would stand to reason that there should be evidence of such a stress and addiction relationship among all behavioral, emotional, cognitive, and physiological parameters of stress specific to gambling. This is, indeed, the case for many of these parameters. For example, reported psychosocial stresses such as divorce, marital strife, and a history of childhood abuse are more prevalent in samples of pathological gamblers (PGs) (Black et al., 2012). Higher life stress at the time of treatment is one of the strongest predicts PG relapse at 4 months posttreatment (Gomes and Pascual-Leone, 2014). Although baseline cortisol may not be elevated with gambling disorder, there are negative correlations with the length of pathological gambling and cortisol, total gambling dysfunction, and distress over gambling behavior (Geisel et al., 2015). Further, gambling-related behaviors increase with experimental induction of a stress state, though not with all types of stressors (Steinberg et al., 2011). Early research indicates that stress physiology, as measured by the HPA,

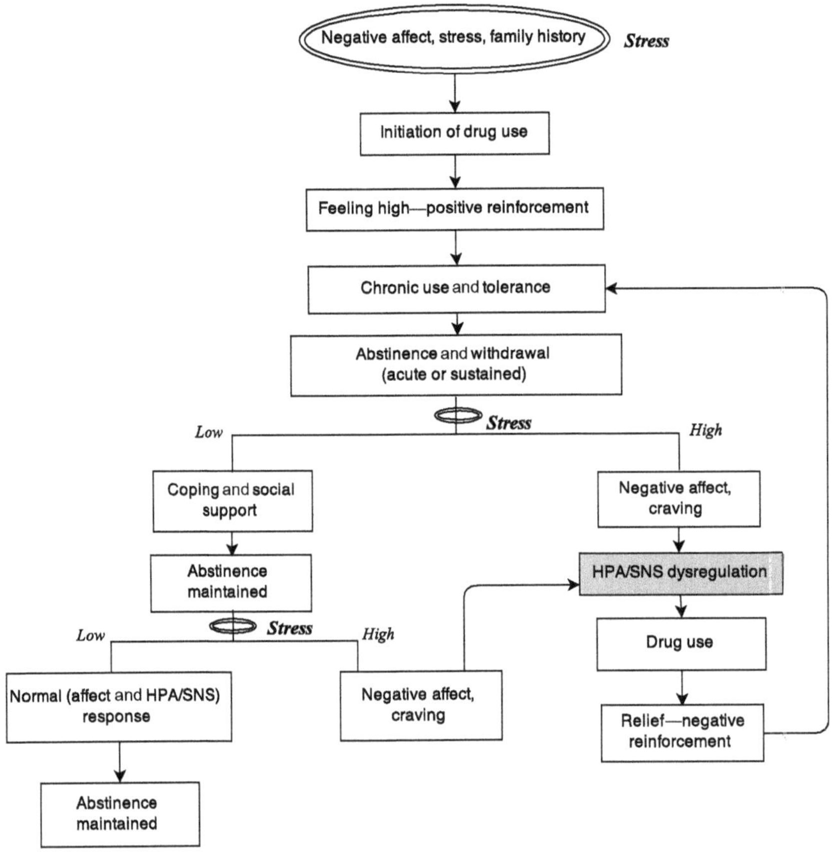

FIGURE 1

A heuristic representation of the multiple roles of stress in the addiction process including initiation, maintenance, and relapse is shown above. As described in the text, stress is a vulnerability factor for both initiation and relapse, and dysregulation of the HPA axis and the sympathetic nervous system work in concert to foster maintenance of addiction.

sympathetic, serotonergic, dopaminergic, and endogenous opioid systems, has been linked to gambling behaviors, maintenance, and relapse (Blanchard et al., 2000; Campbell-Meiklejohn et al., 2011; van den Bos et al., 2009). For PGs, basal circulating NE, EPI, and dopamine are elevated and the act of gambling is an arousal state (Meyer et al., 2004). In contrast to basal levels or gambling behaviors, the cortisol response to gambling cues may be absent for PGs but not recreational gamblers (Paris et al., 2010a,b). Finally, neuroimaging studies indicate that, like alcohol addiction, pathological gambling is associated with abnormalities in the anterior cingulate, ventral striatum, and prefrontal cortices (Koehler et al., 2013).

Other behaviors with addictive qualities (persistent and dysfunctional use or behavior leading to clinically significant impairment or distress) include hypersexuality, internet use disorder, and noneating disordered excessive eating (AKA "food addiction"). While each of these have not yet risen to the level of inclusion in the DSM 5 as a substance use disorder, there is some recognition that each of these have in common an escalating pattern of use leading to dysfunction. Further, early evidence exists of a link between these potential behavioral addictions (purported sex addiction and internet use disorder) and dopamine, self-reported stress, or the HPA axis dysregulation (Farre et al., 2015; Hou et al., 2012). Finally, there is currently a large debate on whether or not there is a "food addiction" or "eating addiction" that is distinct from the traditional eating disorders of anorexia or bulimia exists (Rogers and Smit, 2000). Although still quite controversial, those who support the notion of a food addiction point to its vulnerability to stress and the dopaminergic reward system as supporting evidence of its distinction from other eating disorders (Adam and Epel, 2007; Volkow et al., 2013).

6 MODERATORS OF THE STRESS EFFECTS ON THE ADDICTION CYCLE

6.1 SEX

Differential sex effects are seen across a broad range of addiction types and at all many levels of analysis. It has been proposed that estrogen mediates many of the sex differences observed related to drug cue sensitivity, stress responsivity, and the negative reinforcement properties of drugs (Bobzean et al., 2014). In addition, there are sex differences in the HPA, emotional, and cognitive dysregulations with most, if not all, addictions (Hildebrandt and Greif, 2013). Within the area of nicotine addiction and smoking, elevated salivary cortisol collected in women during the early abstinence phase predicted quicker relapse while the opposite was true for men (al'Absi et al., 2015). Such sex differences do not appear to be exclusive to drug addiction. For example, behavioral addictions such as gambling are also notable for sex differences in HPA functioning. Studies of the HPA axis show hypoarousal in female and hyperarousal in male gamblers (Franco et al., 2010). While both genders show an increase in cortisol in response to gambling cues (film), with actual or mock gambling males show a linear negative relationship between cortisol levels a decline in performance but females have shown a curvilinear relationship where very low responding or high responding (cortisol) leads to declining performance (Paris et al., 2010a,b; van den Bos et al., 2009). There are, however, many unanswered questions, particularly on other sex-dependent cofactors and their effects on other neuromodulators (for an excellent review, see Bisagno and Cadet, 2014).

6.2 GENETICS

The genetics and epigenetics of addiction have progressed in multiple areas in the study of specific addictive substances (Levran et al., 2012; McCarthy et al., 2012; Ponomarev, 2013) and the common behaviors span all addictions such as craving, impulsivity, conditioning, or reward (Nestler, 2014). In parallel, the study of stress-related genes or epigenetic factors has also made significant strides related to early life stress and later risk of psychopathology (Lewis and Olive, 2014), stress across the developmental spectrum, the regulation of the stress response, cognition, or reproductive behavior (Gudsnuk and Champagne, 2012), social adversity and multiple indicators of health and well-being, and stress in learning (Mifsud et al., 2011). In contrast, limited studies have simultaneously examined the specific interaction between stress and addiction. Notable exceptions include the study of transcription factors such as Nur which is capable of acting as intermediate-early genes (Campos-Melo et al., 2013), nuclear factor kappa-B (NFκB) induction of innate immune genes (Crews et al., 2011), and cAMP response element-binding protein (CREB) target genes, specifically those that regulate brain-derived neurotrophic factor, CRF, and dynorphin (Briand and Blendy, 2010). For example, both stress and drug abuse increase NFκB transcription results in increases in brain inflammatory mediators which, in turn, have been linked to widespread and persistent alterations in behavioral factors reflective of the neurobiology of addiction such as frontal lobe excitability, negative affect, behavioral flexibility, and loss of behavioral control (Crews et al., 2011). Animal work has shown that Nur transcription factors are activated by stress and that drugs of abuse may induce dopamine- and glutamate-dependent Nur transcription in the striatum and nucleus accumbens, both of which display tolerance and sensitization (Campos-Melo et al., 2013). It has been proposed that CREB may be a critical link controlling the link between reward circuits and the HPA axis and it may therefore help to explain, in part, the complex interaction between stress and drug abuse (Briand and Blendy, 2010). Finally, variants to the μ-opioid receptors have been identified as a potential gene mediating food addiction via reward, craving, and preference. Work in humans has shown that individuals who are AA or GG homozygous for the A118G μ-opioid receptor gene have enhanced hedonic responsiveness relative to heterozygous (GA) individuals which, in turn, predicts greater propensity toward food-related addictive behaviors (Davis and Loxton, 2014).

7 DISCUSSION

As reviewed above, research on nicotine and alcohol addiction clearly indicates that there are important sex differences in the pattern of addictive behaviors, relapse, and stress neurobiology. There is a need for further clarity on sex differences in other areas of addiction as well, particularly the newer areas of study such as gambling disorder, internet use disorder, and hypersexuality. Much of this literature includes males only and those that do include women fail to analyze for sex differences. Likewise, there are very few studies with non-Caucasian minority groups.

Some studies involving a direct comparison of addiction types have shown that not all stressors affect addictions equally. For example, controllable and uncontrollable noise stress increase the desire for alcohol equally across alcoholics, PGs, dually diagnosed participants, and controls while only uncontrollable noise increased the desire to gamble selectively within those with a history of pathological gambling (Steinberg et al., 2011). How stressor characteristics interact with genetic vulnerability is, as yet, understudied. Thus, further studies of the gene by environment interaction specific to stress and addiction are needed. Finally, further study of stress resilience and coping will be important for informing future intervention research.

8 CONCLUSION

The stress response has been shown to increase risk of initiation, maintenance, and relapse of a variety of addictions. Evidence was presented in support of this proposal utilizing all levels of analysis from the macro- (cross-sectional, population research) to the micro-level (molecular, genetic, epigenetic research). The study of reward systems in the brain and their interactions with HPA, sympathetic, eCB, and other systems has taught us a great deal about the relationship between stress and addiction. We propose the psychological stress and the biological stress response acts to moderate the probability of initiation, maintenance, and relapse (see Fig. 1). As an individual encounters stress-inducing environments, there are individual differences that help to determine whether there is increased (high stress, right-hand path on Fig. 1) or decreased (low stress, left-hand path on Fig. 1) probability of problematic substance use. There are, however, other brain systems and moderators to explore. In particular, we have only just begun to understand the gene by environment interactions that occur within the context of stress and addiction. Likewise, there are a host of behavioral, neuroendocrine, and cognitive differences between men and women that have yet to be explored as potential moderators capable of accounting for sex differences in stress and addiction. Finally, funding sources and the public are increasingly demanding of reliable and efficient interventions, particularly in light of the high treatment failure rates and the role that stress plays in relapse. To address this gap future treatment-specific work, whether they include pharmacological or behavioral interventions, could benefit from further investigation of stress resilience.

REFERENCES

Adam, T.C., Epel, E.S., 2007. Stress, eating and the reward system. Physiol. Behav. 91, 449–458.

Ahmed, S.H., Koob, G.F., 2005. Transition to drug addiction: a negative reinforcement model based on an allostatic decrease in reward function. Psychopharmacology (Berl) 180, 473–490.

Akerstedt, T., 2006. Psychosocial stress and impaired sleep. Scand. J. Work Environ. Health 32, 493–501.

al'Absi, M., 2006. Hypothalamic-pituitary-adrenocortical responses to psychological stress and risk for smoking relapse. Int. J. Psychophysiol. 59, 218–227.

al'Absi, M., Bongard, S., 2006. Neuroendocrine and behavioral mechanisms mediating the relationship between anger expression and cardiovascular risk: assessment considerations and improvements. J. Behav. Med. 29, 573–591.

al'Absi, M., Hatsukami, D., Davis, G.L., Wittmers, L.E., 2004. Prospective examination of effects of smoking abstinence on cortisol and withdrawal symptoms as predictors of early smoking relapse. Drug Alcohol Depend. 73, 267–278.

al'Absi, M., Hatsukami, D., Davis, G.L., 2005. Attenuated adrenocorticotropic responses to psychological stress are associated with early smoking relapse. Psychopharmacology (Berl) 181, 107–117.

al'Absi, M., Carr, S.B., Bongard, S., 2007. Anger and psychobiological changes during smoking abstinence and in response to acute stress: prediction of smoking relapse. Int. J. Psychophysiol. 66, 109–115.

al'Absi, M., Wittmers, L.E., Hatsukami, D., Westra, R., 2008. Blunted opiate modulation of hypothalamic-pituitary-adrenocortical activity in men and women who smoke. Psychosom. Med. 70, 928–935.

al'Absi, M., Nakajima, M., Allen, S., Lemieux, A., Hatsukami, D., 2015. Sex differences in hormonal responses to stress and smoking relapse: a prospective examination. Nicotine Tob. Res. 17, 382–389.

Allolio, B., 2015. Extensive expertise in endocrinology. Adrenal crisis. Eur. J. Endocrinol. 172, R115–R124.

American Psychiatric Association, 2000. Diagnostic and Statistical Manual of Mental Disorders, fourth ed. American Psychiatric Association, Washington, DC, text revision.

American Psychiatric Association, 2014. Diagnostic and Statistical Manual of Mental Disorders. American Psychiatric Association, Arlington, VA.

Baik, J.H., 2013. Dopamine signaling in reward-related behaviors. Front. Neural Circuits 7, 152.

Banducci, A.N., Hoffman, E.M., Lejuez, C.W., Koenen, K.C., 2014. The impact of childhood abuse on inpatient substance users: specific links with risky sex, aggression, and emotion dysregulation. Child Abuse Negl. 38, 928–938.

Barrot, M., Marinelli, M., Abrous, D.N., Rouge-Pont, F., Le Moal, M., Piazza, P.V., 2000. The dopaminergic hyper-responsiveness of the shell of the nucleus accumbens is hormone-dependent. Eur. J. Neurosci. 12, 973–979.

Bisagno, V., Cadet, J.L., 2014. Stress, sex, and addiction: potential roles of corticotropin-releasing factor, oxytocin, and arginine-vasopressin. Behav. Pharmacol. 25, 445–457.

Black, D.W., Shaw, M.C., McCormick, B.A., Allen, J., 2012. Marital status, childhood maltreatment, and family dysfunction: a controlled study of pathological gambling. J. Clin. Psychiatry 73, 1293–1297.

Blanchard, E.B., Wulfert, E., Freidenberg, B.M., Malta, L.S., 2000. Psychophysiological assessment of compulsive gamblers' arousal to gambling cues: a pilot study. Appl. Psychophysiol. Biofeedback 25, 155–165.

Bobzean, S.A., DeNobrega, A.K., Perrotti, L.I., 2014. Sex differences in the neurobiology of drug addiction. Exp. Neurol. 259, 64–74.

Boyson, C.O., Holly, E.N., Shimamoto, A., Albrechet-Souza, L., Weiner, L.A., DeBold, J.F., Miczek, K.A., 2014. Social stress and CRF-dopamine interactions in the VTA: role in long-term escalation of cocaine self-administration. J. Neurosci. 34, 6659–6667.

Brenhouse, H.C., Lukkes, J.L., Andersen, S.L., 2013. Early life adversity alters the developmental profiles of addiction-related prefrontal cortex circuitry. Brain Sci. 3, 143–158.

Brensilver, M., Heinzerling, K.G., Swanson, A.N., Telesca, D., Furst, B.A., Shoptaw, S.J., 2013. Cigarette smoking as a target for potentiating outcomes for methamphetamine abuse treatment. Drug Alcohol Rev. 32, 96–99.

Briand, L.A., Blendy, J.A., 2010. Molecular and genetic substrates linking stress and addiction. Brain Res. 1314, 219–234.

Buss, C., Davis, E.P., Shahbaba, B., Pruessner, J.C., Head, K., Sandman, C.A., 2012. Maternal cortisol over the course of pregnancy and subsequent child amygdala and hippocampus volumes and affective problems. Proc. Natl. Acad. Sci. USA 109, E1312–E1319.

Campbell, J.C., Szumlinski, K.K., Kippin, T.E., 2009. Contribution of early environmental stress to alcoholism vulnerability. Alcohol 43, 547–554.

Campbell-Meiklejohn, D., Wakeley, J., Herbert, V., Cook, J., Scollo, P., Ray, M.K., Selvaraj, S., Passingham, R.E., Cowen, P., Rogers, R.D., 2011. Serotonin and dopamine play complementary roles in gambling to recover losses. Neuropsychopharmacology 36, 402–410.

Campos-Melo, D., Galleguillos, D., Sanchez, N., Gysling, K., Andres, M.E., 2013. Nur transcription factors in stress and addiction. Front. Mol. Neurosci. 6, 44.

Carboni, E., Barros, V.G., Ibba, M., Silvagni, A., Mura, C., Antonelli, M.C., 2010. Prenatal restraint stress: an in vivo microdialysis study on catecholamine release in the rat prefrontal cortex. Neuroscience 168, 156–166.

Ceballos, N.A., al'Absi, M., 2006. Dehydroepiandrosterone sulfate, cortisol, mood state and smoking cessation: relationship to relapse status at 4-week follow-up. Pharmacol. Biochem. Behav. 85, 23–28.

Chaijale, N.N., Curtis, A.L., Wood, S.K., Zhang, X.Y., Bhatnagar, S., Reyes, B.A., Van Bockstaele, E.J., Valentino, R.J., 2013. Social stress engages opioid regulation of locus coeruleus norepinephrine neurons and induces a state of cellular and physical opiate dependence. Neuropsychopharmacology 38, 1833–1843.

Chandler, D.J., Waterhouse, B.D., Gao, W.J., 2014. New perspectives on catecholaminergic regulation of executive circuits: evidence for independent modulation of prefrontal functions by midbrain dopaminergic and noradrenergic neurons. Front. Neural Circuits 8, 53.

Chida, Y., Hamer, M., 2008. Chronic psychosocial factors and acute physiological responses to laboratory-induced stress in healthy populations: a quantitative review of 30 years of investigations. Psychol. Bull. 134, 829–885.

Cohen, S., Lichtenstein, E., 1990. Perceived stress, quitting smoking, and smoking relapse. Health Psychol. 9, 466–478.

Coplan, J.D., Abdallah, C.G., Kaufman, J., Gelernter, J., Smith, E.L., Perera, T.D., Dwork, A.J., Kaffman, A., Gorman, J.M., Rosenblum, L.A., Owens, M.J., Nemeroff, C.B., 2011. Early-life stress, corticotropin-releasing factor, and serotonin transporter gene: a pilot study. Psychoneuroendocrinology 36, 289–293.

Crews, F.T., Zou, J., Qin, L., 2011. Induction of innate immune genes in brain create the neurobiology of addiction. Brain Behav. Immun. 25 (Suppl. 1), S4–S12.

Curtis, A.L., Leiser, S.C., Snyder, K., Valentino, R.J., 2012. Predator stress engages corticotropin-releasing factor and opioid systems to alter the operating mode of locus coeruleus norepinephrine neurons. Neuropharmacology 62, 1737–1745.

Dai, X., Thavundayil, J., Santella, S., Gianoulakis, C., 2007. Response of the HPA-axis to alcohol and stress as a function of alcohol dependence and family history of alcoholism. Psychoneuroendocrinology 32, 293–305.

Davis, C., Loxton, N.J., 2014. A psycho-genetic study of hedonic responsiveness in relation to "food addiction" Nutrients 6, 4338–4353.

Di Marzo, V., De Petrocellis, L., Bisogno, T., 2005. The biosynthesis, fate and pharmacological properties of endocannabinoids. Handb. Exp. Pharmacol. 168, 147–185.
Dlugos, A., Childs, E., Stuhr, K.L., Hillard, C.J., de Wit, H., 2012. Acute stress increases circulating anandamide and other N-acylethanolamines in healthy humans. Neuropsychopharmacology 37, 2416–2427.
Dunn, A.J., Swiergiel, A.H., Palamarchouk, V., 2004. Brain circuits involved in corticotropin-releasing factor-norepinephrine interactions during stress. Ann. N. Y. Acad. Sci. 1018, 25–34.
Enoch, M.A., 2011. The role of early life stress as a predictor for alcohol and drug dependence. Psychopharmacology (Berl) 214, 17–31.
Erb, S., 2007. Neurobiology of stress and risk for relapse. In: al'Absi, M. (Ed.), Stress and Addiction: Biological and Psychological Mechanisms. Academic Press/Elsevier, London, pp. 147–169.
Evren, C., Cinar, O., Evren, B., Ulku, M., Karabulut, V., Umut, G., 2013. The mediator roles of trait anxiety, hostility, and impulsivity in the association between childhood trauma and dissociation in male substance-dependent inpatients. Compr. Psychiatry 54, 158–166.
Farre, J.M., Fernandez-Aranda, F., Granero, R., Aragay, N., Mallorqui-Bague, N., Ferrer, V., More, A., Bouman, W.P., Arcelus, J., Savvidou, L.G., Penelo, E., Aymami, M.N., Gomez-Pena, M., Gunnard, K., Romaguera, A., Menchon, J.M., Valles, V., Jimenez-Murcia, S., 2015. Sex addiction and gambling disorder: similarities and differences. Compr. Psychiatry 56, 59–68.
Fatseas, M., Denis, C., Massida, Z., Verger, M., Franques-Reneric, P., Auriacombe, M., 2011. Cue-induced reactivity, cortisol response and substance use outcome in treated heroin dependent individuals. Biol. Psychiatr. 70 (8), 720–727.
Fox, H.C., Garcia Jr., M., Kemp, K., Milivojevic, V., Kreek, M.J., Sinha, R., 2006. Gender differences in cardiovascular and corticoadrenal response to stress and drug cues in cocaine dependent individuals. Psychopharmacology (Berl) 185, 348–357.
Fox, H.C., Tuit, K.L., Sinha, R., 2013. Stress system changes associated with marijuana dependence may increase craving for alcohol and cocaine. Hum. Psychopharmacol. 28, 40–53.
Franco, C., Paris, J., Wulfert, E., Frye, C., 2010. Male gamblers have significantly greater salivary cortisol before and after betting on a horse race, than do female gamblers. Physiol. Behav. 99, 225–229.
Fride, E., Gobshtis, N., Dahan, H., Weller, A., Giuffrida, A., Ben-Shabat, S., 2009. The endocannabinoid system during development: emphasis on perinatal events and delayed effects. Vitam. Horm. 81, 139–158.
Geisel, O., Panneck, P., Hellweg, R., Wiedemann, K., Muller, C.A., 2015. Hypothalamic-pituitary-adrenal axis activity in patients with pathological gambling and internet use disorder. Psychiatry Res. 226, 97–102.
Gianoulakis, C., Dai, X., Thavundayil, J., Brown, T., 2005. Levels and circadian rhythmicity of plasma ACTH, cortisol, and beta-endorphin as a function of family history of alcoholism. Psychopharmacology (Berl) 181, 437–444.
Goldstein, D.S., Kopin, I.J., 2008. Adrenomedullary, adrenocortical, and sympathoneural responses to stressors: a meta-analysis. Endocr. Regul. 42, 111–119.
Gomes, K., Pascual-Leone, A., 2014. A resource model of change: client factors that influence problem gambling treatment outcomes. J. Gambl. Stud. Epub ahead of print, PMID: 25112220.

Graf, E.N., Wheeler, R.A., Baker, D.A., Ebben, A.L., Hill, J.E., McReynolds, J.R., Robble, M.A., Vranjkovic, O., Wheeler, D.S., Mantsch, J.R., Gasser, P.J., 2013. Corticosterone acts in the nucleus accumbens to enhance dopamine signaling and potentiate reinstatement of cocaine seeking. J. Neurosci. 33, 11800–11810.

Gudsnuk, K., Champagne, F.A., 2012. Epigenetic influence of stress and the social environment. ILAR J. 53, 279–288.

Hamidovic, A., Childs, E., Conrad, M., King, A., de Wit, H., 2010. Stress-induced changes in mood and cortisol release predict mood effects of amphetamine. Drug Alcohol Depend. 109, 175–180.

Hasin, D.S., O'Brien, C.P., Auriacombe, M., Borges, G., Bucholz, K., Budney, A., Compton, W.M., Crowley, T., Ling, W., Petry, N.M., Schuckit, M., Grant, B.F., 2013. DSM-5 criteria for substance use disorders: recommendations and rationale. Am. J. Psychiatry 170, 834–851.

Hausknecht, K., Haj-Dahmane, S., Shen, R.Y., 2013. Prenatal stress exposure increases the excitation of dopamine neurons in the ventral tegmental area and alters their responses to psychostimulants. Neuropsychopharmacology 38, 293–301.

Herman, J.P., Tasker, J.G., Ziegler, D.R., Cullinan, W.E., 2002. Local circuit regulation of paraventricular nucleus stress integration: glutamate-GABA connections. Pharmacol. Biochem. Behav. 71, 457–468.

Herman, J.P., Mueller, N.K., Figueiredo, H., 2004. Role of GABA and glutamate circuitry in hypothalamo-pituitary-adrenocortical stress integration. Ann. N. Y. Acad. Sci. 1018, 35–45.

Heron, J., Barker, E.D., Joinson, C., Lewis, G., Hickman, M., Munafo, M., Macleod, J., 2013. Childhood conduct disorder trajectories, prior risk factors and cannabis use at age 16: birth cohort study. Addiction 108, 2129–2138.

Het, S., Ramlow, G., Wolf, O.T., 2005. A meta-analytic review of the effects of acute cortisol administration on human memory. Psychoneuroendocrinology 30, 771–784.

Higley, A.E., Crane, N.A., Spadoni, A.D., Quello, S.B., Goodell, V., Mason, B.J., 2011. Craving in response to stress induction in a human laboratory paradigm predicts treatment outcome in alcohol-dependent individuals. Psychopharmacology (Berl) 218, 121–129.

Hildebrandt, T., Greif, R., 2013. Stress and addiction. Psychoneuroendocrinology 38, 1923–1927.

Hill, M.N., McEwen, B.S., 2010. Involvement of the endocannabinoid system in the neurobehavioural effects of stress and glucocorticoids. Prog. Neuropsychopharmacol. Biol. Psychiatry 34, 791–797.

Hinkle Jr., L.E., 1987. Stress and disease: the concept after 50 years. Soc. Sci. Med. 25, 561–566.

Hou, H., Jia, S., Hu, S., Fan, R., Sun, W., Sun, T., Zhang, H., 2012. Reduced striatal dopamine transporters in people with internet addiction disorder. J. Biomed. Biotechnol. 2012, 854524.

Hughes, J.R., Hatsukami, D., 1986. Signs and symptoms of tobacco withdrawal. Arch. Gen. Psychiatry 43, 289–294.

Hyman, S.M., Garcia, M., Sinha, R., 2006. Gender specific associations between types of childhood maltreatment and the onset, escalation and severity of substance use in cocaine dependent adults. Am. J. Drug Alcohol Abuse 32, 655–664.

Kassel, J.D., Veilleux, J.C., Wardle, M.C., Yates, M.C., Greenstein, J.E., Evatt, D.P., Roesch, L.L., 2007. Negative affect and addiction. In: al'Absi, M. (Ed.), Stress and Addiction. Academic Press, London, pp. 171–190.

Kavalali, E.T., 2015. The mechanisms and functions of spontaneous neurotransmitter release. Nat. Rev. Neurosci. 16, 5–16.

Kim, S.W., Kang, H.J., Kim, S.Y., Kim, J.M., Yoon, J.S., Jung, S.W., Lee, M.S., Yim, H.W., Jun, T.Y., 2013. Impact of childhood adversity on the course and suicidality of depressive disorders: the CRESCEND study. Depress. Anxiety 30, 965–974.

Kippin, T.E., Szumlinski, K.K., Kapasova, Z., Rezner, B., See, R.E., 2008. Prenatal stress enhances responsiveness to cocaine. Neuropsychopharmacology 33, 769–782.

Kirby, L.G., Zeeb, F.D., Winstanley, C.A., 2011. Contributions of serotonin in addiction vulnerability. Neuropharmacology 61, 421–432.

Koehler, S., Ovadia-Caro, S., van der Meer, E., Villringer, A., Heinz, A., Romanczuk-Seiferth, N., Margulies, D.S., 2013. Increased functional connectivity between prefrontal cortex and reward system in pathological gambling. PLoS One 8, e84565.

Kondo, F., Tachi, M., Gosho, M., Fukayama, M., Yoshikawa, K., Okada, S., 2015. Changes in hypothalamic neurotransmitter and prostanoid levels in response to NMDA, CRF, and GLP-1 stimulation. Anal. Bioanal. Chem. 407, 5261–5272. doi:http://dx.doi.org/10.1007/s00216-015-8496-6. PMID:25633219.

Korte, S.M., Van Duin, S., Bouws, G.A., Koolhaas, J.M., Bohus, B., 1991. Involvement of hypothalamic serotonin in activation of the sympathoadrenomedullary system and hypothalamo-pituitary-adrenocortical axis in male Wistar rats. Eur. J. Pharmacol. 197, 225–228.

Kringelbach, M.L., Stein, A., van Hartevelt, T.J., 2012. The functional human neuroanatomy of food pleasure cycles. Physiol. Behav. 106, 307–316.

Lakehayli, S., Said, N., Battas, O., Hakkou, F., Tazi, A., 2015. Prenatal stress alters sensitivity to benzodiazepines in adult rats. Neurosci. Lett. 591, 187–191.

Lawrence, A.D., Brooks, D.J., 2014. Ventral striatal dopamine synthesis capacity is associated with individual differences in behavioral disinhibition. Front. Behav. Neurosci. 8, 86.

Levran, O., Yuferov, V., Kreek, M.J., 2012. The genetics of the opioid system and specific drug addictions. Hum. Genet. 131, 823–842.

Lewis, C.R., Olive, M.F., 2014. Early-life stress interactions with the epigenome: potential mechanisms driving vulnerability toward psychiatric illness. Behav. Pharmacol. 25, 341–351.

Lovallo, W.R., 2013. Early life adversity reduces stress reactivity and enhances impulsive behavior: implications for health behaviors. Int. J. Psychophysiol. 90, 8–16.

Lupien, S.J., McEwen, B.S., Gunnar, M.R., Heim, C., 2009. Effects of stress throughout the lifespan on the brain, behaviour and cognition. Nat. Rev. Neurosci. 10, 434–445.

Macleod, J., Hickman, M., Jones, H.E., Copeland, L., McKenzie, J., De Angelis, D., Kimber, J., Robertson, J.R., 2013. Early life influences on the risk of injecting drug use: case control study based on the Edinburgh Addiction Cohort. Addiction 108, 743–750.

Marinelli, M., 2007. Dopaminergic reward pathways and effects of stress. In: Al'Absi, M. (Ed.), Stress and Addiction: Biological and Psychological Mechanisms. Elsevier, San Diego, CA, pp. 41–83.

Martens, A., Greenberg, J., Allen, J.J., 2008. Self-esteem and autonomic physiology: parallels between self-esteem and cardiac vagal tone as buffers of threat. Pers. Soc. Psychol. Rev. 12, 370–389.

Mason, W.A., Spoth, R.L., 2012. Sequence of alcohol involvement from early onset to young adult alcohol abuse: differential predictors and moderation by family-focused preventive intervention. Addiction 107, 2137–2148.

McCarthy, D.M., Brown, A.N., Bhide, P.G., 2012. Regulation of BDNF expression by cocaine. Yale J. Biol. Med. 85, 437–446.

Meerlo, P., Sgoifo, A., Suchecki, D., 2008. Restricted and disrupted sleep: effects on autonomic function, neuroendocrine stress systems and stress responsivity. Sleep Med. Rev. 12, 197–210.

Meneses, A., 2013. 5-HT systems: emergent targets for memory formation and memory alterations. Rev. Neurosci. 24, 629–664.

Meyer, G., Schwertfeger, J., Exton, M.S., Janssen, O.E., Knapp, W., Stadler, M.A., Schedlowski, M., Kruger, T.H., 2004. Neuroendocrine response to casino gambling in problem gamblers. Psychoneuroendocrinology 29, 1272–1280.

Mick, I., Spring, K., Uhr, M., Zimmermann, U.S., 2013. Alcohol administration attenuates hypothalamic-pituitary-adrenal (HPA) activity in healthy men at low genetic risk for alcoholism, but not in high-risk subjects. Addict. Biol. 18, 863–871.

Mifsud, K.R., Gutierrez-Mecinas, M., Trollope, A.F., Collins, A., Saunderson, E.A., Reul, J.M., 2011. Epigenetic mechanisms in stress and adaptation. Brain Behav. Immun. 25, 1305–1315.

Naqvi, N.H., Bechara, A., 2010. The insula and drug addiction: an interoceptive view of pleasure, urges, and decision-making. Brain Struct. Funct. 214, 435–450.

Nestler, E.J., 2014. Epigenetic mechanisms of drug addiction. Neuropharmacology 76 (Pt. B), 259–268.

Nylander, I., Roman, E., 2013. Is the rodent maternal separation model a valid and effective model for studies on the early-life impact on ethanol consumption? Psychopharmacology (Berl) 229, 555–569.

Paris, J.J., Franco, C., Sodano, R., Freidenberg, B., Gordis, E., Anderson, D.A., Forsyth, J.P., Wulfert, E., Frye, C.A., 2010a. Sex differences in salivary cortisol in response to acute stressors among healthy participants, in recreational or pathological gamblers, and in those with posttraumatic stress disorder. Horm. Behav. 57, 35–45.

Paris, J.J., Franco, C., Sodano, R., Frye, C.A., Wulfert, E., 2010b. Gambling pathology is associated with dampened cortisol response among men and women. Physiol. Behav. 99, 230–233.

Ponomarev, I., 2013. Epigenetic control of gene expression in the alcoholic brain. Alcohol Res. 35, 69–76.

Reynaert, M.L., Marrocco, J., Gatta, E., Mairesse, J., Van Camp, G., Fagioli, F., Maccari, S., Nicoletti, F., Morley-Fletcher, S., 2015. A self-medication hypothesis for increased vulnerability to drug abuse in prenatally restraint stressed rats. Adv. Neurobiol. 10, 101–120.

Rogers, P.J., Smit, H.J., 2000. Food craving and food "addiction": a critical review of the evidence from a biopsychosocial perspective. Pharmacol. Biochem. Behav. 66, 3–14.

Rokyta, R., Yamamotova, A., Slamberova, R., Franek, M., Vaculin, S., Hruba, L., Schutova, B., Pometlova, M., 2008. Prenatal and perinatal factors influencing nociception, addiction and behavior during ontogenetic development. Physiol. Res. 57 (Suppl. 3), S79–S88.

Salamone, J.D., Correa, M., 2013. Dopamine and food addiction: lexicon badly needed. Biol. Psychiatry 73, e15–e24.

Schuckit, M.A., 2014. A brief history of research on the genetics of alcohol and other drug use disorders. J. Stud. Alcohol Drugs Suppl. 75 (Suppl. 17), 59–67.

Shiffman, S., Paty, J.A., Gnys, M., Kassel, J.A., Hickcox, M., 1996. First lapses to smoking: within-subjects analysis of real-time reports. J. Consult. Clin. Psychol. 64, 366–379.

Silvagni, A., Barros, V.G., Mura, C., Antonelli, M.C., Carboni, E., 2008. Prenatal restraint stress differentially modifies basal and stimulated dopamine and noradrenaline release in the nucleus accumbens shell: an 'in vivo' microdialysis study in adolescent and young adult rats. Eur. J. Neurosci. 28, 744–758.

Sinha, R., 2011. New findings on biological factors predicting addiction relapse vulnerability. Curr. Psychiatry Rep. 13, 398–405.

Steinberg, L., Tremblay, A.M., Zack, M., Busto, U.E., Zawertailo, L.A., 2011. Effects of stress and alcohol cues in men with and without problem gambling and alcohol use disorder. Drug Alcohol Depend. 119, 46–55.

Steiner, M.A., Wotjak, C.T., 2008. Role of the endocannabinoid system in regulation of the hypothalamic-pituitary-adrenocortical axis. Prog. Brain Res. 170, 397–432.

Tiffany, S.T., Drobes, D.J., 1991. The development and initial validation of a questionnaire on smoking urges. Br. J. Addict. 86, 1467–1476.

Van Bockstaele, E.J., Valentino, R.J., 2013. Neuropeptide regulation of the locus coeruleus and opiate-induced plasticity of stress responses. In: Eiden, L.E. (Ed.), New Era of Catecholamines in the Laboratory and Clinic, vol. 68. Elsevier, San Diego, CA, pp. 405–420.

van den Bos, R., Harteveld, M., Stoop, H., 2009. Stress and decision-making in humans: performance is related to cortisol reactivity, albeit differently in men and women. Psychoneuroendocrinology 34, 1449–1458.

Van Waes, V., Enache, M., Berton, O., Vinner, E., Lhermitte, M., Maccari, S., Darnaudery, M., 2011. Effect of prenatal stress on alcohol preference and sensitivity to chronic alcohol exposure in male rats. Psychopharmacology (Berl) 214, 197–208.

Volkow, N.D., Wang, G.J., Tomasi, D., Baler, R.D., 2013. The addictive dimensionality of obesity. Biol. Psychiatry 73, 811–818.

Walsh, E.G., Cawthon, S.W., 2014. The mediating role of depressive symptoms in the relationship between adverse childhood experiences and smoking. Addict. Behav. 39, 1471–1476.

Wang, H., Lupica, C.R., 2014. Release of endogenous cannabinoids from ventral tegmental area dopamine neurons and the modulation of synaptic processes. Prog. Neuropsychopharmacol. Biol. Psychiatry 52, 24–27.

CHAPTER

Reinforcement principles for addiction medicine; from recreational drug use to psychiatric disorder

4

Scott Edwards[1]

Department of Physiology, Alcohol and Drug Abuse Center of Excellence, Neuroscience Center of Excellence, Louisiana State University Health Sciences Center, New Orleans, LA, USA
[1]Corresponding author: Tel.: +1-619-2413380; Fax: +1-504-5686158, e-mail address: sedwa5@lsuhsc.edu

Abstract

The transition from recreational drug use to addiction can be conceptualized as a pathological timeline whereby the psychological mechanisms responsible for disordered drug use evolve from positive reinforcement to favor elements of negative reinforcement. Abused substances (ranging from alcohol to psychostimulants) are initially ingested at regular occasions according to their positive reinforcing properties. Importantly, repeated exposure to rewarding substances sets off a chain of secondary reinforcing events, whereby cues and contexts associated with drug use may themselves become reinforcing and thereby contribute to the continued use and possible abuse of the substance(s) of choice. Indeed, the powerful reinforcing efficacy of certain drugs may eclipse that of competing social rewards (such as career and family) and lead to an aberrant narrowing of behavioral repertoire. In certain vulnerable individuals, escalation of drug use over time is thought to drive specific molecular neuroadaptations that foster the development of addiction. Research has identified neurobiological elements of altered reinforcement following excessive drug use that comprise within-circuit and between-circuit neuroadaptations, both of which contribute to addiction. Central to this process is the eventual potentiation of negative reinforcement mechanisms that may represent the final definitive criterion locking vulnerable individuals into a persistent state of addiction. Targeting the neural substrates of reinforcement likely represents our best chances for therapeutic intervention for this devastating disease.

Keywords

Addiction, Alcoholism, Amygdala, Dopamine, Nucleus accumbens, Reward, Reinforcement, Self-administration, Substance use disorder

1 INTRODUCTION

Drug addiction is a chronic disease that is often characterized by escalation of drug intake over time and a pathological and compulsive drug-seeking behavior. Emerging conceptualizations of drug addiction (or substance use disorder; DSM-5, 2013) have included a constantly evolving set of defining characteristics (Wise and Koob, 2014). However, the behavioral principle of reinforcement has remained as a central feature of multiple diagnostic criteria. As such, a fundamental understanding of reinforcement principles and protocols is essential to model both the psychological and neurobiological processes separating initial drug use from the eventual development of addiction in vulnerable individuals. An ongoing marriage of behavioral and molecular techniques has driven a steady evolution of the "black box" working model of human behavior by early psychologists toward the revelation of neurophysiological mechanisms linking stimulus processing to action. While drugs of abuse represent powerful pharmacological agents that produce significant neuroadaptations in the short-term, further clarification of the neuroadaptations that functionally drive specific elements (e.g., positive vs. negative reinforcement) within the addiction framework is crucial (Kalivas, 2005). Additionally, since addiction can be described as an evolving disease whereby recreational drug sampling can quickly transition to dependence, determining the precise brain circuitry changes associated with each stage would prove beneficial for the development of new and more effective therapeutic strategies. A central question is whether reinforcement processes and underlying neurobiological changes caused by early drug use are simply potentiated in the addicted state, or whether there is a separate recruitment and potentiation of additional reinforcement elements and circuitry that is indispensable for the addiction process.

2 POSITIVE AND NEGATIVE REINFORCEMENT

The process of reinforcement is most commonly associated with *positive* reinforcement, whereby the presence of a motivational stimulus augments or strengthens a particular behavioral response. Although the antecedent stimulus can be generally described as subjectively rewarding or hedonically positive, this is not a necessary requirement for a stimulus to reinforce a response. Likewise, stimuli that are subjectively rewarding are not automatically reinforcing. Consequently, it is important to distinguish reward from reinforcement as psychological constructs and to investigate their neurobiological mechanisms separately. Stimuli that do lead to response strengthening (i.e., reinforcers) may produce either an increase in frequency or duration of responding or a decrease in latency to respond, and an emphasis on measuring such objective criteria is crucial to the study of reinforcement as originally conceptualized by Skinner (1938). Neuroanatomical and neurochemical substrates underlying the positive reinforcing effects of abused drugs have been elucidated over the past few decades, with mesolimbic dopamine and endogenous opioid systems that mediate natural reward playing a central role (Corbett and Wise, 1980; Wise and Bozarth, 1981; Yokel and Wise, 1975). *Negative* reinforcement is a special condition associated with a strengthening of behavioral responses that terminate some ongoing (presumably aversive) stimulus. In this case we can define a negative reinforcer as a motivational stimulus that strengthens such an "escape" response. Historically, in relation to drug addiction, this phenomenon has been consistently observed in humans whereby drugs of abuse are self-administered to quench a motivational need in the state of withdrawal (Wikler, 1952). It's worth reiterating that both positive and negative reinforcement always leads to an increase in organized behavior, and this relationship stands in contrast to the concept of *punishment*, whereby behavioral action is suppressed or terminated by a stimulus. Importantly, relationships between stimulus and response are further complicated by timing and contingency. A stimulus that acts as a negative reinforcer if presented before an action might be expected to act as a punisher if presented at the same time as (i.e., made contingent on) the response. As one example, electrical shock may act as an antecedent negative reinforcer that facilitates a subsequent behavioral action to terminate shock delivery (if this option is available), whereas any ongoing behavioral action conducted at the exact time of delivery of the foot shock would be expected to be suppressed.

3 SECONDARY AND CONDITIONED REINFORCEMENT

An important dimension of reinforcement highly relevant to the addiction process (and particularly relapse) is secondary reinforcement (Stewart, 1992). Secondary reinforcers (in many cases also considered conditioned reinforcers) likely drive the majority of reinforcement processes in humans. In the specific case of drug addition, cues and contexts that are intimately and repeatedly associated with drug use will often themselves become reinforcing, leading some to conceptualize addiction as conditions of disordered reward learning and memory (Hyman et al., 2006; White, 1996). In this regard conditioned reinforcers may act to precipitate or magnify craving in withdrawal states, or alternatively to function at a subconscious level to drive habitual relapse processes (Weiss, 2005; Wikler et al., 1971). A fundamental piece of Robinson and Berridge's incentive-sensitization theory of addiction posits that the incentive value or attractive nature of such secondary reinforcement processes, in addition to the primary reinforcers themselves, may persist and even become sensitized over time in league with the development of drug addiction (Robinson and Berridge, 1993). Indeed, this phenomenon may hold true for both positive and negative reinforcers (Goldberg and Schuster, 1970; Schulteis and Koob, 1996), and is likely associated with strengthened activity within specific subcortical circuitry including the amygdala, ventral striatum, prefrontal cortex, and insular cortex (Everitt et al., 1999; Paulus and Stewart, 2014; Peters et al., 2013; White and Milner, 1992). Subjective craving precipitated by reexposure to (secondary) reinforcing cues and contexts do not diminish but are in fact magnified over the course of abstinence in human drug addicts (Gawin and Kleber, 1986), possibly in concert with the discounting of more noble and socially acceptable pursuits and secondary reinforcers such as those associated with family and career development. Such discounting is reflected by multiple DSM-5 substance use disorder criteria (e.g., "important social, occupational, or recreational activities are given up or reduced because of substance use") and has been described as a narrowing of behavioral repertoire. This further underscores the need for preclinical development and testing of novel therapeutics targeting secondary reinforcement mechanisms associated with drug addiction.

4 MEASURING REINFORCEMENT IN ANIMAL MODELS

The foundations of behavioral analysis as a quantitative science began during Skinner's time as a graduate student at Harvard University where he originated the operant conditioning chamber in the early 1930s to study response rate as a dependent variable. Since then, operant self-administration protocols that mimic multiple elements of the addiction process and timeline (including drug discrimination, intake escalation, and reinstatement of drug seeking) have been developed. In 1954, James Olds and Peter Milner at McGill University also found that rats would readily press a lever to obtain very small amounts of electric current directly into the brain (Olds and Milner, 1954). Stimulation of specific brain areas (most notably the lateral hypothalamus and medial forebrain bundle) supports response rates of thousands of presses per hour. The reinforcing properties of electrical stimulation (termed intracranial self-stimulation, ICSS) outweigh virtually all "natural" rewards such as food, water, and sex, while fluctuations in ICSS thresholds are now believed to reflect temporary or persistent adjustments in hedonic or reward set point. For example, reinforcing stimuli are entirely capable of substituting for electrical current in this procedure, such that ICSS thresholds are lowered following acute administration of various drugs of abuse ranging from cocaine to heroin (Simon et al., 1979). In addition, acute abstinence from drugs of abuse, particularly following a history of excessive drug exposure, reveals elevations in ICSS thresholds thought to represent hedonic deficits associated with a motivational withdrawal state (Jang et al., 2013; Kenny et al., 2006; Schulteis et al., 1995).

Another valuable extension of self-administration models of reinforcement is the runway model of drug self-administration (Ettenberg, 2009). This procedure combines the traditional operant lever pressing behavior with place conditioning, and is capable of independently accessing positive reinforcement from antecedent motivation. Ettenberg and colleagues have further used this technique to distinguish the naturally composite "approach" versus "avoidance" conflict associated with cocaine as a reinforcing stimulus. Interestingly, animals with a more extensive history of cocaine self-administration exhibit heightened approach behavior and attenuated avoidance of cocaine reward (Ben-Shahar et al., 2008).

Animal models of protracted and excessive exposure to drugs and alcohol that display high construct and predictive validity for the human condition have been developed and are being intensively investigated. These models primarily employ extended access to intravenous drug self-administration (Ahmed and Koob, 1998) or chronic, intermittent exposure to dependence-inducing alcohol vapors (Gilpin et al., 2008; Vendruscolo and Roberts, 2014). Such procedures have demonstrated significant associations between excessive drug taking, somatic and motivational indices of withdrawal, and persistent drug seeking (e.g., Ahmed et al., 2000), all critical DSM-5 criteria for substance dependence. The powerful ability of contextual cues to elicit drug-seeking (or relapse-like) behavior can be readily modeled in animals by a simple reexposure to the specific environment where drugs or alcohol were self-administered on previous occasions (Stewart et al., 1984). In this design, and consistent with the reports of human craving described above, drug seeking in animals increases time-dependently as the abstinence period progresses (Grimm et al., 2001; Tran-Nguyen et al., 1998) a phenomenon that likely prolongs and potentiates relapse propensity. Intermittent schedules of repeated abstinence between drinking sessions also significantly increases alcohol self-administration in rodents (Simms et al., 2008; Spanagel et al., 1996). In composite, behavioral methodologies to study reinforcement and addiction in animals arguably represents the most valid preclinical model of any psychiatric disorder available to neuroscientists (Edwards and Koob, 2012). Data derived from these models have underscored the notion that both drug exposure and withdrawal represent dynamic states characterized by profound neuroadaptations that can exacerbate the progression of addiction but can also be targeted for therapeutic intervention.

5 DRUGS OF ABUSE AND EMERGENT WITHDRAWAL SYMPTOMS AS REINFORCERS

Research over the past several decades has successfully uncovered the neuroanatomical sites and signaling processes in the brain where various drugs of abuse act to facilitate their positive reinforcing effects. An abundance of work has implicated ascending monoamine signaling, with a particular emphasis on dopamine neurotransmission, although many nondopamine systems also contribute directly or indirectly to primary reward and reinforcement. Most notably, mesolimbic dopamine system activation attributes incentive salience to environmental stimuli and facilitates goal-directed behavior (Everitt et al., 2008; Robinson and Berridge, 1993; Self and Nestler, 1995) with important contributions from the ventral tegmental area, nucleus accumbens, and amygdala. Psychostimulant drugs (cocaine, amphetamines) lead to a particularly strong activation of mesolimbic dopamine circuitry, while dopamine-independent processes and neural substrates are additionally implicated in opioid and alcohol reinforcement (Shippenberg and Koob, 2002). However, the neurochemical mechanisms underlying the reinforcing effects of acute and/or limited drug and alcohol exposure contrast with the additional motivational circuitry recruited following escalated drug use that most significantly manifests during withdrawal to promote negative reinforcement (Edwards and Koob, 2013; Fig. 1).

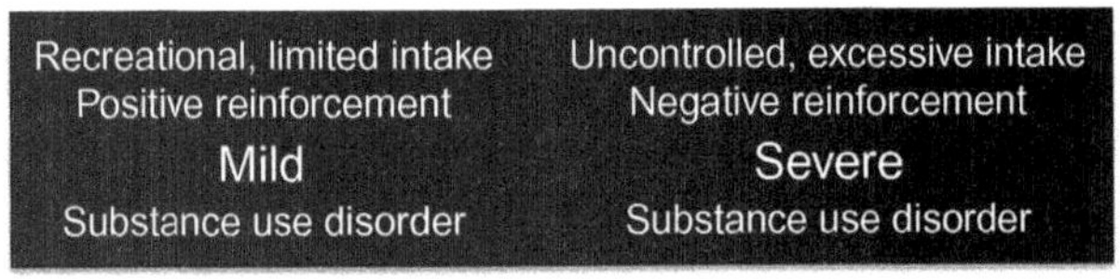

FIGURE 1

Reinforcement principles impact virtually all behavioral and neurobiological processes along the addiction timeline, from initial use to escalation of intake and relapse behavior. However, the transition from mild to severe substance use disorder is hypothesized to result from a recruitment and potentiation of negative reinforcement mechanisms in particular.

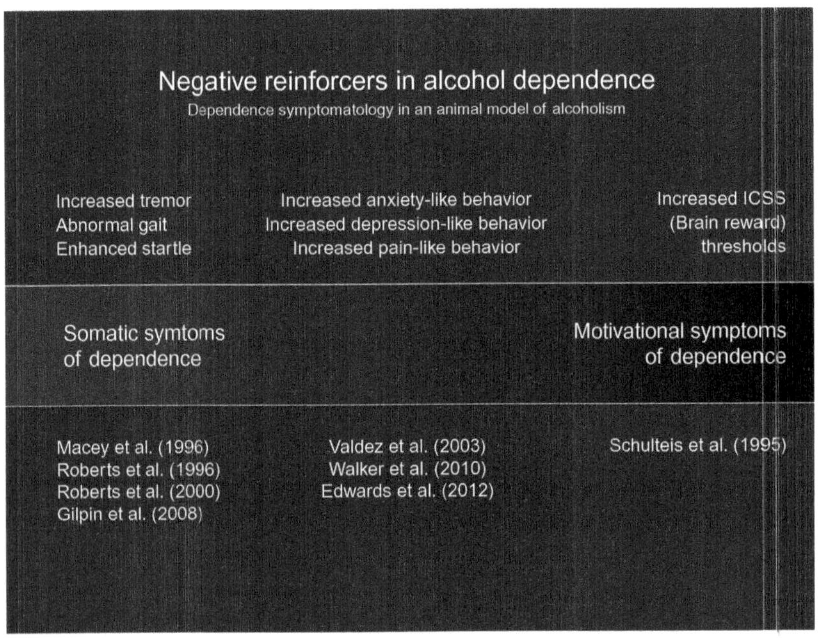

FIGURE 2

Somatic and motivational withdrawal symptoms that manifest during alcohol dependence can represent potent negative reinforcers that promote excessive alcohol drinking.

Indeed, a universal response to both acute withdrawal and attempted protracted abstinence from virtually all drugs of abuse in humans is the display of negative affective symptomatology, and numerous animal models have reliably discovered similar responses during drug and alcohol withdrawal, including increases in anxiety-, depression-, and hyperalgesia-like behaviors (Edwards and Koob, 2012; Edwards et al., 2012; Fig. 2).

6 NEUROADAPTATIONAL INTERSECTIONS OF REINFORCEMENT PROCESSES IN ADDICTION

The interplay of positive and negative reinforcement endemic to chronic drug exposure and withdrawal has led to the development of an allostatic load model of brain reinforcement systems to explain the more persistent changes in motivation associated with the transition to addiction (Koob and Le Moal, 2001). Here, addiction is considered to result from chronic, intermittent cycles of signaling changes in brain reward and antireward (or stress) circuitry that ultimately results in the potentiation of negative emotional states (chronic irritability, somatic and emotional pain, dysphoria, and anhedonia; Egli et al., 2012) that serve as profound negative reinforcers. This imbalance comes as a result of failed homeostatic mechanisms that normally act to buffer reward and stress function. As such, this theory posits that the magnification of negative reinforcement processes takes a predominating position over positive reinforcement in driving the compulsive seeking and use of drugs. At the neuroanatomical level, the dysregulation of reinforcement systems is hypothesized to be regulated by two distinct processes, contrasted as "within-system" neuroadaptations versus "between-system" neuroadaptations (Koob and Le Moal, 2008).

Within-system neuroadaptations often act as opponent processes to a reinforcing stimulus such as ingestion of a drug. For example, the primary signaling cascade linked to the drug of abuse might itself adapt to limit the drug's effects, while persistence of such counterbalancing effects after termination of drug use would produce a significant somatic and/or motivational withdrawal response. Such a within-system opponent neuroadaptation within a given motivational brain region or circuit would therefore result in decreased reward neurotransmitter function over time. For example, excessive drug use (in contrast to limited drug exposure) leads to acute withdrawal deficits in dopamine levels and diminished reward function in regions such as the ventral striatum that mediate the initial positive reinforcing effects of abused drugs (Melis et al., 2005; Weiss et al., 1992). In a between-system neuroadaptation, brain regions or systems other than those involved in the initial positive reinforcing effects of drugs are recruited (and often dysregulated) following chronic activation. In this case, another distinct neural substrate is activated and exerts opposing actions to limit primary reward function. For example, the transition to escalated drug and alcohol use in animal models is associated with the recruitment and potentiation of numerous and diverse central brain stress systems that may magnify negative reinforcement processes (Koob et al., 2014; Muschamp and Carlezon, 2013), particularly in combination with the reward deficits described above.

Indeed, a pathological interaction of brain stress and reward systems may represent a neurophysiological point of intersection between positive and negative reinforcement circuitry acting together to promote or maintain drug addiction. While compromised mesolimbic dopamine levels are associated with anhedonia during drug withdrawal, this condition may also set the stage for enhanced sensitivity to reinforcing stimuli that elevate dopamine levels within this circuitry, including stress

(Kalivas and Stewart, 1991; Lammel et al., 2014; Sorg and Kalivas, 1991). One key mediator of central stress signaling, corticotropin-releasing factor (CRF), also directly activates dopamine terminals in the nucleus accumbens (Pan et al., 1995), while drug-induced dopamine release in this area is reduced by CRF receptor blockade (Lodge and Grace, 2005). Conversely, mesolimbic dopamine promotes CRF release (Kash et al., 2008) and CRF-driven behaviors within extended amygdala areas (Meloni et al., 2006). Moreover, animal models of stress-induced relapse have identified a circuitry link between the central amygdala and ventral tegmental area (VTA) that is key to this phenomenon (McFarland et al., 2004). Adding to its signaling properties to regulate amygdala function, CRF also drives VTA activity to promote stress-primed reinstatement in cocaine-experienced animals (Wang et al., 2005). Similar interactions between glucocorticoid and dopamine signaling exist within reinforcement circuitry (Piazza et al., 1996), and corticosterone itself has been demonstrated to be reinforcing (Piazza et al., 1993). Synergistic relationships between stress and dopamine systems may make individuals more sensitive to the reinforcing effects of drugs as well as expend more effort to obtain the drug of choice (Ambroggi et al., 2009).

7 CONCLUSION AND FUTURE DIRECTIONS

According with the hypothesized transition from positive to negative reinforcement processes driving excessive drug use (Fig. 1), attenuation of brain stress system overactivation in individuals suffering from addiction may represent a valid therapeutic strategy to reduce the impact of aberrant reinforcement on behavior. For example, systemic and intracerebral administration of CRF receptor antagonists significantly reduces drug withdrawal-induced behaviors, including conditioned behaviors and withdrawal symptoms manifest during protracted abstinence (e.g., Heinrichs et al., 1995; Overstreet et al., 2004; Stinus et al., 2005; Valdez et al., 2003). These data strongly suggest that blockade of CRF signaling would prevent the ability of negative affective conditions in withdrawal to serve as negative reinforcers promoting relapse and reescalation of drug intake. At the preclinical level, CRF receptor antagonists are therapeutically effective in animal models of excessive drug use, where they block increased drug intake associated with extended access to intravenous self-administration of cocaine, heroin, and nicotine (George et al., 2007; Greenwell et al., 2009; Park et al., 2015; Specio et al., 2008). CRF receptor antagonists also reduce excessive drinking in alcohol-dependent rats (Funk et al., 2007), while further evidence at the clinical level suggests that antagonists with longer receptor residence rates might exhibit greater efficacy in reducing drinking (Kwako et al., 2015). Importantly, CRF antagonism does not appear to alter drug and alcohol intake in nondependent animals or the intake of natural rewards such as sucrose or water. As such, these results provide substantial evidence for a specific role of CRF receptors in the escalation of drug intake associated with negative reinforcement. This selectivity would suggest a robust and specific efficacy of this drug class to treat addictive disorders, with a critical feature lacking in existing medications such as naltrexone that suffers from substantial compliance issues (Swift et al., 2011) due to its primary actions to reduce positive reinforcement mechanisms via opioid receptor blockade.

Recent data have implicated additional central stress-regulated neuropeptides in promoting escalation of drug intake and addiction-related symptomatology, such as the endogenous dynorphin/kappa-opioid receptor system (Butelman et al., 2012). In contrast to other endogeonous opioids, dynorphin likely plays a central role in negative reinforcement, as kappa-opioid receptor stimulation produces aversion and anhedonia in animal models as well as negative affect in humans (Mucha and Herz, 1985; Pfeiffer et al., 1986; Todtenkopf et al., 2004). Dynorphin inhibits dopamine release and function, both at the origins and within the terminals of the mesolimbic dopamine system, representing a within-system neuroadaptation that underlies the aversive properties of dynorphin (Margolis et al., 2003; Spanagel et al., 1990). Dynorphin signaling is potentiated by various drugs of abuse (e.g., Lindholm et al., 2000; Spangler et al., 1993), and in turn, blockade of kappa-opioid receptors by the antagonist norbinaltorphimine reduces drug and alcohol intake escalation selectively in dependent (but not nondependent) animals (Walker et al., 2012; Wee and Koob, 2010).

Finally, given its ability to attenuate the efficacy of conditioned reinforcers, the employment of extinction therapy may represent a useful behavioral strategy to combat relapse to drug and alcohol use in abstinent individuals. Extinction is the process by which previously conditioned cues and/or contexts gradually lose their motivational impact following repeated presentation in the absence of the reinforcing stimulus. Unfortunately, a metaanalysis of nine previous extinction therapy trials for drug addiction failed to uncover a substantial or consistent benefit (Conklin and Tiffany, 2002), although the authors suggested that future trials should strongly consider basing their methodologies on what has been recently learned from preclinical models of reinforcement learning. Additional design or treatment variables might incorporate elements associated with known neuroadaptations resulting from extinction training that are entirely distinct from the withdrawal process itself (Self et al., 2004). For example, combinations of extinction therapy and traditional pharmaceutical approaches targeting endogenous opioid signaling (O'Brien et al., 1984) or glutamate receptor systems (Myers et al., 2011) would appear to be one promising avenue for maximizing treatment efficacy. Ultimately, future medical interventions for addiction will need to factor in the inextricable link between the addiction process and maladaptive changes in underlying positive and negative reinforcement mechanisms.

ACKNOWLEDGMENTS

Preparation of this work was generously supported by the LSU Health Sciences Center School of Medicine and the National Institute on Alcohol Abuse and Alcoholism (AA020839, S.E.).

REFERENCES

Ahmed, S.H., Koob, G.F., 1998. Transition from moderate to excessive drug intake: change in hedonic set point. Science 282, 298–300.

Ahmed, S.H., Walker, J.R., Koob, G.F., 2000. Persistent increase in the motivation to take heroin in rats with a history of drug escalation. Neuropsychopharmacology 22, 413–421.

Ambroggi, F., Turiault, M., Milet, A., Deroche-Gamonet, V., Parnaudeau, S., Balado, E., Barik, J., Van Der Veen, R., Maroteaux, G., Lemberger, T., Schutz, G., Lazar, M., Marinelli, M., Piazza, P.V., Tronche, F., 2009. Stress and addiction: glucocorticoid receptor in dopaminoceptive neurons facilitates cocaine seeking. Nat. Neurosci. 12, 247–249.

Ben-Shahar, O., Posthumus, E.J., Waldroup, S.A., Ettenberg, A., 2008. Heightened drug-seeking motivation following extended daily access to self-administered cocaine. Prog. Neuropsychopharmacol. Biol. Psychiatry 32, 863–869.

Butelman, E.R., Yuferov, V., Kreek, M.J., 2012. kappa-opioid receptor/dynorphin system: genetic and pharmacotherapeutic implications for addiction. Trends Neurosci. 35, 587–596.

Conklin, C.A., Tiffany, S.T., 2002. Applying extinction research and theory to cue-exposure addiction treatments. Addiction 97, 155–167.

Corbett, D., Wise, R.A., 1980. Intracranial self-stimulation in relation to the ascending dopaminergic systems of the midbrain: a moveable electrode mapping study. Brain Res. 185, 1–15.

DSM-5, 2013. American Psychiatric Association: Diagnostic and Statistical Manual of Mental Disorders, fifth ed. American Psychiatric Association, Arlington, VA.

Edwards, S., Koob, G.F., 2012. Experimental psychiatric illness and drug abuse models: from human to animal, an overview. Methods Mol. Biol. 829, 31–48.

Edwards, S., Koob, G.F., 2013. Escalation of drug self-administration as a hallmark of persistent addiction liability. Behav. Pharmacol. 24, 356–362.

Edwards, S., Vendruscolo, L.F., Schlosburg, J.E., Misra, K.K., Wee, S., Park, P.E., Schulteis, G., Koob, G.F., 2012. Development of mechanical hypersensitivity in rats during heroin and ethanol dependence: alleviation by CRF(1) receptor antagonism. Neuropharmacology 62, 1142–1151.

Egli, M., Koob, G.F., Edwards, S., 2012. Alcohol dependence as a chronic pain disorder. Neurosci. Biobehav. Rev. 36, 2179–2192.

Ettenberg, A., 2009. The runway model of drug self-administration. Pharmacol. Biochem. Behav. 91, 271–277.

Everitt, B.J., Parkinson, J.A., Olmstead, M.C., Arroyo, M., Robledo, P., Robbins, T.W., 1999. Associative processes in addiction and reward. The role of amygdala-ventral striatal subsystems. Ann. N. Y. Acad. Sci. 877, 412–438.

Everitt, B.J., Belin, D., Economidou, D., Pelloux, Y., Dalley, J.W., Robbins, T.W., 2008. Review. Neural mechanisms underlying the vulnerability to develop compulsive drug-seeking habits and addiction. Philos. Trans. R. Soc. Lond. B Biol. Sci. 363, 3125–3135.

Funk, C.K., Zorrilla, E.P., Lee, M.J., Rice, K.C., Koob, G.F., 2007. Corticotropin-releasing factor 1 antagonists selectively reduce ethanol self-administration in ethanol-dependent rats. Biol. Psychiatry 61, 78–86.

Gawin, F.H., Kleber, H.D., 1986. Abstinence symptomatology and psychiatric diagnosis in cocaine abusers. Clinical observations. Arch. Gen. Psychiatry 43, 107–113.

George, O., Ghozland, S., Azar, M.R., Cottone, P., Zorrilla, E.P., Parsons, L.H., O'Dell, L.E., Richardson, H.N., Koob, G.F., 2007. CRF-CRF1 system activation mediates withdrawal-induced increases in nicotine self-administration in nicotine-dependent rats. Proc. Natl. Acad. Sci. U. S. A. 104, 17198–17203.

Gilpin, N.W., Richardson, H.N., Cole, M., Koob, G.F., 2008. Vapor inhalation of alcohol in rats. Curr. Protocol. Neurosci. Chapter 9, Unit 9 29.

Goldberg, S.R., Schuster, C.R., 1970. Conditioned nalorphine-induced abstinence changes: persistence in post morphine-dependent monkeys. J. Exp. Anal. Behav. 14, 33–46.

Greenwell, T.N., Funk, C.K., Cottone, P., Richardson, H.N., Chen, S.A., Rice, K.C., Zorrilla, E.P., Koob, G.F., 2009. Corticotropin-releasing factor-1 receptor antagonists decrease heroin self-administration in long- but not short-access rats. Addict. Biol. 14, 130–143.

Grimm, J.W., Hope, B.T., Wise, R.A., Shaham, Y., 2001. Neuroadaptation. Incubation of cocaine craving after withdrawal. Nature 412, 141–142.

Heinrichs, S.C., Menzaghi, F., Schulteis, G., Koob, G.F., Stinus, L., 1995. Suppression of corticotropin-releasing factor in the amygdala attenuates aversive consequences of morphine withdrawal. Behav. Pharmacol. 6, 74–80.

Hyman, S.E., Malenka, R.C., Nestler, E.J., 2006. Neural mechanisms of addiction: the role of reward-related learning and memory. Annu. Rev. Neurosci. 29, 565–598.

Jang, C.G., Whitfield, T., Schulteis, G., Koob, G.F., Wee, S., 2013. A dysphoric-like state during early withdrawal from extended access to methamphetamine self-administration in rats. Psychopharmacology (Berl.) 225, 753–763.

Kalivas, P.W., 2005. How do we determine which drug-induced neuroplastic changes are important? Nat. Neurosci. 8, 1440–1441.

Kalivas, P.W., Stewart, J., 1991. Dopamine transmission in the initiation and expression of drug- and stress-induced sensitization of motor activity. Brain Res. Brain Res. Rev. 16, 223–244.

Kash, T.L., Nobis, W.P., Matthews, R.T., Winder, D.G., 2008. Dopamine enhances fast excitatory synaptic transmission in the extended amygdala by a CRF-R1-dependent process. J. Neurosci. 28, 13856–13865.

Kenny, P.J., Chen, S.A., Kitamura, O., Markou, A., Koob, G.F., 2006. Conditioned withdrawal drives heroin consumption and decreases reward sensitivity. J. Neurosci. 26, 5894–5900.

Koob, G.F., Le Moal, M., 2001. Drug addiction, dysregulation of reward, and allostasis. Neuropsychopharmacology 24, 97–129.

Koob, G.F., Le Moal, M., 2008. Addiction and the brain antireward system. Annu. Rev. Psychol. 59, 29–53.

Koob, G.F., Buck, C.L., Cohen, A., Edwards, S., Park, P.E., Schlosburg, J.E., Schmeichel, B., Vendruscolo, L.F., Wade, C.L., Whitfield Jr., T.W., George, O., 2014. Addiction as a stress surfeit disorder. Neuropharmacology 76 (Pt B), 370–382.

Kwako, L.E., Spagnolo, P.A., Schwandt, M.L., Thorsell, A., George, D.T., Momenan, R., Rio, D.E., Huestis, M., Anizan, S., Concheiro, M., Sinha, R., Heilig, M., 2015. The corticotropin releasing hormone-1 (CRH1) receptor antagonist pexacerfont in alcohol dependence: a randomized controlled experimental medicine study. Neuropsychopharmacology 40, 1053–1063.

Lammel, S., Lim, B.K., Malenka, R.C., 2014. Reward and aversion in a heterogeneous midbrain dopamine system. Neuropharmacology 76 (Pt B), 351–359.

Lindholm, S., Ploj, K., Franck, J., Nylander, I., 2000. Repeated ethanol administration induces short- and long-term changes in enkephalin and dynorphin tissue concentrations in rat brain. Alcohol 22, 165–171.

Lodge, D.J., Grace, A.A., 2005. Acute and chronic corticotropin-releasing factor 1 receptor blockade inhibits cocaine-induced dopamine release: correlation with dopamine neuron activity. J. Pharmacol. Exp. Ther. 314, 201–206.

Macey, D.J., Schulteis, G., Heinrichs, S.C., Koob, G.F., 1996. Time-dependent quantifiable withdrawal from ethanol in the rat: effect of method of dependence induction. Alcohol 13, 163–170.

Margolis, E.B., Hjelmstad, G.O., Bonci, A., Fields, H.L., 2003. Kappa-opioid agonists directly inhibit midbrain dopaminergic neurons. J. Neurosci. 23, 9981–9986.

McFarland, K., Davidge, S.B., Lapish, C.C., Kalivas, P.W., 2004. Limbic and motor circuitry underlying footshock-induced reinstatement of cocaine-seeking behavior. J. Neurosci. 24, 1551–1560.

Melis, M., Spiga, S., Diana, M., 2005. The dopamine hypothesis of drug addiction: hypodopaminergic state. Int. Rev. Neurobiol. 63, 101–154.

Meloni, E.G., Gerety, L.P., Knoll, A.T., Cohen, B.M., Carlezon Jr., W.A., 2006. Behavioral and anatomical interactions between dopamine and corticotropin-releasing factor in the rat. J. Neurosci. 26, 3855–3863.

Mucha, R.F., Herz, A., 1985. Motivational properties of kappa and mu opioid receptor agonists studied with place and taste preference conditioning. Psychopharmacology (Berl.) 86, 274–280.

Muschamp, J.W., Carlezon Jr., W.A., 2013. Roles of nucleus accumbens CREB and dynorphin in dysregulation of motivation. Cold Spring Harbor Perspect. Med. 3, a012005.

Myers, K.M., Carlezon Jr., W.A., DAVIS, M., 2011. Glutamate receptors in extinction and extinction-based therapies for psychiatric illness. Neuropsychopharmacology 36, 274–293.

O'Brien, C.P., Childress, A.R., McLellan, A.T., Ternes, J., Ehrman, R.N., 1984. Use of naltrexone to extinguish opioid-conditioned responses. J. Clin. Psychiatry 45, 53–56.

Olds, J., Milner, P., 1954. Positive reinforcement produced by electrical stimulation of septal area and other regions of rat brain. J. Comp. Physiol. Psychol. 47, 419–427.

Overstreet, D.H., Knapp, D.J., Breese, G.R., 2004. Modulation of multiple ethanol withdrawal-induced anxiety-like behavior by CRF and CRF1 receptors. Pharmacol. Biochem. Behav. 77, 405–413.

Pan, J.T., Lookingland, K.J., Moore, K.E., 1995. Differential effects of corticotropin-releasing hormone on central dopaminergic and noradrenergic neurons. J. Biomed. Sci. 2, 50–56.

Park, P.E., Schlosburg, J.E., Vendruscolo, L.F., Schulteis, G., Edwards, S., Koob, G.F., 2015. Chronic CRF1 receptor blockade reduces heroin intake escalation and dependence-induced hyperalgesia. Addict. Biol. 20, 275–284.

Paulus, M.P., Stewart, J.L., 2014. Interoception and drug addiction. Neuropharmacology 76 (Pt B), 342–350.

Peters, G.J., David, C.N., Marcus, M.D., Smith, D.M., 2013. The medial prefrontal cortex is critical for memory retrieval and resolving interference. Learn. Mem. 20, 201–209.

Pfeiffer, A., Brantl, V., Herz, A., Emrich, H.M., 1986. Psychotomimesis mediated by kappa opiate receptors. Science 233, 774–776.

Piazza, P.V., Deroche, V., Deminiere, J.M., Maccari, S., Le Moal, M., Simon, H., 1993. Corticosterone in the range of stress-induced levels possesses reinforcing properties: implications for sensation-seeking behaviors. Proc. Natl. Acad. Sci. U. S. A. 90, 11738–11742.

Piazza, P.V., Rouge-Pont, F., Deroche, V., Maccari, S., Simon, H., Le Moal, M., 1996. Glucocorticoids have state-dependent stimulant effects on the mesencephalic dopaminergic transmission. Proc. Natl. Acad. Sci. U. S. A. 93, 8716–8720.

Roberts, A.J., Cole, M., Koob, G.F., 1996. Intra-amygdala muscimol decreases operant ethanol self-administration in dependent rats. Alcohol. Clin. Exp. Res. 20, 1289–1298.

Roberts, A.J., Heyser, C.J., Cole, M., Griffin, P., Koob, G.F., 2000. Excessive ethanol drinking following a history of dependence: animal model of allostasis. Neuropsychopharmacology 22, 581–594.

Robinson, T.E., Berridge, K.C., 1993. The neural basis of drug craving: an incentive-sensitization theory of addiction. Brain Res. Brain Res. Rev. 18, 247–291.

Schulteis, G., Koob, G.F., 1996. Reinforcement processes in opiate addiction: a homeostatic model. Neurochem. Res. 21, 1437–1454.

Schulteis, G., Markou, A., Cole, M., Koob, G.F., 1995. Decreased brain reward produced by ethanol withdrawal. Proc. Natl. Acad. Sci. U. S. A. 92, 5880–5884.

Self, D.W., Nestler, E.J., 1995. Molecular mechanisms of drug reinforcement and addiction. Annu. Rev. Neurosci. 18, 463–495.

Self, D.W., Choi, K.H., Simmons, D., Walker, J.R., Smagula, C.S., 2004. Extinction training regulates neuroadaptive responses to withdrawal from chronic cocaine self-administration. Learn. Mem. 11, 648–657.

Shippenberg, T.S., Koob, G.F., 2002. Recent advances in animal models of drug addiction and alcoholism. In: Davis, K.L., Charney, D., Coyle, J.T., Nemeroff, C. (Eds.), Neuropsychopharmacology: The Fifth Generation of Progress. Lippincott Williams and Wilkins, Philadelphia, PA, pp. 1381–1397.

Simms, J.A., Steensland, P., Medina, B., Abernathy, K.E., Chandler, L.J., Wise, R., Bartlett, S.E., 2008. Intermittent access to 20% ethanol induces high ethanol consumption in Long-Evans and Wistar rats. Alcohol. Clin. Exp. Res. 32, 1816–1823.

Simon, H., Stinus, L., Tassin, J.P., Lavielle, S., Blanc, G., Thierry, A.M., Glowinski, J., Le Moal, M., 1979. Is the dopaminergic mesocorticolimbic system necessary for intracranial self-stimulation? Biochemical and behavioral studies from A10 cell bodies and terminals. Behav. Neural Biol. 27, 125–145.

Skinner, B.F., 1938. The Behavior of Organisms. Appleton-Century-Crofts, New York, NY.

Sorg, B.A., Kalivas, P.W., 1991. Effects of cocaine and footshock stress on extracellular dopamine levels in the ventral striatum. Brain Res. 559, 29–36.

Spanagel, R., Herz, A., Shippenberg, T.S., 1990. The effects of opioid peptides on dopamine release in the nucleus accumbens: an in vivo microdialysis study. J. Neurochem. 55, 1734–1740.

Spanagel, R., Holter, S.M., Allingham, K., Landgraf, R., Zieglgansberger, W., 1996. Acamprosate and alcohol: I. Effects on alcohol intake following alcohol deprivation in the rat. Eur. J. Pharmacol. 305, 39–44.

Spangler, R., Unterwald, E.M., Kreek, M.J., 1993. 'Binge' cocaine administration induces a sustained increase of prodynorphin mRNA in rat caudate-putamen. Brain Res. Mol. Brain Res. 19, 323–327.

Specio, S.E., Wee, S., O'Dell, L.E., Boutrel, B., Zorrilla, E.P., Koob, G.F., 2008. CRF(1) receptor antagonists attenuate escalated cocaine self-administration in rats. Psychopharmacology (Berl.) 196, 473–482.

Stewart, J., 1992. Neurobiology of conditioning to drugs of abuse. Ann. N. Y. Acad. Sci. 654, 335–346.

Stewart, J., De Wit, H., Eikelboom, R., 1984. Role of unconditioned and conditioned drug effects in the self-administration of opiates and stimulants. Psychol. Rev. 91, 251–268.

Stinus, L., Cador, M., Zorrilla, E.P., Koob, G.F., 2005. Buprenorphine and a CRF1 antagonist block the acquisition of opiate withdrawal-induced conditioned place aversion in rats. Neuropsychopharmacology 30, 90–98.

Swift, R., Oslin, D.W., Alexander, M., Forman, R., 2011. Adherence monitoring in naltrexone pharmacotherapy trials: a systematic review. J. Stud. Alcohol Drugs 72, 1012–1018.
Todtenkopf, M.S., Marcus, J.F., Portoghese, P.S., Carlezon Jr., W.A., 2004. Effects of kappa-opioid receptor ligands on intracranial self-stimulation in rats. Psychopharmacology (Berl.) 172, 463–470.
Tran-Nguyen, L.T., Fuchs, R.A., Coffey, G.P., Baker, D.A., O'Dell, L.E., Neisewander, J.L., 1998. Time-dependent changes in cocaine-seeking behavior and extracellular dopamine levels in the amygdala during cocaine withdrawal. Neuropsychopharmacology 19, 48–59.
Valdez, G.R., Zorrilla, E.P., Roberts, A.J., Koob, G.F., 2003. Antagonism of corticotropin-releasing factor attenuates the enhanced responsiveness to stress observed during protracted ethanol abstinence. Alcohol 29, 55–60.
Vendruscolo, L.F., Roberts, A.J., 2014. Operant alcohol self-administration in dependent rats: focus on the vapor model. Alcohol 48, 277–286.
Walker, B.M., Drimmer, D.A., Walker, J.L., Liu, T., Mathe, A.A., Ehlers, C.L., 2010. Effects of prolonged ethanol vapor exposure on forced swim behavior, and neuropeptide Y and corticotropin-releasing factor levels in rat brains. Alcohol 44, 487–493.
Walker, B.M., Valdez, G.R., McLaughlin, J.P., Bakalkin, G., 2012. Targeting dynorphin/kappa opioid receptor systems to treat alcohol abuse and dependence. Alcohol 46, 359–370.
Wang, B., Shaham, Y., Zitzman, D., Azari, S., Wise, R.A., You, Z.B., 2005. Cocaine experience establishes control of midbrain glutamate and dopamine by corticotropin-releasing factor: a role in stress-induced relapse to drug seeking. J. Neurosci. 25, 5389–5396.
Wee, S., Koob, G.F., 2010. The role of the dynorphin-kappa opioid system in the reinforcing effects of drugs of abuse. Psychopharmacology (Berl.) 210, 121–135.
Weiss, F., 2005. Neurobiology of craving, conditioned reward and relapse. Curr. Opin. Pharmacol. 5, 9–19.
Weiss, F., Markou, A., Lorang, M.T., Koob, G.F., 1992. Basal extracellular dopamine levels in the nucleus accumbens are decreased during cocaine withdrawal after unlimited-access self-administration. Brain Res. 593, 314–318.
White, N.M., 1996. Addictive drugs as reinforcers: multiple partial actions on memory systems. Addiction 91, 921–949. discussion 951–965.
White, N.M., Milner, P.M., 1992. The psychobiology of reinforcers. Annu. Rev. Psychol. 43, 443–471.
Wikler, A., 1952. A psychodynamic study of a patient during experimental self-regulated re-addiction to morphine. Psychiatr. Q. 26, 270–293.
Wikler, A., Pescor, F.T., Miller, D., Norrell, H., 1971. Persistent potency of a secondary (conditioned) reinforcer following withdrawal of morphine from physically dependent rats. Psychopharmacologia 20, 103–117.
Wise, R.A., Bozarth, M.A., 1981. Brain substrates for reinforcement and drug self-administration. Prog. Neuropsychopharmacol. 5, 467–474.
Wise, R.A., Koob, G.F., 2014. The development and maintenance of drug addiction. Neuropsychopharmacology 39, 254–262.
Yokel, R.A., Wise, R.A., 1975. Increased lever pressing for amphetamine after pimozide in rats: implications for a dopamine theory of reward. Science 187, 547–549.

CHAPTER

Neuroscience of attentional processes for addiction medicine: from brain mechanisms to practical considerations

5

Javad Salehi Fadardi[*,†,‡,1], **W. Miles Cox**[†], **Arash Rahmani**[§]

[*]*Ferdowsi University of Mashhad, Mashhad, Iran*

[†]*Bangor University, Bangor, UK*

[‡]*Addiction Research Centre, Mashhad University of Medical Sciences, Mashhad, Iran*

[§]*Iranian National Center for Addiction Studies, Tehran University of Medical Sciences, Tehran, Iran*

[1]*Corresponding author: Tel.: +988138805867; Fax: +988138805867, e-mail address: j.s.fadardi@um.ac.ir*

Abstract

The present chapter first argues how having a goal for procuring alcohol or other substances leads to the development of a time-binding, dynamic, and goal oriented motivational state termed *current concern*, as the origin of substance-related attentional bias. Next, it discusses the importance of attentional bias in the development, continuation of, and relapsing to substance abuse. It further proceeds with a review of selective evidence from cognitive psychology that helps account for making decisions about using an addictive substance or refraining from using it. A discussion on the various brain loci that are involved in attentional bias and other kinds of cue reactivity is followed by presenting findings from neurocognitive research. Finally, from an interdisciplinary perspective, the chapter presents new trends and ideas that can be applied to addiction-related cognitive measurement and training.

Keywords

Motivation, Current concern, Attentional bias, Brain, Attention retraining, Implicit cognition, Technology

1 INTRODUCTION

Attentional bias means that a person selectively attends to a certain category or certain categories of stimuli in the environment while tending to overlook, ignore, or disregard other kinds of stimuli. For example, one person might selectively attend to stimuli related to food (particularly food that is perceived to be particularly delicious). Sexual stimuli might be particularly distracting for another person; fashion-related stimuli might capture another person's attention. Of greater relevance for the present chapter is that certain other people are distracted by addiction-related stimuli. The latter stimuli might be ones related to alcohol, another substance of abuse, or some other form of addiction (e.g., gambling) (Fig. 1).

What do these different kinds of attentional bias mean, and how do they arise? Often—if not always—the attentional bias is motivationally relevant, i.e., it is related to the person's goal-directed behavior. Formally, we define motivation as "the internal states of the organism that lead to the instigation, persistence, energy, and direction of behavior towards a goal" (Klinger and Cox, 2011, p. 4). How do such states come about, and how do they give rise to attentional bias?

Motivation starts when a person is aware of incentives that he or she finds attractive. An incentive is defined as any object or event that could potentially change a person's affect in a positive way, either by enhancing positive affect or by reducing negative affect. In addition to *incentive*, *affect* and *affective change* are, therefore, key motivational constructs. *Affect*—which can be either positive or negative—is the subjective component of an emotional response. *Affective change* is a change in affect from its present state; it is the essence of what people are motivated to achieve. People want to feel better than they currently do, either by increasing their *positive affect* (e.g., joy, happiness, or satisfaction) or by decreasing their *negative affect* (e.g., fear, boredom, or depression).

Of the many positive incentives that could potentially enhance a person's positive affect and the many negative incentives that could reduce the person's negative affect if they were removed, each person might set a goal of acquiring only a subset of the positive incentives or a goal of getting rid of only a subset of the negative incentives. In either case, a goal is formed from the moment that the person makes a commitment to either obtain or to get rid of an incentive.

Why are some incentives but not others transformed into goals? Motivational psychologists often rely on Value X Expectancy Theory (Bundorf et al., 2013; Cox et al., 2015; Morone and Morone, 2014) to explain this outcome. According to this theory, two primary variables determine whether an incentive is transformed into a goal. They are (a) the value that the person attributes to the incentive (how valuable to the person the affective change that the incentive would produce would be) and (b) the person's expected likelihood of actually being able to obtain the incentive if he or she puts forth the effort. Because the relationship between *value* and *expectancy* is multiplicative, if either of them is zero (or near zero), a goal to pursue the incentive will not be formed. For example, a person might imagine that becoming a millionaire would bring about a very positive affective change, but the person

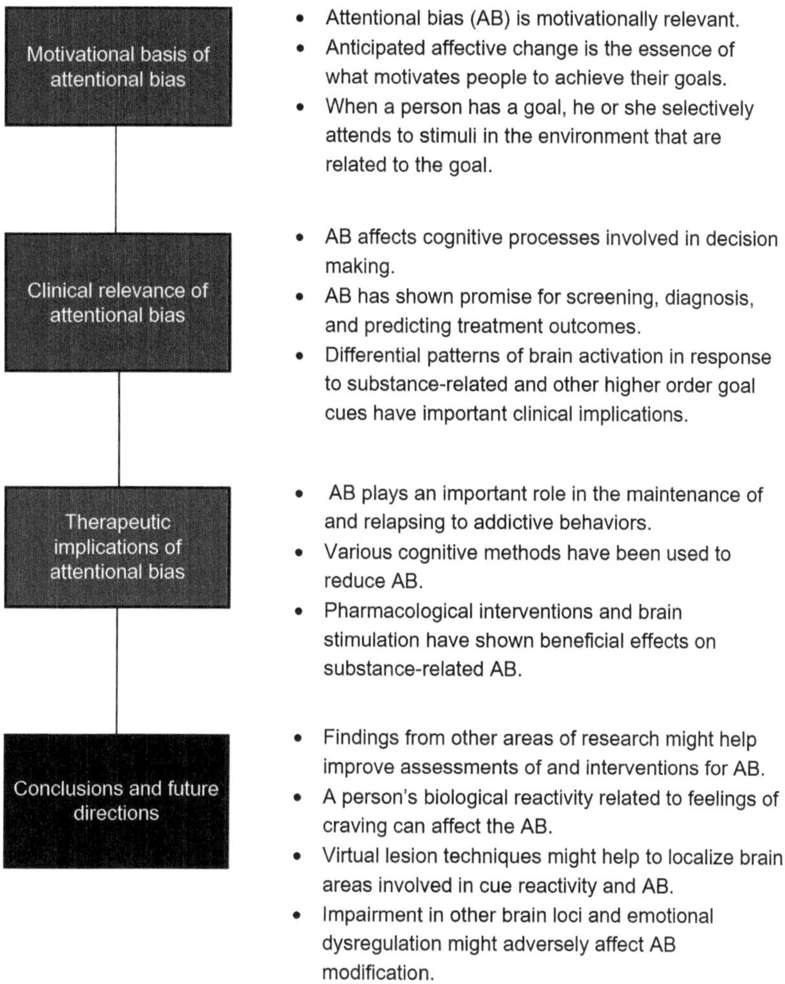

FIGURE 1

Summary of the main topics that have been discussed in the present chapter.

might not actually try to become a millionaire because the expected chances of being able to do so are virtually nil.

People, of course, vary widely with regard to the value that they attribute to using addictive substances and their actual use of them. What are the factors that determine the value of using these substances? To answer this question, we will use alcohol consumption as an example; nevertheless, much of the discussion can be generalized to use of other kinds of addictive substances.

One kind of variable that affects the value of drinking alcohol is each person's own biochemical reaction to alcohol (e.g., Dickson et al., 2006). Pharmacologically, some people react positively to drinking alcohol, and they experience few negative consequences. Other people experience primarily unpleasant reactions, such as facial flushing and nausea. Sociocultural and environmental factors also affect the value of drinking alcohol (e.g., Dantzer et al., 2006). Societies differ widely in how they view drinking alcohol and the extent to which they condone or prohibit it; thus, people living in a particular society will be overtly or subtly reinforced for drinking in the same manner as other people living in that society. Within each society, additional environmental factors—such as advertising alcohol and taxation on alcohol and the extent to which drinking alcohol is promoted in a particular situation (e.g., Hollingworth et al., 2006; Huckle et al., 2008; Paschall et al., 2014)—also affect the value that people attribute to drinking alcohol.

Other characteristics of individuals themselves (other than each person's biochemical reactivity to alcohol) might affect the degree to which they value drinking. Notable among these is each person's personality characteristics (e.g., see Vrieze et al., 2014) and the degree to which they are feeling stressed because of frustrations in other areas of their lives (Demirbas et al., 2012).

As in the case of all goal pursuits, a person will form a goal of drinking alcohol or using another addictive substance when both (a) the value that the person attributes to using the substance (the expected desirable affective change) is high and (b) the person's expected chances of being able to actually achieve the desired change in affect is high. When the goal is formed, the person is in a distinctive motivational state—called a *current concern* (Cox et al., 2015; Klinger and Cox, 2011). This state lasts from the moment that the commitment to the goal pursuit is first made until either the goal is reached or the pursuit of the goal is relinquished. During this period, the goal striving is reflected in the person cognitive processes (e.g., his or her thoughts, memories, attention, and even dreams). The construct *current concern* is presumed to refer to latent goal-related brain processes, and recent neuroscientific research has identified clues about how these processes are represented in the brain (Berkman and Lieberman, 2009; Klinger and Cox, 2011; Kouneiher et al., 2009). For example, expected satisfaction from goal attainments are mainly processed in amygdala and with interactions with orbital prefrontal and anterior cingulate cortex; the interactions among these structures help to determine anticipated goal outcomes, cue reactivity, and response selections (Baxter et al., 2000; Murray, 2007). Later in the chapter, we discuss how anterior cingulate cortex plays an important role in determining attentional bias for addiction-related cues. In any case when a person has a goal of drinking alcohol, the person selectively attends to stimuli in the environment that are related to procuring and imbibing alcohol. This process facilitates the goal striving by increasing the person's motivation to drink and his or her actual consumption of alcohol.

Extensive research has been conducted on alcohol and other substance-related attentional bias (e.g., Cox et al., 2006) and other kinds of cognitive biases (e.g., automatic action tendencies; Wiers et al., 2011) related to people's goal of drinking alcohol. In the following sections, we (a) briefly review this research, (b) describe how dual process models help to account for decisions about whether or not to use an addictive substance, (c) discuss how various brain loci are involved in attentional bias and other kinds of cue reactivity, and (d) suggest how findings from neurocognitive research can be applied to cognitive training and future research.

2 CLINICAL RELEVANCE OF ATTENTIONAL BIAS: IS ATTENTIONAL BIAS AND ARTIFACT OF IMPAIRED EXECUTIVE CONTROL?

Substance abuse has adverse effects on the entire body, including the brain. It can impair the brain's prefrontal and frontal loci that are mainly responsible for executive cognitive functions (ECF), including inhibition, task switching, sustained attention, decision making, and planning (de la Monte and Kril, 2014; Terrett et al., 2014; Vonmoos et al., 2014). Performance on measures of attentional bias and other implicit cognition tasks requires the involvement of ECFs and working memory. For example, on an addiction Stroop test (Cox et al., 2006), which measures attentional bias for substance-related stimuli, the individual needs to *inhibit* the distracting feature of a stimulus (e.g., alcohol) and mentally switch to the experimental task (responding to the color of the stimulus) and keep this task requirement continuously in mind. As mentioned earlier, alcohol and other drugs of abuse adversely affect higher order processes; therefore, the question is whether we can attribute longer latencies on measures of attentional bias to the substance-related characteristics of the stimuli. Or could longer latencies on addiction-related tasks simply result from impaired ECFs. Additionally, there are studies suggesting that other characteristics of abusers, such as their impulsivity or poor inhibitory control, can also affect their performance on attentional bias tasks (Liu et al., 2011).

Fadardi and Cox (2006) found that longer latencies on measures of attentional bias were not a result only of poorer ECF. They tested alcohol abusers using general cognitive measures and an alcohol-Stroop test. The results suggested that excessive drinking sensitized the abusers' attentional responsiveness to alcohol-related stimuli to a degree that exceeded the drinkers' ECF. In other words, the addiction-related attentional bias was not an artifact of cognitive impairment that either preceded or followed the alcohol abuse. A clinically valuable method would be able to (a) screen clinical samples, (b) differentiate among various levels of problem severity, and (c) predict treatment outcome. In fact, several reviews suggest that implicit measures of addiction-related cognitive biases have diagnostic and prognostic validity (Cox et al., 2006; Field and Cox, 2008; Leeman et al., 2014).

In a recent study, a cocaine version of the Stroop test and neuroimaging were used to differentiate recreational users from abusers. Cocaine-dependent individuals showed significantly larger attentional bias for cocaine words, but nonusers and recreational users did not show the bias. Moreover, recreational users showed a distinctive pattern of underactivation in prefrontal cortices, including orbitofrontal and anterior cingulate during task performance compared to both dependent and control participants (Smith et al., 2014). Janes et al. (2010) used *f*MRI to assess brain activity related to attentional bias for smoking-related stimuli on a Stroop test. They reported that both measures predicted treatment outcome during an 8-week smoking cessation period.

3 ATTENTION RETRAINING: A MULTIMODAL ACTIVITY

Evidence from cognitive neuroscience suggests that ignoring addiction-related stimuli requires increased activation of control networks and reduced processing in emotion and reward regions of the brain. Although during attentional bias drug abusers typically show increased activity in the regions associated with cognitive control (i.e., lateral prefrontal and dorsal anterior cingulate), such increases seem to be insufficient to overcome simultaneous increases in processing in the emotion/reward regions (i.e., amygdala, insula, and striatum) (Hester and Luijten, 2014). In another study (Janes et al., 2010), when heavy smokers were exposed to smoking-related words, increased activation in the brain's bilateral anterior insula and dorsal anterior cingulate cortex were observed.

There are similarities in the brain regions that react to drug-related and drug-unrelated stimuli. For example, Lorenz et al. (2013) reported that gamblers showed stronger brain responses in medial prefrontal cortex (MPFC) during short presentations of gambling stimuli (i.e., attentional bias), but stronger brain responses in anterior cingulate gyrus and in lingual gyrus during long presentations of gambling stimuli (i.e., cue reactivity). The authors also reported that in long presentations of gambling-related stimuli (when participants' task was to ignore them), stronger brain responses were observed in right inferior frontal gyrus (associated with inhibition processing), left orbitofrontal cortex, and ventral striatum.

What actually happens when there is increased activity in certain loci of the brain in response to addiction-related stimuli? According to Franken (2003), a conditioned drug stimulus produces increase in dopamine in the corticostriatal circuit, particularly in the anterior cingulate gyrus, amygdala, and nucleus accumbens; the increase serves to draw the person's attention toward a drug-related stimulus. The process also results in motor preparation and hyper-attentiveness for drug-related stimuli, which ultimately promotes craving and perhaps relapse. An implication for treatment would be to frequently expose abusers to dopamine-releasing stimuli but to prevent them from taking the substance. This could gradually attenuate the dopamine-releasing characteristic of the addictive stimuli.

It is interesting to note that such conditioned brain responses in abusive or heavy drinkers are not limited to the brain's maladaptive neural responses to addiction-related stimuli; rather, anomalies can also be observed in the brain's reactivity to other kinds of stimuli. In an *f*MRI study, Ihssen et al. (2011) exposed heavy drinkers to visual cues that were related to alcohol. Similar to the findings of other studies, parts of the brain that are involved in emotional processing (i.e., insular cortex) and reward circuitry (i.e., ventral striatum) showed greater activity in heavy drinkers than light drinkers. When heavy drinkers were presented with visual stimuli that were related to their higher order life goals (such as those related to family, health, and finances) they showed weaker responses in frontal areas than the light drinkers. They also showed reduced activity in the cingulate cortex when exposed to attractive food stimuli. These findings suggest that heavy drinkers have difficulty forming socially desirable goals as an alternative to drinking alcohol. The combination of overactivation to cues related to alcohol but underactivation to cues related to alternative, alcohol-unrelated goals may have important clinical implications. These findings suggest that interventions that decrease alcohol abusers' hyperreactivity to alcohol-related stimuli but increase their sensitivity to alternative goals might be beneficial.

4 ATTENTIONAL BIAS: THERAPEUTIC IMPLICATIONS

Given the potential role of attentional bias in continuation of and relapsing to addictive behaviors, various methods have been used to reduce the bias (Field et al., 2014). However, there are few double-blind clinical trials that report the effects of attentional bias retraining on treatment outcomes or relapse prevention (Begh et al., 2013). Nevertheless, among other methods for targeting addictive cognitions, such as implicit interpretations or approach-avoidance response tendencies, attentional bias modification has been the focus of much research (see Cox et al., 2014). As Cox et al. (2014) discussed, practical applications of attentional retraining are being developed; for example, a mobile phone application has been created to help people with various types of addictive behaviors to overcome their attentional bias. Wiers et al. (2015) used a modified version of the Alcohol Attention Control Training Program for delivery over the Internet to help alcohol abusers decrease their attentional bias. Finally, in an ongoing research project funded by the European Commission called *BRAINTRAIN* and using *f*MRI technology (Linden et al., 2015), alcohol-dependent participants are being trained to downregulate their brain reactions to alcohol-related pictorial stimuli. In turn, the short- and long-term effects of the training on participants' urges to drink alcohol and their alcohol consumption are being assessed. The general principle used in various types of attentional retraining programs is to help abusers improve their brain's inhibitory processes so that they can more easily disengage their attention from environmental stimuli related to alcohol or other substances of abuse. The training programs help trainees to divert their attention to alternative (usually neutral) stimuli while they try to ignore the emotionally salient stimuli (e.g., the drug-related ones). However, could there be other more nondirective methods that do not require much effort by the person to desensitize his/her addiction-related attentional system?

5 PHARMACOLOGICAL INTERVENTIONS

Several studies have investigated the effectiveness of different kinds of medication for substance abusers (for a review, see Luijten et al., 2013). Some of these studies have employed *f*MRI to monitor neuronal activation in response to drug stimuli, which could provide information about specific brain–behavior modulations. Machielsen et al. (2014) conducted a randomized controlled trial to compare the effectiveness of Clozapine with Risperidone in reducing subjective craving, attentional bias, and brain activation in patients with both schizophrenia and cannabis use disorders (CUDs). Although no differences between CUD and non-CUD patients' behavior were observed, Clozapine-treated CUD participants showed greater reductions in their subjective craving and in the activation of insula during the completion of a cannabis-related Stroop test.

There are likely multiple benefits of pharmacological interventions for substance abuse (e.g., Nalteroxon, nicotine patches). There is evidence, for example, that an intervention that helps to reduce alcohol or drug use increases in the abuser's sense of control (Leeman et al., 2014). Similarly, there is evidence that craving for a substance of abuse can be reduced by improving a person's sense of control, whether through a cognitive-behavioral intervention (e.g., Shamloo and Cox, 2014) or stimulation of the brain's dorsolateral prefrontal cortex (DLPC) using repetitive transcranial magnetic stimulation (rTMS) (Mishra et al., 2010) and transcranial direct current stimulation (tDCS) (Naylor et al., 2014).

6 NONINVASIVE BRAIN STIMULATION

There is consensus that stimulation of DLPC is associated with increases in cognitive control (Leeman et al., 2014) and enhanced processing of emotional stimuli. Brunoni et al. (2014) conducted a study with depressed participants who completed an emotional Stroop test containing neutral, positive, and negative emotional words while receiving active or sham tDCS. The results showed that participants receiving active tDCS were better able to suppress their attentional bias for negative emotional words. In a recent study, Clarke et al. (2014) used a combination of active or sham left tDCS and attentional bias modification. They reported that the combination of active tDCS and attentional bias modification led to greater reduction in attentional bias.

From Jansen et al.'s (2013) review of 17 studies using repetitive rTSM and tDCS on left or right stimulation of DLPFC, the authors concluded that both methods yielded a medium effect size for reducing craving for substances of abuse and for palatable food. Considering the prior established association between craving and attentional bias (Nikolaou et al., 2013), it would also be expected that noninvasive brain stimulation (NIBS) would be effective in reducing attentional bias. Indeed, preliminary findings suggest that using NIBS to reduce attentional bias and substance use is effective. For instance, Meng et al. (2014) found that bilateral cathodal stimulation of frontoparietotemporal areas significantly decreased both attention to smoking-related cues and cigarette use. Additionally, Hoppner et al. (2011) found preliminary support for the effectiveness of rTMS in reducing alcohol-related attentional bias.

7 CONCLUSIONS AND FUTURE DIRECTIONS

Evidence suggests that attentional bias plays an important role in sustaining addictive behaviors. However, incorporating new findings from other areas of research might help to improve assessment of attentional bias and interventions for it. Field et al. (2014) reviewed the literature on attentional bias and concluded that it is associated with subjective craving, and that moment-by-moment fluctuations in attentional bias might precede relapse to substance use. One suggestion for future research would be to investigate events and feelings that precede subjective craving and attentional bias for substance-related stimuli. For example, portable devices might allow substance abusers to measure and record momentary changes in their attentional bias under various conditions and in various situations. The precision of the measurement could be improved by recording the person's biological reactivity prior to, during, and after feelings of craving. To this end, there would be various methods to employ, e.g., *experience sampling method*, *ecological momentary assessment*, *ambulatory assessment*, and *day reconstruction method* (see Trull et al., 2009).

Choi et al. (2015) used neuronavigation-guided rTMS to induce virtual lesions in selected areas of the brain. With this technique, they reported that posterior temporal lobe plays an important role in lexical decision making. It would be possible to use the same technique to induce virtual lesions in the areas of the brain that presumably play a role in attentional bias. This technique could have important implications for precisely localizing the brain areas involved in cue reactivity and attentional bias, and it might help in the development of new clinical interventions to reduce unwanted attentional bias.

Each person's attentional bias varies across time, and it might be difficult to generalize directly from laboratory data (e.g., obtained with *f*MRI) to applied situations (e.g., when the person is experiencing craving or is about to relapse). Therefore, in future research it would be more clinically relevant to assess attentional bias in natural settings using ecologically valid methods (e.g., experience sampling techniques) (Field et al., 2014).

Although epidemiological studies have pointed to an increase in methamphetamine use in recent decades (UNODC, 2012), the majority of studies of attentional bias and substance use have focused on alcohol or nicotine; a few studies have addressed opiates and cocaine use; but methamphetamine use has not been considered. It would be beneficial in future research to research a wide range of substances and their clinical implications.

Converging evidence from animal studies (George et al., 2012; Holmes and Wellman, 2009) suggests that (a) impairment of prefrontal cortex (mPFC), which is responsible for ECFs, and (b) overactivation of the central nucleus of the amygdala is associated with excessive drinking and alcohol dependence. Similar findings have been reported from neurological studies with alcohol abusers (Koob, 2009), indicating that impairment of the brain's systems that are responsible for emotional dysregulation are associated with poorer treatment outcomes in alcohol abuse.

Therefore, attentional bias retraining under laboratory conditions might not benefit alcohol-dependent people when they are in stressful situations. In other words, those dependent on alcohol should perhaps receive some type of brain intervention prior to receiving attentional bias modification.

Given the important role of attentional bias in addictive behaviors and the need for interventions (both pharmaceutical and cognitive behavioral) to counteract the bias, understanding the neural basis for successfully reducing the bias remains an important, but as yet unanswered, question. In future research, methods such as QEEG, ERP, or *f*MRI might be employed for brain mapping of cue reactivity to help identify addictive dispositions at preclinical stages and to clarify the mechanisms that underlie the development and maintenance of alcohol dependence or other types of addictive behaviors.

REFERENCES

Baxter, M.G., Parker, A., Lindner, C.C., Izquierdo, A.D., Murray, E.A., 2000. Control of response selection by reinforcer value requires interaction of amygdala and orbital prefrontal cortex. J. Neurosci. 20, 4311–4319.

Begh, R., Munafo, M.R., Shiffman, S., Ferguson, S.G., Nichols, L., Mohammed, M.A., Holder, R.L., Sutton, S., Aveyard, P., 2013. Attentional bias retraining in cigarette smokers attempting smoking cessation (ARTS): study protocol for a double blind randomised controlled trial. BMC Public Health 13, 1176.

Berkman, E.T., Lieberman, M.D., 2009. The neuroscience of goal pursuit: bridging gaps between theory and data. In: Moskowitz, G., Grant, H. (Eds.), The Psychology of Goals. Guilford Press, New York, NY, pp. 98–126.

Brunoni, A.R., Zanao, T.A., Vanderhasselt, M.A., Valiengo, L., De Oliveira, J.F., Boggio, P.S., Lotufo, P.A., Bensenor, I.M., Fregni, F., 2014. Enhancement of affective processing induced by bifrontal transcranial direct current stimulation in patients with major depression. Neuromodulation 17, 138–142.

Bundorf, M.K., Mata, R., Schoenbaum, M., Bhattacharya, J., 2013. Are prescription drug insurance choices consistent with expected utility theory? Health Psychol. 32, 986–994.

Clarke, P.J., Browning, M., Hammond, G., Notebaert, L., Macleod, C., 2014. The causal role of the dorsolateral prefrontal cortex in the modification of attentional bias: evidence from transcranial direct current stimulation. Biol. Psychiatry 76, 946–952.

Cox, W.M., Fadardi, J.S., Pothos, E.M., 2006. The addiction-stroop test: theoretical considerations and procedural recommendations. Psychol. Bull. 132, 443–476.

Cox, W.M., Fadardi, J.S., Intriligator, J.M., Klinger, E., 2014. Attentional bias modification for addictive behaviors: clinical implications. CNS Spectr. 19 (3), 215–224. http://dx.doi.org/10.1017/S1092852914000091.

Cox, W.M., Klinger, E., Fadardi, J.S., 2015. Nonconscious motivational influences on cognitive processes in addictive behaviors. In: Heather, N., Segal, G. (Eds.), Addiction and Choice. Oxford University Press, Oxford, UK. in press.

Choi, Y.H., Park, H.K., Paik, N.J., 2015. Role of the posterior temporal lobe during language tasks: a virtual lesion study using repetitive transcranial magnetic stimulation. Neuroreport 26 (6), 314–319. http://dx.doi.org/10.1097/WNR.0000000000000339.

Dantzer, C., Wardle, J., Fuller, R., Pampalone, S.Z., Steptoe, A., 2006. International study of heavy drinking: attitudes and sociodemographic factors in university students. J. Am. Coll. Health 55, 83–89.

De La Monte, S.M., Kril, J.J., 2014. Human alcohol-related neuropathology. Acta Neuropathol. 127, 71–90.

Demirbas, H., Ilhan, I.O., Dogan, Y.B., 2012. Ways of problem solving as predictors of relapse in alcohol dependent male inpatients. Addict. Behav. 37, 131–134.

Dickson, P.A., James, M.R., Heath, A.C., Montgomery, G.W., Martin, N.G., et al., 2006. Effects of variation at the ALDH2 locus on alcohol metabolism, sensitivity, consumption, and dependence in Europeans. Alcohol. Clin. Exp. Res. 30, 1093–1100.

Fadardi, J.S., Cox, W.M., 2006. Alcohol attentional bias: drinking salience or cognitive impairment? Psychopharmacology (Berl) 185, 169–178.

Field, M., Cox, W.M., 2008. Attentional bias in addictive behaviors: A review of its development, causes, and consequences. Drug Alcohol. Depend. 97 (1–2), 1–20.

Field, M., Marhe, R., Franken, I.H.A., 2014. The clinical relevance of attentional bias in substance use disorders. CNS Spectr. 19, 225–230.

Franken, I.H., 2003. Drug craving and addiction: integrating psychological and neuropsychopharmacological approaches. Prog. Neuropsychopharmacol. Biol. Psychiatry 27, 563–579.

George, O., Sanders, C., Freiling, J., Grigoryan, E., VU, S., Allen, C.D., Crawford, E., Mandyam, C.D., Koob, G.F., 2012. Recruitment of medial prefrontal cortex neurons during alcohol withdrawal predicts cognitive impairment and excessive alcohol drinking. Proc. Natl. Acad. Sci. USA 109, 18156–18161.

Hester, R., Luijten, M., 2014. Neural correlates of attentional bias in addiction. CNS Spectr. 19, 231–238.

Hollingworth, W., Ebel, B.E., McCarty, C.A., Garrison, M.M., Christakis, D.A., et al., 2006. Prevention of deaths from harmful drinking in the United States: the potential effects of tax increases and advertising bans on young drinkers. J. Stud. Alcohol 67, 300–308.

Holmes, A., Wellman, C.L., 2009. Stress-induced prefrontal reorganization and executive dysfunction in rodents. Neurosci. Biobehav. Rev. 33, 773–783.

Hoppner, J., Broese, T., Wendler, L., Berger, C., Thome, J., 2011. Repetitive transcranial magnetic stimulation (rTMS) for treatment of alcohol dependence. World J. Biol. Psychiatry 12 (Suppl. 1), 57–62.

Huckle, T., Huakau, J., Sweetsur, P., Huisman, O., Casswell, S., 2008. Density of alcohol outlets and teenage drinking: living in an alcogenic environment is associated with higher consumption in a metropolitan setting. Addiction 103, 1614–1621.

Ihssen, N., Cox, W.M., Wiggett, A., Fadardi, J.S., Linden, D.E., 2011. Differentiating heavy from light drinkers by neural responses to visual alcohol cues and other motivational stimuli. Cereb. Cortex 21, 1408–1415.

Janes, A.C., et al., 2010. Brain reactivity to smoking cues prior to smoking cessation predicts ability to maintain tobacco abstinence. Biol. Psychiatry 67, 722–729.

Jansen, J.M., Daams, J.G., Koeter, M.W., Veltman, D.J., Van Den Brink, W., Goudriaan, A.E., 2013. Effects of non-invasive neurostimulation on craving: a meta-analysis. Neurosci. Biobehav. Rev. 37, 2472–2480.

Klinger, E., Cox, W.M., 2011. Motivation and the goal theory of current concerns. In: Cox, W.M., Klinger, E. (Eds.), Handbook of Motivational Counseling, Second ed. Wiley-Blackwell, Chichester, UK, pp. 3–47.

Koob, G.F., 2009. Brain stress systems in the amygdala and addiction. Brain Res. 1293, 61–75.

Kouneiher, F., Charron, S., Koechlin, E., 2009. Motivation and cognitive control in the human prefrontal cortex. Nat. Neurosci. 12, 939–945.

Leeman, R.F., Bogart, D., Fucito, L.M., Boettiger, C.A., 2014. "Killing Two Birds with One Stone": alcohol use reduction interventions with potential efficacy in enhancing self-control. Curr. Addict. Rep. 1, 41–52.

Linden, D.E.J., et al., 2015. BRAINTRAIN. Taking imaging into the therapeutic domain: self-regulation of brain systems for mental disorders. Ongoing research funded by the European Commission. Grant agreement no: 602186.

Liu, S., Lane, S.D., Schmitz, J.M., Waters, A.J., Cunningham, K.A., Moeller, F.G., 2011. Relationship between attentional bias to cocaine-related stimuli and impulsivity in cocaine-dependent subjects. Am. J. Drug Alcohol Abuse 37, 117–122.

Lorenz, R.C., Kruger, J.K., Neumann, B., Schott, B.H., Kaufmann, C., Heinz, A., Wustenberg, T., 2013. Cue reactivity and its inhibition in pathological computer game players. Addict. Biol. 18, 134–146.

Luijten, M., Field, M., Franken, I.H.a., 2013. Pharmacological interventions to modulate attentional bias in addiction. CNS Spectr. 19, 239–246.

Machielsen, M.W., et al., 2014. The effect of clozapine and risperidone on attentional bias in patients with schizophrenia and a cannabis use disorder: an fMRI study. J. Psychopharmacol. (Oxford, England) 28, 633–642.

Meng, Z., et al., 2014. Transcranial direct current stimulation of the frontal-parietal-temporal area attenuates smoking behavior. J. Psychiatr. Res. 54, 19–25.

Mishra, B.R., Nizamie, S.H., Das, B., Praharaj, S.K., 2010. Efficacy of repetitive transcranial magnetic stimulation in alcohol dependence: a sham-controlled study. Addiction 105, 49–55.

Morone, A., Morone, P., 2014. Estimating individual and group preference functionals using experimental data. Theor. Decis. 77, 403–422.

Murray, E.A., 2007. The amygdala, reward and emotion. Trends Cogn. Sci. 11, 489–497.

Naylor, J.C., Borckardt, J.J., Marx, C.E., Hamer, R.M., Fredrich, S., Reeves, S.T., George, M.S., 2014. Cathodal and anodal left prefrontal tDCS and the perception of control over pain. Clin. J. Pain 30, 693–700.

Nikolaou, K., et al., 2013. Acute alcohol effects on attentional bias are mediated by subcortical areas associated with arousal and salience attribution. Neuropsychopharmacology 38, 1365–1373.

Paschall, M.J., Lipperman-Kreda, S., Grube, J.W., 2014. Effects of the performance: a role analysis. Psychol. Rev. 76 (574), 591.

Shamloo, Z.S., Cox, W.M., 2014. Information-enhancement and goal setting techniques for increasing adaptive motivation and decreasing urges to drink alcohol. Addict. Behav. 39, 1205–1213.

Smith, D.G., et al., 2014. Enhanced orbitofrontal cortex function and lack of attentional bias to cocaine cues in recreational stimulant users. Biol. Psychiatry 75, 124–131.

Terrett, G., Mclennan, S.N., Henry, J.D., Biernacki, K., Mercuri, K., Curran, H.V., Rendell, P.G., 2014. Prospective memory impairment in long-term opiate users. Psychopharmacology (Berl) 231, 2623–2632.

Trull, T.J., Ebner-Priemer, U.W., 2009. Using experience sampling methods/ecological momentary assessment (ESM/EMA) in clinical assessment and clinical research: introduction to the special section. Psychol. Assess. 21 (4), 457–462. http://dx.doi.org/10.1037/a0017653.

UNODC, 2012. World Drug Report. 2012. United Nations Publication, Sales No. E.12.XI.1.

Vonmoos, M., Hulka, L.M., Preller, K.H., Minder, F., Baumgartner, M.R., Quednow, B.B., 2014. Cognitive impairment in cocaine users is drug-induced but partially reversible: evidence from a longitudinal study. Neuropsychopharmacology 39, 2200–2210.

Vrieze, S.I., Vaidyanathan, U., Hicks, B.M., Iacono, W.G., McGue, M., 2014. The role of constraint in the development of nicotine, marijuana, and alcohol dependence in young adulthood. Behav. Genet. 44, 14–24.

Wiers, R.W., Eberl, C., Rinck, M., Becker, E.S., Lindenmeyer, J., 2011. Retraining automatic action tendencies changes alcoholic patients' approach bias for alcohol and improves treatment outcome. Psychol. Sci. 22, 490–497.

Wiers, R.W., Houben, K., Fadardi, J.S., Van Beek, P., Rhemtulla, M., Cox, W.M., 2015. Alcohol cognitive bias modification training for problem drinkers over the web. Addict. Behav. 40, 21–26.

CHAPTER

Neuroscience of learning and memory for addiction medicine: from habit formation to memory reconsolidation

6

Mary M. Torregrossa*[,1], **Jane R. Taylor**[†]

*Department of Psychiatry, University of Pittsburgh, Pittsburgh, PA, USA

[†]Department of Psychiatry, Yale University, New Haven, CT, USA

[1]Corresponding authors: Tel.: +1-412-6245723, e-mail address: torregrossam@upmc.edu

Abstract

Identifying effective pharmacological treatments for addictive disorders has remained an elusive goal. Many different classes of drugs have shown some efficacy in preclinical models, but the number of effective clinical therapeutics has remained stubbornly low. The persistence of drug use and the high frequency of relapse is at least partly attributable to the enduring ability of environmental stimuli associated with drug use to maintain behavioral patterns of drug use and induce craving during abstinence. We propose that stimuli associated with drug use exert such powerful control over behavior through the development of abnormally strong memories, and their ability to initiate subconscious sequences of motor actions (habits) that promote uncontrolled drug use. In this chapter, we will review the evidence suggesting that drugs of abuse strengthen associations with cues in the environment and facilitate habit formation. We will also discuss potential mechanisms for disrupting memories associated with drug use to help improve treatments for addiction.

Keywords

Memory, Cues, Reconsolidation, Habit, Extinction, Relapse, Self-administration, Goal-directed action

Progress in Brain Research, Volume 223, ISSN 0079-6123, http://dx.doi.org/10.1016/bs.pbr.2015.07.006

1 INTRODUCTION

1.1 ROLE OF LEARNING AND MEMORY IN ADDICTION

Our brains are designed to learn. Learning about our environments and retrieving accurate information from memory storage is critical to survival. All organisms to some degree or another must learn where to find food, how to prepare it, how to find shelter, and what predators to avoid. A variety of factors can influence how effectively an individual acquires and stores new information, including availability of attentional resources, motivation, and arousal. Drugs of abuse (at certain doses) can positively affect all of these factors, increasing attention, motivation, and arousal, so that the behaviors leading to drug use and the stimuli encountered during drug exposure are strongly encoded. The ability of drugs of abuse to increase dopamine signaling and the signaling of other neurotransmitter systems involved in learning and memory accounts for this oft-cited ability of drugs to "hijack" normal brain systems (Abel and Lattal, 2001; Gipson et al., 2013; Milton and Everitt, 2012; Nestler, 2013; Torregrossa et al., 2011).

During this period of initial drug use, memories surrounding drug use may be forming and becoming exceptionally strong, but the drugs are used in a controlled manner, and consumption is initiated with the goal of experiencing the reinforcing effects of the drug. At this stage, individuals can readily refrain from drug use as needed. However, with repeated use, the behaviors associated with consumption can become automatic or habitual. Indeed, many experienced drug users describe their drug-taking actions as "ritualistic." Eventually, behaviors oriented towards drug use become compulsive, such that use continues even if the individual no longer feels that the drug is reinforcing or experiences adverse consequences.

Nevertheless, the majority of individuals suffering from an addiction eventually recognize that they have a problem and will quit and abstain from use for some time. Unfortunately, most individuals eventually relapse and ultimately endure multiple cycles of abstinence and relapse throughout their lives. At this point, the strong memories about the people, places, and things (i.e., cues) associated with drug use can induce craving or drug "wanting" that promotes relapse (Bossert et al., 2013; Epstein et al., 2006; Torregrossa et al., 2011). Moreover, these cues may also initiate the subconscious habitual behaviors associated with obtaining and taking drugs that further increases the likelihood of relapse (Everitt and Robbins, 2005).

Therefore, the successful treatment of addictive disorders may require an approach that addresses neurobiological changes to learning and memory systems that occurs as part of the addictive process. In this review, we will go over evidence supporting the theory that drugs of abuse alter learning and memory, including Pavlovian associative learning and habit formation. In addition, we will discuss factors that may increase risk for drug-induced alterations in learning and memory. Finally, we will describe possible approaches for treating addictive disorders by manipulating learning and memory systems, and potential caveats to those approaches.

2 DRUG-INDUCED ALTERATIONS TO LEARNING AND MEMORY

2.1 ASSOCIATIVE LEARNING

2.1.1 Clinical studies

One characteristic of addictive disorders is that substance use becomes more important than relationships, work, or other activities. It has been hypothesized that this shift in choice for drugs over natural rewards may be partially mediated by the competing strength of drug-associated memories that overshadows memories associated with other rewards. Therefore, one would predict that drug-associated memories are either encoded more strongly or are retrieved more readily than other forms of memory. However, whether or not a new drug association forms a stronger memory than a new memory for a natural reward has not been directly tested clinically. Nevertheless, several pieces of indirect evidence exist to suggest that this may true. For example, in a study comparing brain response of cocaine-using individuals to controls when confronted with a sexually arousing stimulus versus a cocaine stimulus the cocaine-using individuals showed a greatly diminished response to the sexual stimulus relative to controls, but showed robust brain activation when presented with a cocaine-associated stimulus (Garavan et al., 2000). Thus, cocaine-associated memories may be so strong that they overshadow those of natural rewards.

Further support for the ability of drug-cue memories to overshadow other memories comes from studies examining the "overshadowing effect" in drug users. Overshadowing is a psychological phenomenon that occurs when learning about one cue overshadows learning about another cue even when both cues are equally predictive of reward. Drug users have been shown to exhibit overshadowing of neutral cues by drug-associated cues, indicating that drug cues can interfere with learning about alternative reinforcers (Freeman et al., 2012a,b). In addition, craving has been associated with the degree of attentional bias to drug-associated cues (Hogarth et al., 2003a,b, 2006).

2.1.2 Preclinical studies

The clinical studies described above strongly support the hypothesis that drugs of abuse engage associative learning processes that lead to the over valuation of drug-associated cues. Substantial evidence also exists to suggest that similar phenomena occur in animal models of addiction.

First, prior exposure to various drugs of abuse can facilitate subsequent associative learning about cues predictive of natural rewards, using Pavlovian approach, conditioned reinforcement, and stimulus discrimination procedures (Hankosky et al., 2013; Harmer and Phillips, 1998; Olausson et al., 2003, 2004a,b; Shiflett, 2012; Taylor and Horger, 1999; Taylor and Jentsch, 2001). In addition, drugs like nicotine can enhance motivation to respond for presentation of visual cues (cf., Liu et al., 2007; Palmatier et al., 2007).

Moreover, individual differences in innate associative learning ability also associate with risk for addiction. For example, in previous studies we have found that mice with high approach to a magazine delivering food reward specifically during presentation of a predictive cue (high Pavlovian approach mice), have an impaired ability to stop responding for alcohol when alcohol is no longer present (impaired extinction), suggesting that these mice were innately more likely to form strong memories about drug use (Barker et al., 2012). In addition, several studies from Flagel and colleagues have demonstrated that rats that allocate their behavior towards cues associated with food reward delivery (sign-trackers), as opposed to spending their time interacting with the place of food delivery (goal-trackers), are at increased risk for addiction-like behaviors (Flagel et al., 2008, 2009, 2010). Therefore, individual differences in appetitive associative learning may predispose individuals to form strong drug-associated memories that promote the development of addiction.

2.2 TRANSLATING MEMORY TO ACTION

Both individual differences in associative learning and drug-induced enhancement of associative learning circuits may promote the compulsive use and propensity to relapse that characterizes addiction. However, even with the formation of strong drug-cue memories, increased attentional bias and craving, ultimately the ability of cues to stimulate drug seeking *actions* is critical to maintaining drug abuse. In other words, it may be possible that environmental cues elicit strong drug memories, but with some cognitive control it may be possible to inhibit drug taking. However, what if individual traits or chronic drug exposure increase the translation of memories into drug-taking actions? Such effects would make abstaining from use even more difficult.

The ability of associative (i.e., Pavlovian) memories to invigorate actions aimed at obtaining a reinforcer is studied experimentally using the Pavlovian-to-Instrumental Transfer (PIT) procedure. Similar to associative memory formation, preclinical studies suggest that exposure to drugs of abuse promotes the expression of PIT. For example, several studies have found that prior cocaine exposure, particularly self-administered cocaine, can promote PIT for natural reinforcers, and that cocaine cues can promote cocaine seeking (LeBlanc et al., 2012, 2013, 2014; Ostlund et al., 2014). In addition, prior amphetamine exposure can also promote PIT (Shiflett, 2012; Wyvell and Berridge, 2001).

Fewer clinical studies have investigated whether or not PIT is enhanced in chronic drug users. One recent report suggests that alcohol-dependent subjects demonstrate greater PIT than healthy controls (Garbusow et al., 2014), and presentation of cigarette packs can induce PIT in smokers (Hogarth et al., 2015). Interestingly, plain, nonbranded packs, were not able to induce PIT in this study, suggesting that features of a cue need to be consistent with their predictive relationship with drug experience to induce PIT (Hogarth et al., 2015).

Additional clinical research is needed to fully elucidate the role of PIT-related processes in the development of addiction. Moreover, we need to increase our understanding of how individual differences in propensity for PIT may interact with drug exposure to promote addiction. For example, we have found that mice with high trait PIT have facilitated habit formation and are more likely to prefer an alcohol-paired environment even after being shocked in that environment (Barker et al., 2014). These data suggest that innate differences in PIT may predict one's propensity to develop habitual or compulsive behavior. Further research is needed to determine if innate traits and/or drug exposure promotes PIT and to determine the neural substrates mediating the abilities of cues to potentiate drug-taking actions.

2.3 HABIT FORMATION

Drugs of abuse not only promote associative learning processes, and the ability of cues to invigorate actions, but may also promote the transition from goal-directed actions to stimulus-response habits (Belin et al., 2013; Everitt and Robbins, 2005). Initially, learning to perform any action is guided by the outcome of that action, which is often a reinforcer. However, with repetition or practice, behavior can become habitual, where responses are induced by antecedent environmental stimuli and are independent of their association with the outcome. Thus, habitual behaviors continue even if the reinforcer is no longer valued, the action no longer predicts reinforcer delivery, or if the action prevents reinforcer delivery (Balleine and O'Doherty, 2010). Thus, drug addiction is often conceptualized as a "bad habit" because drug use persists in the face of adverse consequences and is often triggered by environmental stimuli (Barker and Taylor, 2014; Belin et al., 2013; Everitt and Robbins, 2005; O'Tousa and Grahame, 2014; Robbins and Everitt, 1999). Moreover, drugs of abuse stimulate dopamine release in critical prefrontal cortical and dorsal striatal brain regions that are necessary for habit formation (Balleine and O'Doherty, 2010; Barker et al., 2013; Coutureau and Killcross, 2003; Gourley et al., 2013; Yin and Knowlton, 2006; Yin et al., 2006), allowing for the possibility that drugs of abuse will strongly regulate both goal-directed actions and habitual response strategies.

Indeed there is substantial preclinical evidence to suggest that drugs of abuse promote the formation of habit. For example, Dickinson and colleagues demonstrated that a habitual response for an alcohol reinforcer or oral cocaine formed more quickly than for a food reinforcer (Dickinson et al., 2002; Miles et al., 2003). More recently, a study by Corbit and colleagues also found that alcohol habits form more rapidly than a food habit (Corbit et al., 2012). In addition, chronic cocaine or amphetamine exposure can promote habit formation for a nondrug reinforcer (Corbit et al., 2014; LeBlanc et al., 2013; Nelson and Killcross, 2006, 2013; Nordquist et al., 2007). Even brief exposure to cocaine in adolescence can promote the formation of habits later in life (Hinton et al., 2014). Finally, studies have shown that while responding for cocaine is initially goal-directed, cocaine seeking becomes habitual with sufficient training (Olmstead et al., 2001; Zapata et al., 2010). However, it should be noted that all of these studies have been conducted in males, and more research on the effects of drugs of abuse on the circuitry regulating habits in females is warranted.

For example, using a mouse model that allows dissociation of genetic versus gonadal sex, we have found that genetic males are more likely to form habits for alcohol than genetic females (Barker et al., 2010), but females are more likely than males to form habits for food (Quinn et al., 2007).

In addition, other genetic and environmental factors may also influence habit formation, and thus the progression to addiction. For example, studies in rodent models have found that prior chronic stress or glucocorticoid exposure results in a more rapid progression to habitual responding for food (Dias-Ferreira et al., 2009; Gourley et al., 2012). Moreover, acute stress or acute glucocorticoid exposure + noradrenergic stimulation in human subjects biases behavior toward habitual response strategies (Schwabe and Wolf, 2009, 2010; Schwabe et al., 2010). Due to the fact that prior stress exposure is a risk factor for developing addictive disorders, and that acute stress is a strong driver of continued use and relapse after abstinence, it may be that stress mediates increased risk and propensity to relapse through actions on corticostriatal habit circuitry (Guenzel et al., 2014; See and Waters, 2010; Sinha et al., 2011).

On the other hand, we have recently found that adolescents are resistant to forming ethanol-seeking habits relative to adults, but nevertheless consume much more ethanol (Serlin and Torregrossa, 2014). Adolescence is a known period of vulnerability to developing substance use disorders, but our results suggest that this is not due to premature habit development, but rather due to goal-directed reward seeking. Nevertheless, exposure to drugs of abuse may promote habit development in adulthood, as has been shown for juvenile cocaine exposure (Hinton et al., 2014). Interestingly, in the studies examining sex differences in habit formation mentioned above, while we observed that genetic males formed ethanol-seeking habits faster than genetic females, females consumed more ethanol in free access conditions (Barker et al., 2010), indicating females may have had more goal-directed reward seeking.

In addition, rats with high trait impulsivity that have been shown to exhibit many addiction-like qualities, including compulsive responding for psychostimulants (Belin et al., 2008; Dalley et al., 2007), nevertheless demonstrate a delay in the development of behavioral control by dorsal lateral striatum dopamine systems (Murray et al., 2014). Therefore, these animals either use different neural systems to regulate compulsive drug seeking, or alternatively are highly goal-directed in their drug seeking behavior. Indeed, drug seeking may involve both habitual and goal-directed components, as one study has shown that cocaine seeking induced by weakly associated contextual stimuli was habitual, but that cocaine seeking became goal-directed in the presence of discrete cues strongly associated cocaine delivery (Root et al., 2009). Similarly, we have found that habitual food seeking responses (lever presses) become habitual with extended training, while food approach responses (food delivery port entries) remained goal-directed (Kimchi et al., 2009). Moreover, one study found that an alcohol-paired context was sufficient to revert otherwise goal-directed food seeking to habitual control, further indicating that drug exposure and drug-associated cues can promote habitual response strategies (Ostlund et al., 2010).

Based on our current state of knowledge, drug addiction most likely involves both highly goal-directed components involved in drug seeking, but that relapse after abstinence or loss of control over use, such as in a binge, is due to reversion to habitual behavior. Therefore, treatment strategies that can either reduce the ability of drug-paired stimuli to induce habitual response patterns, or treatments that help individuals regain cognitive control over their drug seeking may improve addiction therapy.

3 LEARNING AND MEMORY SYSTEMS AS ADDICTION MEDICINE

3.1 OVERVIEW

As described above, cues associated with drugs of abuse can become strong drivers of behavior and contribute to the development of addiction. Therefore, identifying approaches that reduce the strength of drug-associated memories could aid in the treatment of addiction at multiple levels. Interfering with the associative memories directly, that is disrupting the associations between environmental cues and drug use, is one strategy that has received increased attention over the past several years.

Traditionally, acquisition, consolidation, and retrieval have been considered the three cornerstones of the learning and memory process. However, more recently, research has examined what happens to memories during and after retrieval. First, several studies have established that in the process of retrieving information from long-term storage the memory becomes "destabilized." That is, the molecular mechanisms supporting the memory are reactivated and the memory becomes labile and subject to disruption. Once the memory is destabilized, it is then restored, or "restabilized" in long-term memory in a process termed reconsolidation, which requires many of the same molecular mechanisms essential for initial consolidation (Duvarci and Nader, 2004; Finnie and Nader, 2012; Nader et al., 2000; Taylor et al., 2009; Tronson and Taylor, 2007). Every time a drug-associated memory is retrieved and reconsolidated into long term memory the reconsolidation process may also be stronger resulting in progressively more invasive drug-associated memories with repeated use (Lee, 2008; Sara, 2000; Sorg, 2012; Tronson and Taylor, 2007; Tronson et al., 2006). Ultimately, this may make drug-associated memories particularly difficult to disrupt, but amenable to pharmacological methods for inhibiting the reconsolidation process. In addition, exposure to cues in the absence of drug, either repeatedly or for a prolonged period of time, can lead to formation of a new "extinction" memory. An extinction memory is a new memory that a cue is not associated with drug use, and this new memory interferes with the expression behaviors associated with the original memory (Holmes and Quirk, 2010; Peters et al., 2009; Quirk and Mueller, 2008; Taylor et al., 2009; Torregrossa and Taylor, 2012).

Pharmacological manipulations can affect all phases of memory, including extinction and reconsolidation processes, to either enhance or strengthen the memory or to make it weaker or even forgotten. In the next sections, we will discuss some of the behavioral and pharmacological approaches that show promise for potentially improving the treatment of addictive disorders.

3.2 RECONSOLIDATION

Several reviews have discussed the molecular mechanisms underlying the reconsolidation of memories, including memories associated with exposure to drugs of abuse (Lattal and Wood, 2013; Milton and Everitt, 2010; Sorg, 2012; Taylor et al., 2009; Torregrossa and Taylor, 2012). In this section, we will focus on clinically available pharmacological agents with substantial supporting preclinical data supporting their development as memory-based addiction treatments.

One of the most widely studied manipulations of reconsolidation, both preclinically and clinically, is the inhibition of β-adrenergic receptors (βARs). Notably, the clinically available βAR antagonist propranolol, administered systemically in animal models, can interfere with drug memory reconsolidation (Bernardi et al., 2006; Milton et al., 2008). Therefore, propranolol has the potential to be a clinically viable treatment for addiction. Indeed, one pilot clinical study in cocaine-dependent subjects indicates that under certain conditions postretrieval propranolol can reduce subsequent cocaine craving (Saladin et al., 2013). Several other clinical studies have supported the potential utility of propranolol-induced disruption of reconsolidation in fear-associated disorders, such as posttraumatic stress disorder (Soeter and Kindt, 2011).

In addition to βAR antagonists, several other classes of compounds are approved for use in humans and have demonstrated some ability to disrupt the reconsolidation of drug-associated memories in rodent models. For example, the anxiolytic GABAa receptor agonist midazolam can disrupt reconsolidation of a morphine conditioned place preference memory (Robinson and Franklin, 2010; Robinson et al., 2011). However, the efficacy of midazolam seems to be highly dependent on the strength and age of the memory, and the physiological state of the animal, including being morphine dependent or under stress (Bustos et al., 2010; Robinson and Franklin, 2010; Robinson et al., 2011). Therefore, midazolam may only be effective for very weak and recent memories that are easily destabilized. Further, preclinical research is needed to determine if midazolam is effective in drug self-administration models and in clinically relevant situations, such as disrupting very strong and well-established memories associated with drug use.

Finally, several studies have pointed toward the effectiveness of inhibiting the mTOR signaling pathway with rapamycin to interfere with drug memory reconsolidation (Barak et al., 2013; Lin et al., 2014; Shi et al., 2014). mTOR signaling regulates protein translation, cell growth, and mitochondrial activity. Targeting components of the mTOR signaling pathway has been of great clinical interest, and rapamycin has been approved for use in humans for a number of years (Lamming, 2014). Importantly, rapamycin can be given systemically postmemory reactivation to disrupt reconsolidation allowing the possibility of using rapamycin as a targeted treatment for maladaptive memory-based disorders (Glover et al., 2010; Mac Callum et al., 2013).

As with any reconsolidation-based treatment for addiction, rapamycin would have to be given after reactivation of the specific memories to be targeted in a controlled environment. To date, neither rapamycin, midazolam, or any other compound proposed to disrupt reconsolidation has been tested clinically in the treatment of addiction, but we anticipate a proliferation of these studies in the future given the availability of clinically approved compounds.

3.3 EXTINCTION

In addition to reconsolidation disruption, drug-associated memories can also be weakened through the process of extinction. Clinical studies have demonstrated that extinction can reduce some of the conditioned physiological effects induced by drug cues and reduces reported subjective levels of craving (Foltin and Haney, 2000; O'Brien et al., 1993). However, clinical cue extinction approaches have generally been ineffective (Conklin and Tiffany, 2002). Therefore, more recent research has focused on potential pharmacological manipulations that might be used in conjunction with extinction to help individuals maintain abstinence. For example, much attention has been given to enhancing the consolidation of extinction learning with the cognitive enhancer D-cycloserine (DCS) (Davis et al., 2006; Myers and Carlezon, 2012).

DCS is a positive modulator of NMDA receptor signaling, acting as a partial agonist at the glycine site of the receptor. DCS can enhance learning in a variety of paradigms, including extinction of drug-associated memories (Botreau et al., 2006; Nic Dhonnchadha et al., 2010; Paolone et al., 2009; Thanos et al., 2009, 2011; Torregrossa et al., 2010). DCS is safe to give clinically and has been assessed in a number of studies in human subjects. DCS has proven effective in some anxiety disorders, including specific phobias (Ressler et al., 2004), and has shown some promise in reducing cue reactivity in smokers (Santa Ana et al., 2009). However, the majority of clinical studies have not found DCS to be an effective adjunct to extinction-based therapy, including studies in cocaine- and alcohol-dependent subjects (Hofmann et al., 2012; Price et al., 2012; Watson et al., 2011). Nevertheless, no clinical studies have yet administered DCS after the extinction sessions to assure that individuals achieve significant levels of extinction learning prior to receiving this memory-enhancing drug. Therefore, it is possible that in these studies competing reconsolidation processes were also being enhanced by DCS treatment, masking any potential benefit. Indeed, depending on the conditions DCS can either enhance reconsolidation or enhance extinction (Lee et al., 2006). Therefore, careful design of clinical studies is warranted. Preferably, other targets will be found for extinction enhancement that do not also possess the potential to strengthen memory reconsolidation.

To date the majority of other pharmacological agents tested for extinction enhancing effects are other cognitive enhancing agents also acting on the glycine site of the NMDA receptor, such as D-serine, or AMPA receptor potentiators. D-serine is more potent than DCS, and has been shown to facilitate extinction in a number of studies (Hafenbreidel et al., 2014; Hammond et al., 2013; Kelamangalath and Wagner,

2010). In addition, the AMPA receptor positive allosteric modulator PEPA also produces cognitive enhancing effects and can facilitate the extinction of instrumental cocaine seeking in a self-administration paradigm (LaLumiere et al., 2010, 2012). In fear conditioning paradigms, the effects of PEPA are selective to extinction, not enhancing the reconsolidation of fear memories (Yamada et al., 2009). Therefore, PEPA may overcome the problem with DCS, that addictive memories may be unintentionally strengthened through reconsolidation mechanisms. However, the ability of PEPA to affect drug memory reconsolidation has not yet been tested.

Finally, another promising class of compounds that may selectively enhance extinction learning are histone deacetylase (HDAC) inhibitors. Gene expression is tightly controlled by the regulation of chromatin in the nucleus. Chromatin consists of DNA wrapped around histone proteins, and modification of histones by acetylation and methylation can condense or relax chromatin around specific genes to allow transcription. Histone modifications are dynamic and are known to regulate learning and memory processes. Histone acetylation generally creates a permissive gene transcription state, and maintaining histone acetylation with HDAC inhibitors can enhance learning and memory (Jarome and Lubin, 2013; Lattal and Wood, 2013; Lubin et al., 2011). Nonspecific HDAC inhibition with sodium butyrate given after cocaine conditioned place preference extinction sessions facilitates extinction and reduces subsequent reinstatement of preference induced by re-exposure to cocaine (Malvaez et al., 2010). However, sodium butyrate can potentially enhance associative learning about cocaine (Itzhak et al., 2013; Raybuck et al., 2013), suggesting that the administration of sodium butyrate would have to be closely monitored in clinical settings. A more specific HDAC3 inhibitor, however, holds promise as a potential extinction enhancing agent that can promote extinction and reduce reinstatement of a conditioned place preference for cocaine (Malvaez et al., 2013). Future studies in self-administration models will provide further evidence to support testing modulators of histone acetylation in clinical studies.

3.4 RESTORING GOAL-DIRECTED BEHAVIOR

Finally, it may be possible to interfere with the expression of habitual and compulsive drug use to help individuals maintain abstinence. Theoretically, reducing the strength of cue-drug associative memories through either reconsolidation or extinction mechanisms will disrupt the ability of these cues to induce habitual patterns of behavior (Fig. 1). However, little research has been conducted to directly test whether or not memory manipulations interfere with habitual behaviors or if habit memories are capable of disruption. To date, the majority of studies have assessed drug place preference or operant self-administration under conditions that do not result in habit formation (Corbit et al., 2012; Economidou et al., 2009; Olmstead et al., 2001; Zapata et al., 2010). Therefore, it is difficult to know if any of the manipulations listed above are effective under conditions of habitual or compulsive drug seeking. In addition, it is only possible to assess habit if the action-outcome contingency or value of the outcome is disrupted. Thus, in animal studies where the value of the

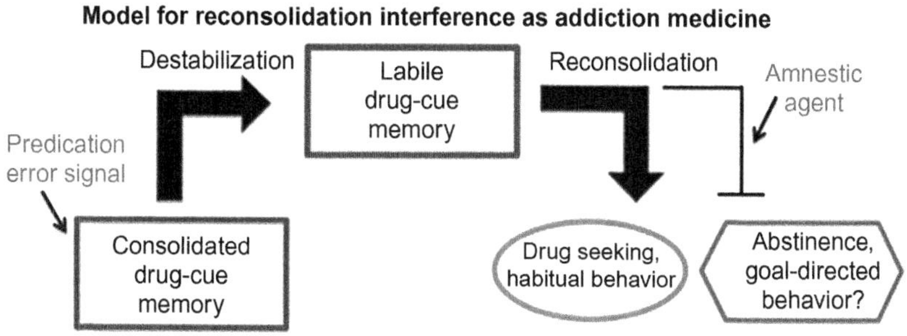

FIGURE 1

The destabilization–restabilization process in cocaine-cue memory reconsolidation. A theoretical framework for the impact of prediction error on destabilization and reconsolidation of a cocaine-cue memory and subsequent effects on drug seeking and expression of habitual behaviors.

drug presumably remains high, restoration of goal-directed behavior might be predicted to increase rather decrease drug seeking. There is, therefore, a need for more animal studies testing whether memory manipulations or manipulations targeting habit circuitry can reduce drug seeking in models where the drug is associated with adverse consequences (e.g., foot shock). One study has found that the cystine prodrug, *n*-acetylcysteine, which can promote extinction of drug seeking (Zhou and Kalivas, 2008), and has demonstrated some clinical efficacy in addiction treatment (cf., LaRowe et al., 2013; Mousavi et al., 2015), can also restore goal-directed food seeking behaviors after cocaine exposure (Corbit et al., 2014). Additional studies are needed to further explore the relationship between memory modulating agents and the expression of habitual behaviors.

Few studies have explicitly tested whether or not restoring goal-directed behavior can reduce drug seeking or treat addiction. However, using a compulsive drug seeking model where rats are periodically punished for drug seeking with a foot shock, Economidou and colleagues demonstrated that treatment with the norepinephrine reuptake inhibitor atomoxetine reduced compulsive cocaine seeking (Economidou et al., 2009). In addition, systemic administration of the dopamine D2 receptor partial agonist aripiprazole, also reduces compulsive cocaine seeking. Thus, activating D2 receptors may help reduce habitual or compulsive drug use and similar effects may also be observed by antagonizing dopamine D1 receptors (Barker et al., 2013; Nelson and Killcross, 2013). Unfortunately, clinical laboratory studies of smokers or cocaine users have not found any therapeutic potential for D2 agonists or D1 antagonists (Haney et al., 1998, 2001, 2011; Lofwall et al., 2014). However, it should be noted that in these studies the value of cocaine or cigarettes was not explicitly manipulated, so if goal-directed behavior was restored, individuals may have maintained or increased goal-directed drug seeking. Therefore, any treatments aimed at interfering with habitual behavior will have to be given in conjunction with explicit behavioral treatment paradigms that emphasize the negative consequences of drug use.

4 SPECIAL CONSIDERATIONS FOR MEMORY MANIPULATIONS IN THE TREATMENT OF ADDICTION

4.1 OVERVIEW

The sections above described many examples of potentially clinically applicable treatments for addictive disorders based on altering learning and memory systems. However, all of these approaches rely on combining some sort of behavioral intervention or laboratory manipulation with the administration of the therapeutic agent. Treatments based on learning and memory cannot be given as a daily pill as is typical for treatments of other chronic disorders. In some circumstances, this may be an advantage of memory-based treatments as concerns about medication compliance and side effects induced by chronic use would be mitigated. However, memory-based strategies do require substantial experimentation to determine the ideal conditions for memory disruption, including duration of cue exposure, timing of medication, frequency of intervention, to name a few. The sections below will briefly describe some of the considerations that need to be addressed for the successful implementation of memory-based interventions as addiction medicine.

4.2 TARGETING DRUG MEMORY DESTABILIZATION MECHANISMS?

An additional possible approach would be to use behavioral methods to selectively enhance memory destabilization to make a drug-cue memory more sensitive to subsequent disruption by amnestic agents. As discussed above, according to theories of reconsolidation, a drug-CS memory may become *destabilized* by a CS reminder (memory reactivation). Destabilization pushes the CS to a labile state, in which the memory is subject to modification prior to being restabilized. Memory destabilization is a fundamental constraint in reconsolidation that has been only studied within the context of nondrug memories. There has been little work to establish the neurobehavioral triggers of cocaine memory destabilization, but conditions that require new information to be integrated into the existing memory appear to be important (cf., Lee, 2009; Sevenster et al., 2013; Tronson and Taylor, 2013). Indeed, we, and others, have recently hypothesized that a prediction error (PE) triggered by a difference between predicted outcomes and the actual outcomes experienced specifically engages *destabilization* (Exton-McGuinness et al., 2015). Expectancy-based error correction processes are common to most theories of associative learning, and are argued to be a factor in drug-associated learning mechanisms (cf., Corlett and Taylor, 2013; Corlett et al., 2009; Tronson and Taylor, 2013). Interestingly, cocaine use disorders have been associated with possible deficits in negative PE signals (Parvaz et al., 2015) and thus manipulating PE directly in addicts may aid in abstinence. If, for example, memory destabilization could be enhanced, amnestic agents could be used to more persistently or effectively block memory restabilization. We hypothesize that behavioral interventions could be used to enhance PEs to render a CS-drug memory labile and subject to modifications to reduce the strength of the memory. We have previously shown that the strength of a cocaine-cue memory can be reduced by administering an amnestic agent following reactivation of the cocaine-cue memory in the absence of the US, which may result in a negative PE (Sanchez et al., 2010; Wan et al., 2014). Therefore, we hypothesize that the ability to behaviorally induce and/or enhance the destabilization of cocaine-cue memories through PE manipulation could be a key factor for the efficacy of subsequent or combined pharmacological therapies that aim to disrupt reconsolidation process and weaken maladaptive drug memories (see Fig. 1).

4.3 MEMORY SPECIFICITY AND BOUNDARY CONDITIONS

One concern often expressed about using reconsolidation inhibitors to treat any disorder is the possibility that other important memories will also be disrupted. This is certainly a valid concern, and any manipulation used to disrupt reconsolidation should be thoroughly tested in animal models for memory specificity. However, in general, reconsolidation disruption seems to be fairly specific to the manipulated memory, with studies in fear conditioning demonstrating that protein synthesis inhibition only disrupts reconsolidation of reactivated conditioned stimulus memories, but not memories for other conditioned stimuli that were not reactivated (Debiec et al., 2006). Moreover, we have seen similar effects for a cocaine-associated memory, where disruption of memory reconsolidation with PKA inhibition, did not affect the ability of the rats to remember to press the lever in a test of cocaine-primed reinstatement (Sanchez et al., 2010). In addition, several studies have indicated remote or strongly encoded memories are harder to destabilize and are less subject to reconsolidation disruption than newer memories (Bustos et al., 2009; Inda et al., 2011; Raybuck and Lattal, 2014). Therefore, established biographical memories, for example, may be unlikely to be destabilized and disrupted by reconsolidation manipulations.

On the flip side, one could argue that memories associated with drugs of abuse are often very strongly encoded, and are remote after years of use. Individuals may also have many memories associated with drug use, such that a more generalized disruption of all drug memories would be desired. Future studies are certainly warranted to address these issues in the addiction field.

4.4 PERSISTENCE OF EFFECTS

A critical potential limitation to reconsolidation and extinction manipulations is that any positive effects that are observed are unlikely to be long lasting. Persistence of effects is certainly a concern for extinction-based therapies, as extinction is known to be a temporary phenomenon with the expression of the original memory returning with time: a process known as spontaneous recovery, and with exposure to a context not associated with extinction: a process known as renewal (Bouton et al., 2006; Torregrossa et al., 2010, 2013). In addition, while reconsolidation disruption is theoretically a more permanent manipulation, the persistence of reconsolidation manipulations is often not assessed. Moreover, memories can easily be re-established or new memories formed with re-exposure to the drug of abuse. Therefore, successful addiction treatment may require multiple intermittent memory manipulation sessions to maintain abstinence, and to potentially address new memories over time. Developing and implementing a maintenance treatment plan is likely to be critical for the long-term success of memory-based therapies for addiction.

4.5 RESTORATION OF GOAL-DIRECTED DRUG SEEKING?

Finally, as mentioned above, efforts to inhibit habitual patterns of drug use or to regain goal-directed control of behavior must be carefully considered. If individuals do not perceive continued drug use as problematic, then enhancing goal-directedness is likely to just maintain or even increase drug use. For example, adolescents responding for alcohol display highly goal-directed motivation for much larger quantities of alcohol than adults expressing habitual behavior (Serlin and Torregrossa, 2014). Therefore, manipulations aimed at disrupting habit in human subjects will likely need to be conducted in conjunction either with agents that reduce the value of drug use or behavioral therapies that emphasize the negative consequences of continued use. More research is needed to determine the effectiveness of these approaches. Nevertheless, reconsolidation- or extinction-based therapies that interfere with stimulus-response associations may still help individuals refrain for initiating use or engaging in uncontrolled use, such as in a binge.

5 CONCLUSIONS AND FUTURE DIRECTIONS

Considering the burden of addictive disorders on society and the scarcity of effective treatments, new approaches for addiction medicine are certainly warranted. As reviewed above, drugs of abuse produce profound effects on the brain's learning and memory systems that control motivated behavior (Jentsch and Taylor, 1999). Identifying manipulations that can reverse the strong hold drug-associated memories have on behavior could greatly improve rates of abstinence and help individuals regain control over intake. Preclinical studies largely support the use of memory-based approaches to treat addiction, and several clinical studies suggest the potential utility of these treatment strategies. We argue that one novel approach would be to enhance memory lability, by triggering memory destabilization, thereby rendering cocaine-cue memories more sensitive to subsequent disruption by amnestic agents, and consequently to persistently reduce relapse triggered by drug-cue memories. Nevertheless, substantial research is still needed to determine the conditions under which memory manipulations are most effective, how long these manipulations maintain effectiveness, and to ensure that unintentional increases in drug memory strength or goal-directed drug taking do not occur.

REFERENCES

Abel, T., Lattal, K.M., 2001. Molecular mechanisms of memory acquisition, consolidation and retrieval. Curr. Opin. Neurobiol. 11, 180–187.

Balleine, B.W., O'Doherty, J.P., 2010. Human and rodent homologies in action control: corticostriatal determinants of goal-directed and habitual action. Neuropsychopharmacology 35, 48–69.

Barak, S., Liu, F., Ben Hamida, S., Yowell, Q.V., Neasta, J., Kharazia, V., Janak, P.H., Ron, D., 2013. Disruption of alcohol-related memories by mTORC1 inhibition prevents relapse. Nat. Neurosci. 16, 1111–1117.

Barker, J.M., Taylor, J.R., 2014. Habitual alcohol seeking: modeling the transition from casual drinking to addiction. Neurosci. Biobehav. Rev. 47, 281–294.

Barker, J.M., Torregrossa, M.M., Arnold, A.P., Taylor, J.R., 2010. Dissociation of genetic and hormonal influences on sex differences in alcoholism-related behaviors. J. Neurosci. 30, 9140–9144.

Barker, J.M., Torregrossa, M.M., Taylor, J.R., 2012. Low prefrontal PSA-NCAM confers risk for alcoholism-related behavior. Nat. Neurosci. 15, 1356–1358.

Barker, J.M., Torregrossa, M.M., Taylor, J.R., 2013. Bidirectional modulation of infralimbic dopamine D1 and D2 receptor activity regulates flexible reward seeking. Front. Neurosci. 7, 126.

Barker, J.M., Zhang, H., Villafane, J.J., Wang, T.L., Torregrossa, M.M., Taylor, J.R., 2014. Epigenetic and pharmacological regulation of 5HT3 receptors controls compulsive ethanol seeking in mice. Eur. J. Neurosci. 39, 999–1008.

Belin, D., Mar, A.C., Dalley, J.W., Robbins, T.W., Everitt, B.J., 2008. High impulsivity predicts the switch to compulsive cocaine-taking. Science 320, 1352–1355.

Belin, D., Belin-Rauscent, A., Murray, J.E., Everitt, B.J., 2013. Addiction: failure of control over maladaptive incentive habits. Curr. Opin. Neurobiol. 23, 564–572.

Bernardi, R.E., Lattal, K.M., Berger, S.P., 2006. Postretrieval propranolol disrupts a cocaine conditioned place preference. Neuroreport 17, 1443–1447.

Bossert, J.M., Marchant, N.J., Calu, D.J., Shaham, Y., 2013. The reinstatement model of drug relapse: recent neurobiological findings, emerging research topics, and translational research. Psychopharmacology (Berl.) 229, 453–476.

Botreau, F., Paolone, G., Stewart, J., 2006. D-Cycloserine facilitates extinction of a cocaine-induced conditioned place preference. Behav. Brain Res. 172, 173–178.

Bouton, M.E., Westbrook, R.F., Corcoran, K.A., Maren, S., 2006. Contextual and temporal modulation of extinction: behavioral and biological mechanisms. Biol. Psychiatry 60, 352–360.

Bustos, S.G., Maldonado, H., Molina, V.A., 2009. Disruptive effect of midazolam on fear memory reconsolidation: decisive influence of reactivation time span and memory age. Neuropsychopharmacology 34, 446–457.

Bustos, S.G., Giachero, M., Maldonado, H., Molina, V.A., 2010. Previous stress attenuates the susceptibility to Midazolam's disruptive effect on fear memory reconsolidation: influence of pre-reactivation D-cycloserine administration. Neuropsychopharmacology 35, 1097–1108.

Conklin, C.A., Tiffany, S.T., 2002. Applying extinction research and theory to cue-exposure addiction treatments. Addiction 97, 155–167.

Corbit, L.H., Nie, H., Janak, P.H., 2012. Habitual alcohol seeking: time course and the contribution of subregions of the dorsal striatum. Biol. Psychiatry 72, 389–395.

Corbit, L.H., Chieng, B.C., Balleine, B.W., 2014. Effects of repeated cocaine exposure on habit learning and reversal by N-acetylcysteine. Neuropsychopharmacology 39, 1893–1901.

Corlett, P.R., Taylor, J.R., 2013. The translational potential of memory reconsolidation. In: Alberini, C. (Ed.), Memory Reconsolidation. Elsevier Press, New York, NY, pp. 273–292.

Corlett, P.R., Krystal, J.H., Taylor, J.R., Fletcher, P.C., 2009. Why do delusions persist? Front. Hum. Neurosci. 3, 12.

Coutureau, E., Killcross, S., 2003. Inactivation of the infralimbic prefrontal cortex reinstates goal-directed responding in overtrained rats. Behav. Brain Res. 146, 167–174.

Dalley, J.W., Fryer, T.D., Brichard, L., Robinson, E.S.J., Theobald, D.E.H., Lääne, K., Peña, Y., Murphy, E.R., Shah, Y., Probst, K., Abakumova, I., Aigbirhio, F.I., Richards, H.K., Hong, Y., Baron, J.-C., Everitt, B.J., Robbins, T.W., 2007. Nucleus accumbens D2/3 receptors predict trait impulsivity and cocaine reinforcement. Science 315, 1267–1270.

Davis, M., Ressler, K., Rothbaum, B.O., Richardson, R., 2006. Effects of D-cycloserine on extinction: translation from preclinical to clinical work. Biol. Psychiatry 60, 369–375.

Debiec, J., Doyère, V., Nader, K., Ledoux, J.E., 2006. Directly reactivated, but not indirectly reactivated, memories undergo reconsolidation in the amygdala. Proc. Natl. Acad. Sci. U. S. A. 103, 3428–3433.

Dias-Ferreira, E., Sousa, J.C., Melo, I., Morgado, P., Mesquita, A.R., Cerqueira, J.J., Costa, R.M., Sousa, N., 2009. Chronic stress causes frontostriatal reorganization and affects decision-making. Science 325, 621–625.

Dickinson, A., Wood, N., Smith, J.W., 2002. Alcohol seeking by rats: action or habit? Q. J. Exp. Psychol. B 55, 331–348.

Duvarci, S., Nader, K., 2004. Characterization of fear memory reconsolidation. J. Neurosci. 24, 9269–9275.

Economidou, D., Pelloux, Y., Robbins, T.W., Dalley, J.W., Everitt, B.J., 2009. High impulsivity predicts relapse to cocaine-seeking after punishment-induced abstinence. Biol. Psychiatry 65, 851–856.

Epstein, D.H., Preston, K.L., Stewart, J., Shaham, Y., 2006. Toward a model of drug relapse: an assessment of the validity of the reinstatement procedure. Psychopharmacology (Berl.) 189, 1–16.

Everitt, B.J., Robbins, T.W., 2005. Neural systems of reinforcement for drug addiction: from actions to habits to compulsion. Nat. Neurosci. 8, 1481–1489.

Exton-McGuinness, M.T.J., Lee, J.L.C., Reichelt, A.C., 2015. Updating memories—the role of prediction errors in memory reconsolidation. Behav. Brain Res. 278, 375–384.

Finnie, P.S.B., Nader, K., 2012. The role of metaplasticity mechanisms in regulating memory destabilization and reconsolidation. Neurosci. Biobehav. Rev. 36, 1667–1707.

Flagel, S.B., Watson, S.J., Akil, H., Robinson, T.E., 2008. Individual differences in the attribution of incentive salience to a reward-related cue: influence on cocaine sensitization. Behav. Brain Res. 186, 48–56.

Flagel, S.B., Akil, H., Robinson, T.E., 2009. Individual differences in the attribution of incentive salience to reward-related cues: implications for addiction. Neuropharmacology 56 (Suppl. 1), 139–148.

Flagel, S.B., Robinson, T.E., Clark, J.J., Clinton, S.M., Watson, S.J., Seeman, P., Phillips, P.E. M., Akil, H., 2010. An animal model of genetic vulnerability to behavioral disinhibition and responsiveness to reward-related cues: implications for addiction. Neuropsychopharmacology 35, 388–400.

Foltin, R.W., Haney, M., 2000. Conditioned effects of environmental stimuli paired with smoked cocaine in humans. Psychopharmacology (Berl.) 149, 24–33.

Freeman, T.P., Morgan, C.J.A., Beesley, T., Curran, H.V., 2012a. Drug cue induced overshadowing: selective disruption of natural reward processing by cigarette cues amongst abstinent but not satiated smokers. Psychol. Med. 42, 161–171.

Freeman, T.P., Morgan, C.J.A., Pepper, F., Howes, O.D., Stone, J.M., Curran, H.V., 2012b. Associative blocking to reward-predicting cues is attenuated in ketamine users but can be modulated by images associated with drug use. Psychopharmacology (Berl.) 225, 41–50.

Garavan, H., Pankiewicz, J., Bloom, A., Cho, J.K., Sperry, L., Ross, T.J., Salmeron, B.J., Risinger, R., Kelley, D., Stein, E.A., 2000. Cue-induced cocaine craving: neuroanatomical specificity for drug users and drug stimuli. Am. J. Psychiatry 157, 1789–1798.

Garbusow, M., Schad, D.J., Sommer, C., Jünger, E., Sebold, M., Friedel, E., Wendt, J., Kathmann, N., Schlagenhauf, F., Zimmermann, U.S., Heinz, A., Huys, Q.J.M., Rapp, M.A., 2014. Pavlovian-to-instrumental transfer in alcohol dependence: a pilot study. Neuropsychobiology 70, 111–121.

Gipson, C.D., Kupchik, Y.M., Kalivas, P.W., 2013. Rapid, transient synaptic plasticity in addiction. Neuropharmacology 76, 276–286.

Glover, E.M., Ressler, K.J., Davis, M., 2010. Differing effects of systemically administered rapamycin on consolidation and reconsolidation of context vs. cued fear memories. Learn. Mem. 17, 577–581.

Gourley, S.L., Swanson, A.M., Jacobs, A.M., Howell, J.L., Mo, M., Dileone, R.J., Koleske, A.J., Taylor, J.R., 2012. Action control is mediated by prefrontal BDNF and glucocorticoid receptor binding. Proc. Natl. Acad. Sci. U. S. A. 109, 20714–20719.

Gourley, S.L., Olevska, A., Gordon, J., Taylor, J.R., 2013. Cytoskeletal determinants of stimulus-response habits. J. Neurosci. 33, 11811–11816.

Guenzel, F.M., Wolf, O.T., Schwabe, L., 2014. Sex differences in stress effects on response and spatial memory formation. Neurobiol. Learn. Mem. 109, 46–55.

Hafenbreidel, M., Rafa Todd, C., Twining, R.C., Tuscher, J.J., Mueller, D., 2014. Bidirectional effects of inhibiting or potentiating NMDA receptors on extinction after cocaine self-administration in rats. Psychopharmacology (Berl.) 231, 4585–4594.

Hammond, S., Seymour, C.M., Burger, A., Wagner, J.J., 2013. D-Serine facilitates the effectiveness of extinction to reduce drug-primed reinstatement of cocaine-induced conditioned place preference. Neuropharmacology 64, 464–471.

Haney, M., Foltin, R.W., Fischman, M.W., 1998. Effects of pergolide on intravenous cocaine self-administration in men and women. Psychopharmacology (Berl.) 137, 15–24.

Haney, M., Ward, A.S., Foltin, R.W., Fischman, M.W., 2001. Effects of ecopipam, a selective dopamine D1 antagonist, on smoked cocaine self-administration by humans. Psychopharmacology (Berl.) 155, 330–337.

Haney, M., Rubin, E., Foltin, R.W., 2011. Aripiprazole maintenance increases smoked cocaine self-administration in humans. Psychopharmacology (Berl.) 216, 379–387.

Hankosky, E.R., Kofsky, N.M., Gulley, J.M., 2013. Age of exposure-dependent effects of amphetamine on behavioral flexibility. Behav. Brain Res. 252, 117–125.

Harmer, C.J., Phillips, G.D., 1998. Enhanced appetitive conditioning following repeated pretreatment with D-amphetamine. Behav. Pharmacol. 9, 299–308.

Hinton, E.A., Wheeler, M.G., Gourley, S.L., 2014. Early-life cocaine interferes with BDNF-mediated behavioral plasticity. Learn. Mem. 21, 253–257.

Hofmann, S.G., Hüweler, R., MacKillop, J., Kantak, K.M., 2012. Effects of D-cycloserine on craving to alcohol cues in problem drinkers: preliminary findings. Am. J. Drug Alcohol Abuse 38, 101–107.

Hogarth, L., Dickinson, A., Duka, T., 2003a. Discriminative stimuli that control instrumental tobacco-seeking by human smokers also command selective attention. Psychopharmacology (Berl.) 168, 435–445.

Hogarth, L.C., Mogg, K., Bradley, B.P., Duka, T., Dickinson, A., 2003b. Attentional orienting towards smoking-related stimuli. Behav. Pharmacol. 14, 153–160.

Hogarth, L., Dickinson, A., Hutton, S.B., Elbers, N., Duka, T., 2006. Drug expectancy is necessary for stimulus control of human attention, instrumental drug-seeking behaviour and subjective pleasure. Psychopharmacology (Berl.) 185, 495–504.

Hogarth, L., Maynard, O.M., Munafò, M.R., 2015. Plain cigarette packs do not exert Pavlovian to instrumental transfer of control over tobacco-seeking. Addiction 110, 174–182.

Holmes, A., Quirk, G.J., 2010. Pharmacological facilitation of fear extinction and the search for adjunct treatments for anxiety disorders—the case of yohimbine. Trends Pharmacol. Sci. 31, 2–7.

Inda, M.C., Muravieva, E.V., Alberini, C.M., 2011. Memory retrieval and the passage of time: from reconsolidation and strengthening to extinction. J. Neurosci. 31, 1635–1643.

Itzhak, Y., Liddie, S., Anderson, K.L., 2013. Sodium butyrate-induced histone acetylation strengthens the expression of cocaine-associated contextual memory. Neurobiol. Learn. Mem. 102, 34–42.

Jarome, T.J., Lubin, F.D., 2013. Histone lysine methylation: critical regulator of memory and behavior. Rev. Neurosci. 24, 375–387.

Jentsch, J.D., Taylor, J.R., 1999. Impulsivity resulting from frontostriatal dysfunction in drug abuse: implications for the control of behavior by reward-related stimuli. Psychopharmacology (Berl.) 146, 373–390.

Kelamangalath, L., Wagner, J.J., 2010. D-serine treatment reduces cocaine-primed reinstatement in rats following extended access to cocaine self-administration. Neuroscience 169, 1127–1135.

Kimchi, E.Y., Torregrossa, M.M., Taylor, J.R., Laubach, M., 2009. Neuronal correlates of instrumental learning in the dorsal striatum. J. Neurophysiol. 102, 475–489.

LaLumiere, R.T., Niehoff, K.E., Kalivas, P.W., 2010. The infralimbic cortex regulates the consolidation of extinction after cocaine self-administration. Learn. Mem. 17, 168–175.

LaLumiere, R.T., Smith, K.C., Kalivas, P.W., 2012. Neural circuit competition in cocaine-seeking: roles of the infralimbic cortex and nucleus accumbens shell. Eur. J. Neurosci. 35, 614–622.

Lamming, D.W., 2014. Diminished mTOR signaling: a common mode of action for endocrine longevity factors. SpringerPlus 3, 735.

LaRowe, S.D., Kalivas, P.W., Nicholas, J.S., Randall, P.K., Mardikian, P.N., Malcolm, R.J., 2013. A double-blind placebo-controlled trial of N-acetylcysteine in the treatment of cocaine dependence. Am. J. Addict. 22, 443–452.

Lattal, K.M., Wood, M.A., 2013. Epigenetics and persistent memory: implications for reconsolidation and silent extinction beyond the zero. Nat. Neurosci. 16, 124–129.

LeBlanc, K.H., Ostlund, S.B., Maidment, N.T., 2012. Pavlovian-to-instrumental transfer in cocaine seeking rats. Behav. Neurosci. 126, 681–689.

LeBlanc, K.H., Maidment, N.T., Ostlund, S.B., 2013. Repeated cocaine exposure facilitates the expression of incentive motivation and induces habitual control in rats. PLoS One 8, e61355.

LeBlanc, K.H., Maidment, N.T., Ostlund, S.B., 2014. Impact of repeated intravenous cocaine administration on incentive motivation depends on mode of drug delivery. Addict. Biol. 19, 965–971.

Lee, J.L.C., 2008. Memory reconsolidation mediates the strengthening of memories by additional learning. Nat. Neurosci. 11, 1264–1266.

Lee, J.L.C., 2009. Reconsolidation: maintaining memory relevance. Trends Neurosci. 32, 413–420.

Lee, J.L.C., Milton, A.L., Everitt, B.J., 2006. Reconsolidation and extinction of conditioned fear: inhibition and potentiation. J. Neurosci. 26, 10051–10056.

Lin, J., Liu, L., Wen, Q., Zheng, C., Gao, Y., Peng, S., Tan, Y., Li, Y., 2014. Rapamycin prevents drug seeking via disrupting reconsolidation of reward memory in rats. Int. J. Neuropsychopharmacol. 17, 127–136.

Liu, X., Palmatier, M.I., Caggiula, A.R., Donny, E.C., Sved, A.F., 2007. Reinforcement enhancing effect of nicotine and its attenuation by nicotinic antagonists in rats. Psychopharmacology (Berl.) 194, 463–473.

Lofwall, M.R., Nuzzo, P.A., Campbell, C., Walsh, S.L., 2014. Aripiprazole effects on self-administration and pharmacodynamics of intravenous cocaine and cigarette smoking in humans. Exp. Clin. Psychopharmacol. 22, 238–247.

Lubin, F.D., Gupta, S., Parrish, R.R., Grissom, N.M., Davis, R.L., 2011. Epigenetic mechanisms: critical contributors to long-term memory formation. Neuroscientist 17, 616–632.

Mac Callum, P.E., Hebert, M., Adamec, R.E., Blundell, J., 2013. Systemic inhibition of mTOR kinase via rapamycin disrupts consolidation and reconsolidation of auditory fear memory. Neurobiol. Learn. Mem. 112, 176–185.

Malvaez, M., Sanchis-Segura, C., Vo, D., Lattal, K.M., Wood, M.A., 2010. Modulation of chromatin modification facilitates extinction of cocaine-induced conditioned place preference. Biol. Psychiatry 67, 36–43.

Malvaez, M., McQuown, S.C., Rogge, G.A., Astarabadi, M., Jacques, V., Carreiro, S., Rusche, J.R., Wood, M.A., 2013. HDAC3-selective inhibitor enhances extinction of cocaine-seeking behavior in a persistent manner. Proc. Natl. Acad. Sci. U. S. A. 110, 2647–2652.

Miles, F.J., Everitt, B.J., Dickinson, A., 2003. Oral cocaine seeking by rats: action or habit? Behav. Neurosci. 117, 927–938.

Milton, A.L., Everitt, B.J., 2010. The psychological and neurochemical mechanisms of drug memory reconsolidation: implications for the treatment of addiction. Eur. J. Neurosci. 31, 2308–2319.

Milton, A.L., Everitt, B.J., 2012. The persistence of maladaptive memory: addiction, drug memories and anti-relapse treatments. Neurosci. Biobehav. Rev. 36, 1119–1139.

Milton, A.L., Lee, J.L.C., Everitt, B.J., 2008. Reconsolidation of appetitive memories for both natural and drug reinforcement is dependent on {beta}-adrenergic receptors. Learn. Mem. 15, 88–92.

Mousavi, S.G., Sharbafchi, M.R., Salehi, M., Peykanpour, M., Karimian Sichani, N., Maracy, M., 2015. The efficacy of N-acetylcysteine in the treatment of methamphetamine dependence: a double-blind controlled, crossover study. Arch. Iran. Med. 18, 28–33.

Murray, J.E., Dilleen, R., Pelloux, Y., Economidou, D., Dalley, J.W., Belin, D., Everitt, B.J., 2014. Increased impulsivity retards the transition to dorsolateral striatal dopamine control of cocaine seeking. Biol. Psychiatry 76, 15–22.

Myers, K.M., Carlezon, W.A., 2012. D-cycloserine effects on extinction of conditioned responses to drug-related cues. Biol. Psychiatry 71, 947–955.

Nader, K., Schafe, G.E., LeDoux, J.E., 2000. The labile nature of consolidation theory. Nat. Rev. Neurosci. 1, 216–219.
Nelson, A., Killcross, S., 2006. Amphetamine exposure enhances habit formation. J. Neurosci. 26, 3805–3812.
Nelson, A.J.D., Killcross, S., 2013. Accelerated habit formation following amphetamine exposure is reversed by D1, but enhanced by D2, receptor antagonists. Front. Neurosci. 7, 76.
Nestler, E.J., 2013. Cellular basis of memory for addiction. Dialogues Clin. Neurosci. 15, 431–443.
Nic Dhonnchadha, B.A., Szalay, J.J., Achat-Mendes, C., Platt, D.M., Otto, M.W., Spealman, R.D., Kantak, K.M., 2010. D-cycloserine deters reacquisition of cocaine self-administration by augmenting extinction learning. Neuropsychopharmacology 35, 357–367.
Nordquist, R.E., Voorn, P., de Mooij-van Malsen, J.G., Joosten, R.N.J.M.A., Pennartz, C.M. A., Vanderschuren, L.J.M.J., 2007. Augmented reinforcer value and accelerated habit formation after repeated amphetamine treatment. Eur. Neuropsychopharmacol. 17, 532–540.
O'Brien, C.P., Childress, A.R., McLellan, A.T., Ehrman, R., 1993. Developing treatments that address classical conditioning. NIDA Res. Monogr. 135, 71–91.
O'Tousa, D., Grahame, N., 2014. Habit formation: implications for alcoholism research. Alcohol 48, 327–335.
Olausson, P., Jentsch, J.D., Taylor, J.R., 2003. Repeated nicotine exposure enhances reward-related learning in the rat. Neuropsychopharmacology 28, 1264–1271.
Olausson, P., Jentsch, J.D., Taylor, J.R., 2004a. Repeated nicotine exposure enhances responding with conditioned reinforcement. Psychopharmacology (Berl.) 173, 98–104.
Olausson, P., Jentsch, J.D., Taylor, J.R., 2004b. Nicotine enhances responding with conditioned reinforcement. Psychopharmacology (Berl.) 171, 173–178.
Olmstead, M.C., Lafond, M.V., Everitt, B.J., Dickinson, A., 2001. Cocaine seeking by rats is a goal-directed action. Behav. Neurosci. 115, 394–402.
Ostlund, S.B., Maidment, N.T., Balleine, B.W., 2010. Alcohol-paired contextual cues produce an immediate and selective loss of goal-directed action in rats. Front. Integr. Neurosci. 4, 19.
Ostlund, S.B., LeBlanc, K.H., Kosheleff, A.R., Wassum, K.M., Maidment, N.T., 2014. Phasic mesolimbic dopamine signaling encodes the facilitation of incentive motivation produced by repeated cocaine exposure. Neuropsychopharmacology 39, 2441–2449.
Palmatier, M.I., Liu, X., Caggiula, A.R., Donny, E.C., Sved, A.F., 2007. The role of nicotinic acetylcholine receptors in the primary reinforcing and reinforcement-enhancing effects of nicotine. Neuropsychopharmacology 32, 1098–1108.
Paolone, G., Botreau, F., Stewart, J., 2009. The facilitative effects of D-cycloserine on extinction of a cocaine-induced conditioned place preference can be long lasting and resistant to reinstatement. Psychopharmacology (Berl.) 202, 403–409.
Parvaz, M.A., Konova, A.B., Proudfit, G.H., Dunning, J.P., Malaker, P., Moeller, S.J., Maloney, T., Alia-Klein, N., Goldstein, R.Z., 2015. Impaired neural response to negative prediction errors in cocaine addiction. J. Neurosci. 35, 1872–1879.
Peters, J., Kalivas, P.W., Quirk, G.J., 2009. Extinction circuits for fear and addiction overlap in prefrontal cortex. Learn. Mem. 16, 279–288.
Price, K.L., Baker, N.L., McRae-Clark, A.L., Saladin, M.E., Desantis, S.M., Santa Ana, E.J., Brady, K.T., 2012. A randomized, placebo-controlled laboratory study of the effects of

D-cycloserine on craving in cocaine-dependent individuals. Psychopharmacology (Berl.) 226 (4), 739–746.

Quinn, J.J., Hitchcott, P.K., Umeda, E.A., Arnold, A.P., Taylor, J.R., 2007. Sex chromosome complement regulates habit formation. Nat. Neurosci. 10, 1398–1400.

Quirk, G.J., Mueller, D., 2008. Neural mechanisms of extinction learning and retrieval. Neuropsychopharmacology 33, 56–72.

Raybuck, J.D., Lattal, K.M., 2014. Differential effects of dorsal hippocampal inactivation on expression of recent and remote drug and fear memory. Neurosci. Lett. 569, 1–5.

Raybuck, J.D., McCleery, E.J., Cunningham, C.L., Wood, M.A., Lattal, K.M., 2013. The histone deacetylase inhibitor sodium butyrate modulates acquisition and extinction of cocaine-induced conditioned place preference. Pharmacol. Biochem. Behav. 106, 109–116.

Ressler, K.J., Rothbaum, B.O., Tannenbaum, L., Anderson, P., Graap, K., Zimand, E., Hodges, L., Davis, M., 2004. Cognitive enhancers as adjuncts to psychotherapy: use of D-cycloserine in phobic individuals to facilitate extinction of fear. Arch. Gen. Psychiatry 61, 1136–1144.

Robbins, T.W., Everitt, B.J., 1999. Drug addiction: bad habits add up. Nature 398, 567–570.

Robinson, M.J.F., Franklin, K.B.J., 2010. Reconsolidation of a morphine place preference: impact of the strength and age of memory on disruption by propranolol and midazolam. Behav. Brain Res. 213, 201–207.

Robinson, M.J.F., Armson, M., Franklin, K.B.J., 2011. The effect of propranolol and midazolam on the reconsolidation of a morphine place preference in chronically treated rats. Front. Behav. Neurosci. 5, 42.

Root, D.H., Fabbricatore, A.T., Barker, D.J., Ma, S., Pawlak, A.P., West, M.O., 2009. Evidence for habitual and goal-directed behavior following devaluation of cocaine: a multifaceted interpretation of relapse. PLoS One 4 (9), e7170.

Saladin, M.E., Gray, K.M., McRae-Clark, A.L., Larowe, S.D., Yeatts, S.D., Baker, N.L., Hartwell, K.J., Brady, K.T., 2013. A double blind, placebo-controlled study of the effects of post-retrieval propranolol on reconsolidation of memory for craving and cue reactivity in cocaine dependent humans. Psychopharmacology (Berl.) 226, 721–737.

Sanchez, H., Quinn, J.J., Torregrossa, M.M., Taylor, J.R., 2010. Reconsolidation of a cocaine-associated stimulus requires amygdalar protein kinase A. J. Neurosci. 30, 4401–4407.

Santa Ana, E.J., Rounsaville, B.J., Frankforter, T.L., Nich, C., Babuscio, T., Poling, J., Gonsai, K., Hill, K.P., Carroll, K.M., 2009. D-Cycloserine attenuates reactivity to smoking cues in nicotine dependent smokers: a pilot investigation. Drug Alcohol Depend. 104, 220–227.

Sara, S.J., 2000. Retrieval and reconsolidation: toward a neurobiology of remembering. Learn. Mem. 7, 73–84.

Schwabe, L., Wolf, O.T., 2009. Stress prompts habit behavior in humans. J. Neurosci. 29, 7191–7198.

Schwabe, L., Wolf, O.T., 2010. Socially evaluated cold pressor stress after instrumental learning favors habits over goal-directed action. Psychoneuroendocrinology 35, 977–986.

Schwabe, L., Tegenthoff, M., Höffken, O., Wolf, O.T., 2010. Concurrent glucocorticoid and noradrenergic activity shifts instrumental behavior from goal-directed to habitual control. J. Neurosci. 30, 8190–8196.

See, R.E., Waters, R.P., 2010. Pharmacologically-induced stress: a cross-species probe for translational research in drug addiction and relapse. Am. J. Transl. Res. 3, 81–89.

Serlin, H., Torregrossa, M.M., 2014. Adolescent rats are resistant to forming ethanol seeking habits. Dev. Cogn. Neurosci. http://dx.doi.org/10.1016/j.dcn.2014.12.002

Sevenster, D., Beckers, T., Kindt, M., 2013. Prediction error governs pharmacologically induced amnesia for learned fear. Science 339, 830–833.

Shi, X., Miller, J.S., Harper, L.J., Poole, R.L., Gould, T.J., Unterwald, E.M., 2014. Reactivation of cocaine reward memory engages the Akt/GSK3/mTOR signaling pathway and can be disrupted by GSK3 inhibition. Psychopharmacology (Berl.) 231 (16), 3109–3118.

Shiflett, M.W., 2012. The effects of amphetamine exposure on outcome-selective Pavlovian-instrumental transfer in rats. Psychopharmacology (Berl.) 223, 361–370.

Sinha, R., Shaham, Y., Heilig, M., 2011. Translational and reverse translational research on the role of stress in drug craving and relapse. Psychopharmacology (Berl.) 218, 69–82.

Soeter, M., Kindt, M., 2011. Disrupting reconsolidation: pharmacological and behavioral manipulations. Learn. Mem. 18, 357–366.

Sorg, B.A., 2012. Reconsolidation of drug memories. Neurosci. Biobehav. Rev. 36, 1400–1417.

Taylor, J.R., Horger, B.A., 1999. Enhanced responding for conditioned reward produced by intra-accumbens amphetamine is potentiated after cocaine sensitization. Psychopharmacology (Berl.) 142, 31–40.

Taylor, J.R., Jentsch, J.D., 2001. Repeated intermittent administration of psychomotor stimulant drugs alters the acquisition of Pavlovian approach behavior in rats: differential effects of cocaine, D-amphetamine and 3,4-methylenedioxymethamphetamine ("Ecstasy"). Biol. Psychiatry 50, 137–143.

Taylor, J.R., Olausson, P., Quinn, J.J., Torregrossa, M.M., 2009. Targeting extinction and reconsolidation mechanisms to combat the impact of drug cues on addiction. Neuropharmacology 56 (Suppl. 1), 186–195.

Thanos, P.K., Bermeo, C., Wang, G.-J., Volkow, N.D., 2009. D-cycloserine accelerates the extinction of cocaine-induced conditioned place preference in C57bL/c mice. Behav. Brain Res. 199, 345–349.

Thanos, P.K., Bermeo, C., Wang, G.-J., Volkow, N.D., 2011. D-cycloserine facilitates extinction of cocaine self-administration in rats. Synapse 65, 938–944.

Torregrossa, M.M., Taylor, J.R., 2012. Learning to forget: manipulating extinction and reconsolidation processes to treat addiction. Psychopharmacology (Berl.).

Torregrossa, M.M., Sanchez, H., Taylor, J.R., 2010. D-cycloserine reduces the context specificity of pavlovian extinction of cocaine cues through actions in the nucleus accumbens. J. Neurosci. 30, 10526–10533.

Torregrossa, M.M., Corlett, P.R., Taylor, J.R., 2011. Aberrant learning and memory in addiction. Neurobiol. Learn. Mem. 96 (4), 609–623.

Torregrossa, M.M., Gordon, J., Taylor, J.R., 2013. Double dissociation between the anterior cingulate cortex and nucleus accumbens core in encoding the context versus the content of pavlovian cocaine cue extinction. J. Neurosci. 33, 8370–8377.

Tronson, N.C., Taylor, J.R., 2007. Molecular mechanisms of memory reconsolidation. Nat. Rev. Neurosci. 8, 262–275.

Tronson, N.C., Taylor, J.R., 2013. Addiction: a drug-induced disorder of memory reconsolidation. Curr. Opin. Neurobiol. 23, 573–580.

Tronson, N.C., Wiseman, S.L., Olausson, P., Taylor, J.R., 2006. Bidirectional behavioral plasticity of memory reconsolidation depends on amygdalar protein kinase A. Nat. Neurosci. 9, 167–169.

Wan, X., Torregrossa, M.M., Sanchez, H., Nairn, A.C., Taylor, J.R., 2014. Activation of exchange protein activated by cAMP in the rat basolateral amygdala impairs reconsolidation of a memory associated with self-administered cocaine. PLoS One 9, e107359.

Watson, B.J., Wilson, S., Griffin, L., Kalk, N.J., Taylor, L.G., Munafò, M.R., Lingford-Hughes, A.R., Nutt, D.J., 2011. A pilot study of the effectiveness of D-cycloserine during cue-exposure therapy in abstinent alcohol-dependent subjects. Psychopharmacology (Berl.) 216, 121–129.

Wyvell, C.L., Berridge, K.C., 2001. Incentive sensitization by previous amphetamine exposure: increased cue-triggered "wanting" for sucrose reward. J. Neurosci. 21, 7831–7840.

Yamada, D., Zushida, K., Wada, K., Sekiguchi, M., 2009. Pharmacological discrimination of extinction and reconsolidation of contextual fear memory by a potentiator of AMPA receptors. Neuropsychopharmacology 34, 2574–2584.

Yin, H.H., Knowlton, B.J., 2006. The role of the basal ganglia in habit formation. Nat. Rev. Neurosci. 7, 464–476.

Yin, H.H., Knowlton, B.J., Balleine, B.W., 2006. Inactivation of dorsolateral striatum enhances sensitivity to changes in the action-outcome contingency in instrumental conditioning. Behav. Brain Res. 166, 189–196.

Zapata, A., Minney, V.L., Shippenberg, T.S., 2010. Shift from goal-directed to habitual cocaine seeking after prolonged experience in rats. J. Neurosci. 30, 15457–15463.

Zhou, W., Kalivas, P.W., 2008. N-acetylcysteine reduces extinction responding and induces enduring reductions in cue- and heroin-induced drug-seeking. Biol. Psychiatry 63, 338–340.

CHAPTER

Neuroscience of drug craving for addiction medicine: From circuits to therapies

7

Hamed Ekhtiari*,†,‡,1, Padideh Nasseri*,†, Fatemeh Yavari*, Azarkhsh Mokri§, John Monterosso¶,||,#

*Neurocognitive Laboratory, Iranian National Center for Addiction Studies (INCAS), Tehran University of Medical Sciences, Tehran, Iran

†Translational Neuroscience Program, Institute for Cognitive Science Studies (ICSS), Tehran, Iran

‡Research Center for Molecular and Cellular Imaging (RCMCI), Tehran University of Medical Sciences, Tehran, Iran

§Clinical Department, Iranian National Center for Addiction Studies (INCAS), Tehran University of Medical Sciences, Tehran, Iran

¶Neuroscience Graduate Program, University of Southern California, Los Angeles, CA, USA

||Department of Psychology, University of Southern California, Los Angeles, CA, USA

#Brain and Creativity Institute, University of Southern California, Los Angeles, CA, USA

[1]Corresponding author: Tel.: +98-912-1885898; Fax: +98-21-55412232
e-mail address: h_ekhtiari@razi.tums.ac.ir

Abstract

Drug craving is a dynamic neurocognitive emotional–motivational response to a wide range of cues, from internal to external environments and from drug-related to stressful or affective events. The subjective feeling of craving, as an appetitive or compulsive state, could be considered a part of this multidimensional process, with modules in different levels of consciousness and embodiment. The neural correspondence of this dynamic and complex phenomenon may be productively investigated in relation to regional, small-scale networks, large-scale networks, and brain states. Within cognitive neuroscience, this approach has provided a long list of neural and cognitive targets for craving modulations with different cognitive, electrical, or pharmacological interventions. There are new opportunities to integrate different approaches for carving management from environmental, behavioral, psychosocial, cognitive, and neural perspectives. By using cognitive neuroscience models that treat drug craving as a dynamic and multidimensional process, these approaches may yield more effective interventions for addiction medicine.

Progress in Brain Research, Volume 223, ISSN 0079-6123, http://dx.doi.org/10.1016/bs.pbr.2015.10.002

Keywords

Substance use disorders, Drug craving, Functional magnetic resonance imaging, Neurocognitive, Integrative model, Addiction medicine

1 INTRODUCTION

Substance use disorder as a brain disease is characterized by persistent drug-seeking and drug-taking behaviors despite significant negative consequences in physical, emotional, social, and occupational aspects of the individual's life (Shariatirad et al., 2013; Volkow et al., 2011). Drug craving is considered by many researchers to be one of the main driving forces of drug-taking behaviors (Anton, 1999; Kassel and Shiffman, 1992; Robinson and Berridge, 1993). In spite of growing interest in craving which led to its consideration as one of the criteria for substance use disorders in DSM-5, there is a lack of consensuses about its definition (Sayette et al., 2000) and underlying causes (Drummond, 2001). This controversy may arise from insufficient understanding of the phenomenon. Common psychological perspectives describe craving as a subjective motivational state in which an individual experiences an intense urge to engage in a behavior (here drug use), especially when that behavior is not in accord with longer range interests and goals. However, while the subjective state of wanting is important in most conceptions of craving, modern theorists have argued that it is only one component of craving. The complete mental entanglement of craving can include a variety of related phenomena such as memories (specific episodes of past use), positive expectancies, difficulties in concentration (on things other than the target of the craving), heightened attention to substance-related stimuli, particular interpretations of physiological reactions that can occur in response to substance cues, and actual automatic behavior associated with substance use (Maarefvand et al., 2013; Tiffany, 1990).

The ambiguity of craving is not limited to its definition. Different interventions such as mindfulness relapse prevention (Witkiewitz et al., 2005), attentional bias modification (Cox et al., 2014; Field et al., 2014), pharmacological treatments (Dackis et al., 2005; Johansson et al., 2006), and non-invasive brain stimulation (NIBS) (Gorelick et al., 2014; Shahbabaie et al., 2014) designed to manage craving and subsequently prevent relapse, target only some aspects of drug craving. Moreover, there is little evidence regarding interactions between these interventions and particular aspects of drug craving. Identification of these interactions could point a way forward by suggesting complimentary interventions that address distinct components of craving (a difficult task in real-life addiction medicine settings).

In recent years, the methods of cognitive neuroscience are increasingly being used to study cognitive processes underlying different psychiatric disorders including substance use disorders. Neuroimaging techniques, such as functional magnetic resonance imaging (fMRI) and positron emission tomography, have been widely used to identify neural correlates of craving (Ekhtiari et al., 2008; Tabatabaei-Jafari et al., 2014). This naturally raises the question, "Do the advances of neuroscience provide a bases from which to update existing models of drug craving, and to develop more effective neurocognitive treatments for addiction medicine?" (Ekhtiari, 2010).

2 MODELS OF DRUG CRAVING: FROM BEHAVIORAL TO COGNITIVE PERSPECTIVES

Investigators from different perspectives have advanced conceptual accounts of the nature of craving and its associated conditions, as well as its underlying causes. Most of the models can be divided into two main categories: conditioning theories and cognitive theories (Singleton and Gorelick, 1998). Conditioning theories consider craving to be Pavlovian responses that occur as a result of continuous pairings of the effects of a psychoactive substance in the brain with presented environmental cues. These conditioned responses can result from repeated pairing of stimuli with positive effects of substance use (appetitive, substance-like models) (Stewart et al., 1984), or the alleviation of negative effects of withdrawal (compulsive, withdrawal-like models) (Ludwig et al., 1974; Siegel, 1989), or can happen because of incentive motivational features of a substance that sensitize the brain and subsequently change the perception of cues and make the substance wanted (incentive-sensitization model) (Robinson and Berridge, 1993).

In contrast, cognitive theories are based on the assumption that responses to substance-related cues involve various cognitive processes. According to these theories, craving either is mediated by beliefs and expectations about the positive effects of substance use, such as pleasure, relaxation, or relief from negative effects of withdrawal (cognitive – behavioral model) (Larimer et al., 1999), or considered as an automatic cognitive processes elicited when substance use is impeded (cognitive model of drug urges and drug use behavior) (Tiffany, 1990).

Conditioning theories have been influential in the development of cue-exposure treatments, while cognitive theories have been the basis for other cognitive therapies of substance use disorders (Drummond, 2001).

Although all these models have tried to explain the nature of craving, none of them provide a complete explanation of the multidimensional phenomenon of craving. In recent years, neuroscience has identified associations between craving and brain regions, networks, and cognitive functions. Building on these findings, an extended explanation can be provided which takes into account varied components of craving and links neural mechanisms to cognitive and behavioral functions and environmental factors.

3 NEUROCOGNITIVE BASIS OF DRUG CRAVING: REGION-BASED PERSPECTIVE

In recent years, advances in neuroscience methodologies, and particularly in brain-imaging techniques, have allowed better understanding of drug craving by linking the processes involved in craving to certain brain pathways. There are about 200 functional neuroimaging studies, mainly fMRI, which investigated craving with cue-reactivity paradigms among different populations with substance use disorders. Cue-reactivity paradigms generally involve exposure of substance or stress/emotional-related cues in block or event design during image acquisitions.

As it has been shown in Table 1, drug craving could be considered a multidimensional phenomenon associated with a range of neurocognitive functions. The main cognitive processes elicited during cue exposure can be categorized into six different domains spreading from bottom-up perceptual processes to top-down regulatory executive ones (Table 1). These processes involve a wide range of activities in different subcortical and cortical regions and can be activated in different levels based on various individualized factors such as severity and longevity of the disorder and current status of drug use (e.g., recreational user, active user, treatment seeker or abstinent) (Jasinska et al., 2014).

4 NEUROCOGNITIVE BASIS OF DRUG CRAVING: NETWORK-BASED PERSPECTIVE

It has been shown that many cognitive functions result from the coordinated activations of different brain regions as a network, and therefore many neurological and psychiatric diseases likely reflect abnormality in brain networks (Fox et al., 2012; Lerman et al., 2014). Therefore, in order to better understand the neurocognitive basis of substance use disorder (and particularly of drug craving), we need to consider links between craving phenomena and brain network activity. From a networks perspective, the integrity of network interactions and disrupted functional connectivity between them can provide us with clues about the neurocognitive dysfunctions of addiction, and potential targets for effective interventions.

There are various functional brain networks suggested to be involved in different cognitive functions. Six well-known brain networks could explain the main cognitive functions involved in the cue reactivity among drug users as follows (DeWitt et al., 2015; Janes et al., 2015; Lerman et al., 2014; Potvin et al., 2015; Robbins et al., 2008; Seo and Sinha, 2014).

4.1 STRIATAL-LIMBIC NETWORK

A common characteristic of most substance of abuse is their ability to increase extracellular dopamine concentration in ventral striatum, amygdala, hypothalamus, and orbitofrontal cortex which are innervated by dopaminergic projections from the ventral

Table 1 Cognitive Dimensions and Functions Associated with Cue-Elicited Drug Craving and Their Corresponding Brain Areas Among People with Substance Use Disorders Based on Imaging Studies Using Cue-Reactivity Paradigms

Domain	Subdomains	Possible Related Regions	Possible Relations to Cue Reactivity
Perception	Sensory processing and perception	Middle occipital gyrus (Charboneau et al., 2013; Chase et al., 2011), lingual gyrus (Engelmann et al., 2012; Park et al., 2007), cuneus (Charboneau et al., 2013)	Enhanced activity in visual cortex is consistently observed in studies using visual cues. Given top-down influences over visual processing and evidence that higher value targets recruit greater visual activation, it is likely that cue value mediates this effect.
	Object recognition and classification	Inferior temporal cortex (Park et al., 2007), fusiform gyrus (Due et al., 2002)	The users probably acquire an object recognition and classification expertise for cues associated with their substance of abuse. This expertise helps them to have an efficient access to the semantic knowledge, which may in turn enhance processing of substance cues.
	Sensory imagery	Primary and secondary sensory cortices (Kilts et al., 2001)	Imagery carries the emotional qualities of the desired event and also mimics expected pleasure or relief. Continual elaboration of the imagery can elicit craving and trigger mechanisms of substance use.
	Interoception	Insular cortex (Goudriaan et al., 2010)	Interoceptive processes lead to perception of cue-evoked conditioned feeling states, providing the subjective feeling of drug craving.
Action and motor functions	Motor preparation and imagery	Primary motor cortex (Smolka et al., 2006), premotor cortex (Smolka et al., 2006), supplementary motor area (Smolka et al., 2006), precuneus (Engelmann et al., 2012)	Drug cues cause motor cortex activity, because the resulting action is habitually overlearned. Also, anticipation can act as a conditioned stimulus triggering motor mechanisms related to drug administration.
	Motor planning	Inferior parietal lobule (Garavan et al., 2000), superior parietal lobule (Garavan et al., 2000), posterior middle temporal gyrus (Yalachkov et al., 2009), inferior temporal cortex (Yalachkov et al., 2009)	Motor planning mechanisms link perception and action by integrating sensory and motor signals which are the processing of spatial representations for reaching and grasping of drug-related objects and sensory guidance of movements.
	Motor initiation and execution	Dorsal striatum (McClernon et al., 2009), primary motor cortex (McClernon et al., 2009)	Regular exposure to drug-related cues can help the development of automatized actions related to initiation and execution of consuming drugs. These actions can raise the probability of performing drug-taking actions when addicts view drug-related cues.
	Habitual responses	Dorsal striatum (Volkow et al., 2006), cerebellum (Wang et al., 2013)	Habitual response can be operationalized, as stimulus–response links (often contrasted with response–outcome links that serve value-based action) are important in compulsive drug seeking and drug taking.

Continued

Table 1 Cognitive Dimensions and Functions Associated with Cue-Elicited Drug Craving and Their Corresponding Brain Areas Among People with Substance Use Disorders Based on Imaging Studies Using Cue-Reactivity Paradigms—cont'd

Domain	Subdomains	Possible Related Regions	Possible Relations to Cue Reactivity
Learning and memory	Emotional memory retrieval	Amygdala (Grant et al., 1996), hippocampus (Lou et al., 2012)	Drug-related cues can trigger retrieving of drug-associated memories and their affective/motivational value. This increases the degree of vulnerability of drug users to relapse.
	Procedural memory retrieval	Cerebellum (Wang et al., 1999), dorsal striatum (Volkow et al., 2006)	Procedural memories can contribute to learning and formation of drug use behavior, which in turn will be triggered by drug-related cues.
Attention	Sensory attention (mainly visual)	Lingual gyrus (Engelmann et al., 2012), fusiform gyrus (Engelmann et al., 2012), precuneus (Engelmann et al., 2012)	Drug-related cues can have effects like rare targets on brain, which lead to activate brain systems important for highlighting relevant and meaningful stimuli in the environment.
	Attention shifting (bias)	Superior temporal gyrus (Luijten et al., 2011), superior parietal cortex (Luijten et al., 2011), anterior cingulate cortex (Luijten et al., 2011)	Because of an enhanced early processing of drug cues, the sensory representations of drug-related cues activate and trigger robust attentional biases in drug users. These biases can lead to decision-making and motor responses that increase the chance of drug-seeking behavior.
Executive cognitive control	Goal setting, error detection, and monitoring	Anterior cingulate cortex (Goldstein et al., 2007)	Substance users, as a group, exhibit deficits in the goal-based attentional control, which includes checking error behaviors (conflicts) and correcting actions. Goal-setting processes and error detection and monitoring are activated during cue exposure. Disrupted processing can result in impulsive behaviors that lead to drug-taking behaviors.
	Execution of goal-directed behavior	Orbitofrontal cortex (Bonson et al., 2002), dorsolateral prefrontal cortex (McBride et al., 2006)	By integrating sensory inputs that are triggered by drug-related cues, reward expectation and motivation, and homeostatic signals about the current state and needs of the person, the behavior for obtaining desired outcomes will be selected and executed.
	Self-referential processing	Anterior cingulate cortex (Moeller et al., 2014), posterior cingulate cortex (Mel'nikov et al., 2014), precuneus (Mel'nikov et al., 2014)	Substance users have deficits in appraising the motivational significance of errors and self-referential processing. This may reduce concern about negative outcomes of behaviors, which can in turn lead them to greater drug use.
	Response inhibition	Inferior frontal gyrus (Prisciandaro et al., 2014)	Individuals with substance use disorders sometimes try to inhibit their cravings when confronted with drug-related cues.

Emotion	Emotion generation	Amygdala (Courtney et al., 2015), limbic system (Courtney et al., 2015)	Drug-related cues in environment are recognized as immediate sources of pleasure and reward and increase the arousal of substance users, and produce intense positive and or negative emotions related to the appetitive/aversive nature of craving.
	Emotion regulation	Anterior cingulate cortex (Goldstein et al., 2007), inferior frontal gyrus (Goldstein et al., 2007), medial orbitofrontal cortex (Goldstein et al., 2007)	Substance abusers try to control their emotions elicited by exposure to drug cues or internal/external stressors.
Motivation	Prediction and anticipation of reward and punishment	Orbitofrontal cortex (Goldstein et al., 2007), dorsolateral prefrontal cortex (Müller-Oehring et al., 2013), posterior cingulate cortex (Müller-Oehring et al., 2013)	The anticipation of receiving rewards and expectation of availability of drug in the immediate future can be activated in cue exposure and contribute to feelings of craving.
	Appetitive motivation	Ventral striatum (David et al., 2007), amygdala (Brody et al., 2002), orbitofrontal cortex (Brody et al., 2002)	Appetitive motivation toward drugs is created by classical conditioning. Through associative (classical) conditioning, abusers recognize drug cues as pleasant stimuli. This enhanced motivation encourages addicts to procure drugs and results in craving sensations and drug-seeking responses.

tegmental area (Hyman et al., 2006). This circuit originally responds to natural rewards such as food or sex, but drugs of abuse can also activate reward processing in these regions, establish reward-related memories, and promote the actions leading to the reward (Everitt and Robbins, 2005). Also, it has been demonstrated that most midbrain dopaminergic neurons show phasic firing following conditioned visual, auditory, and somatosensory reward-predicting stimuli (Schultz, 2007) and the activity of the neurons is briefly depressed by stimuli predicting the explicit absence of reward. Therefore, the activity of this network in the presence of substance-related cues is in accordance with classical conditioning theories of addiction. Activity within this network triggers a pattern of physiological (e.g., increase in heart rate) and cognitive (e.g., increase in attention) reactions, which can lead to drug seeking and taking.

4.2 SALIENCE NETWORK

The brain receives a continuous flow of exogenous and endogenous information. Success in performing tasks which are beneficial for the individual's survival and well-being depends on the capacity to identify which stimuli merit the very limited resources associated with attention. "Salience" refers to the outcome of this identification process (and is therefore intimately connected with attention, see below). Increased prefrontal norepinephrine outflow has been reported in response to both rewarding and aversive unconditioned stimuli with high motivational salience in animals (Sara and Segal, 1991). In humans, the salience network (SN), which includes the anterior cingulate cortex (ACC) and insula, has been suggested to play a critical role in assessing the relevance of internal and external stimuli and integrating this information to guide behavior (Seeley et al., 2007).

ACC has been shown to be activated by salient stimuli including reward-related stimuli and also stimuli that elicit pain or negative affect (Shackman et al., 2011). Furthermore, the insula integrates information from several sensory modalities and is important for translating emotionally salient stimuli information into a guideline for goal-directed behavior (Cloutman et al., 2012).

In regard to drug dependence, drug-related cues and the internal states that are generated by emotional/stressful cues tend to be experienced as highly salient "internal and external stimuli," as evidenced by the activity within the SN (Janes et al., 2015). It has been shown that greater insula–ACC coupling at rest was significantly correlated with enhanced smoking cue reactivity in brain areas associated with attention and motor preparation, including the visual cortex, right ventral lateral prefrontal cortex, and the dorsal striatum (Janes et al., 2015). Moreover, a triple network connectivity of SN with executive control network (ECN) and default mode network (DMN) has been proposed, suggesting the role of SN in switching activity between DMN and ECN based on attribution of saliency to internal or external stimuli (Menon, 2011). It has been suggested that in substance use disorders, the activity of SN leads to allocation of attentional resources toward drug craving and thereupon to a bias toward enhanced DMN activity and decreased ECN operations (Sutherland et al., 2012).

4.3 ATTENTION NETWORK

It has been shown that goal pursuit makes individual more sensitive to goal-related cues and thus those cues get higher priority in cognitive processing, which entails attentional bias (Hoelscher et al., 1981). In the case of substance use, there are biases in attentional processing of substance-related stimuli because of their incentive salience; hence, they can grab substance abusers' attention. The extent of attentional bias has been proposed to be highly correlated with the intensity of the subjective feeling of craving and with the risk of relapse (Field and Cox, 2008).

From two neuronal networks that are thought to be specifically associated with attentional processing, the ventral and the dorsal attention networks, the dorsal attentional network (DAN) including regions of the dorsal and lateral frontal cortices, and superior parietal lobule (Corbetta and Shulman, 2002) have been shown to be overactive during drug cue exposure (Luijten et al., 2011). Neurophysiological studies indicate that in addition to being recruited for top-down selection (Westerberg et al., 2011), the DAN is modulated by the bottom-up distinctiveness of objects in a visual scene (Corbetta and Shulman, 2002). Thus, the aforementioned hyperactivity during cue exposure could be the result of increased effort for top-down regulation of attentional bias (Lubman et al., 2004). The hyperactivity of DAN which seems to be functionally connected to regions of ECN (Dosenbach et al., 2007) makes use of limited available attentional resources and subsequently interferes with day-to-day life or employment of abstinence-oriented coping skills (Waters et al., 2003).

4.4 DEFAULT MODE NETWORK

The DMN is a set of brain regions that exhibits strong low-frequency oscillations coherent during resting state and is thought to be activated when individuals are focused on their internal mental-state processes, such as self-referential processing, interoception, autobiographical memory retrieval, or imagining future. DMN is deactivated during cognitive task performance. The posterior cingulate cortex, precuneus, medial prefrontal, and inferior parietal cortices are thought to be parts of DMN (Broyd et al., 2009).

Previous studies have suggested that DMN activation is positively correlated with drug craving/withdrawal, while substance consumption suppresses activity of DMN (Cole et al., 2010; Sutherland et al., 2012). Also, as it has been mentioned earlier, DMN–SN–ECN coupling is disrupted in substance use disorders. Alternations in SN–DMN coupling appear to be related to drug craving, which may result in difficulty in disengaging from self-centered thoughts associated with craving (Lerman et al., 2014). Alternations in coupling between DMN–SN–ECN and more specifically failure in suppressing the DMN activity can increase the error rates in cognitive tasks (especially tasks involving inhibitory control) and can interfere with goal-directed behaviors (Eichele et al., 2008). This can lead to higher vulnerability of substance users to relapse in the presence of substance-related cues.

4.5 MEMORY NETWORKS

The level of stimuli's influence on attentional and motivational processes can determine how individuals acquire and encode those stimuli and related information. Drugs can positively modulate and engage attentional and motivational systems of the brain; hence, drug-related information including drug-taking behaviors and drug-related cues are strongly encoded. The activation of memory systems of the brain during cue-induced craving has been observed from the very first imaging studies (Grant et al., 1996). Drug-related memories contribute to the saliency processing of the cues and associated feelings of craving within a spectrum from appetitive (liking) to compulsive (wanting) feelings. Studies shed light on the importance of two types of memory networks in substance abuse: emotional and procedural.

During the period of initial substance use, memories of substance use are formed and become strong, and drug-taking behaviors are initiated with the goal of experiencing the reinforcing effects of the drug. The amygdala and hippocampus play a key role in the conditioned reinforcing effects of drugs of abuse. The amygdala is involved in emotional processing of discrete cues, while the hippocampus has a major role in processing contextual cues and spatial learning (Selden et al., 1991). The interaction of the amygdala and hippocampus is implicated in the mediation of attention toward cues that are highly related to the formation of drug use memories (Floresco et al., 2001), which in return can induce craving for drug of choice.

By contrast, dorsal striatum and cerebellum have been assumed to mediate procedural learning and habit formation in addiction (Miquel et al., 2009; Robbins et al., 2008). Habits are products of stimulus–response learning in which reinforcers primarily strengthen the stimulus–response associations. Constant confrontation with drug-related cues and regular drug consumption facilitate formation of habits and automatized actions. These actions can be triggered by substance-related cues and increase the chance of drug abuse in people with substance use disorders (Yalachkov et al., 2010).

4.6 EXECUTIVE CONTROL NETWORK

The ability to have cognitive control over impulses is a key process in self-regulating healthy behavior. An important characteristic of substance use disorder is an attenuated top-down inhibitory control and an augmented bottom-up signal of appetitive salience for drugs (Bechara, 2005). The ECN includes the lateral prefrontal and parietal cortices and is thought to be involved in goal-directed behavior and cognitive control. The ECN has been shown to have abnormal activity in substance use disorders (Krmpotich et al., 2013; Sutherland et al., 2012). As we mentioned earlier, the decreased ECN activity combined with less suppression of DMN activity in cue-exposure events has been observed in imaging studies. This reduction in ECN activity is reflected in difficulty in executing top-down cognitive control over subjective feeling of craving, which can lead to drug consumption (Lerman et al., 2014).

5 NEUROCOGNITIVE MODEL OF DRUG CRAVING AS A DYNAMIC PROCESS

A schematic of drug craving processing pipeline, beginning from external or internal input stimuli, and ending with drug-taking behavior or successful abstinence, is shown in Fig. 1. As can be seen in the model, the pathway of drug-craving processing is activated by the environmental stimuli (E), which stand in a spectrum from internal/external emotional/drug-related cues. These cues elicit distinct yet collaborative bottom-up and top-down attentional processes that generate attentional allocation (A) to drug-relevant cues (Field and Cox, 2008). The focused attention toward cues and retrieval of previous substance abuse-related memories then triggers saliency evaluation processes (S) (Janes et al., 2015). Sensitized brain circuits attribute pathologically high level of salience to drug-associated cues. This salience can be implicit or explicit, i.e., in the form of "unconscious wanting" or "conscious craving" (Robinson and Berridge, 2008). Saliency evaluation can result in subjective feeling of craving (in a spectrum from positive appetitive to negative compulsive states) which drives ECN to take action for execution of drug-taking behavior or successful effortful executive control that leads to abstinence.

With development of substance use disorder over time, a transition could happen from recreational to habitual use of drugs, i.e., immediate rewards from drug intake with appetitive craving fade and drug-taking behaviors become more compulsive or habitual (Potenza et al., 2012).

Other than the feed-forward connections in this model, there are also several feedback connections; for example, a reciprocal excitatory interaction has been suggested between craving and substance-related attentional bias, i.e., increase in craving makes substance-related stimuli more salient and increase in focused attention toward these

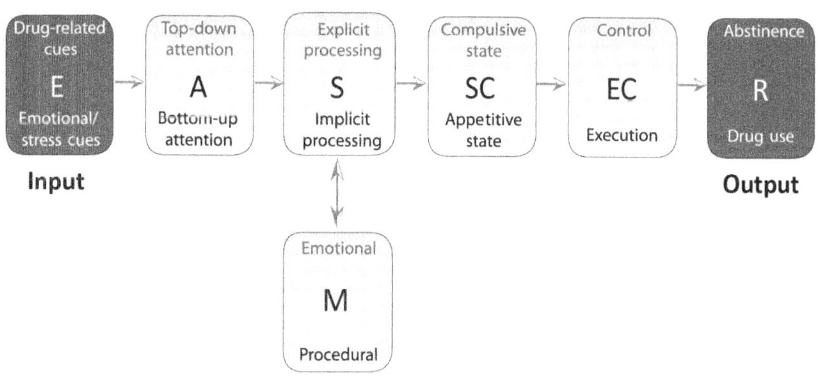

FIGURE 1

A multimodular neurocognitive model for drug craving as a dynamic process. Abbreviations: E, environment (internal or external); A, attentional processing; S, salience processing; SC, subjective craving; EC, executive control; R, response.

cues leads to more boosting of craving (Franken, 2003). Another feedback connection is related to when drug use (e.g., first sips) acts as a multidimensional drug-related cue; in this way, the pathway in Fig. 1 can also be considered as a cycle.

As mentioned before, traditionally craving has been seen as a subjective feeling that could be elicited in external cue-exposure events or internal withdrawal cue experiences. In contrast to traditional viewpoints, in the framework shown in Fig. 1, craving is not just a subjective feeling, but it is a more complex process which involves a dynamic cascade of cognitive functions that begin with exposure to cues and end with behavioral executions. In this extended definition of craving, subjective feeling of craving, from liking to wanting, is only a part of the whole drug-craving process, and not the only player in the scene.

Another important aspect of this model is that each block in this figure could be mainly attributed to a special network in the brain. The striatal-limbic network contributes to bottom-up attentional processing, while dorsal attention network is activated during the top-down one. The saliency evaluation processing is done by SN, which has interactions with memory network through memory retrieval processes. During subjective feeling of craving, the DMN is activated and engages the subject with internal processes including autobiographical drug-related memories and the experience of wanting. Finally, ECN contributes to execution or control of behaviors. Interactions of different networks in this model can also be interpreted in the form of imbalanced interaction between neural reward and control circuitry (Karoly et al., 2013), in which striatal-limbic, default mode, salience and memory networks are correlated with augmented neural reward circuits, while ECN and attention network contribute to top-down control circuit. These interactions can predict the final result of particular cue-exposure events. We will describe different interventions using this network-based model in the next section.

6 NEUROCOGNITIVE INTERVENTIONS IN CRAVING-RELATED PROCESSES

All the theoretical advancements and neuroscience findings in this field will be most useful when they can be integrated and employed in treatment development. Current evidence suggests the crucial need for more effective interventional approaches for addiction therapy. Various interventional strategies for addiction treatment and relapse prevention can be divided into three general categories:

I. *Pharmacological interventions*, which mostly target subcortical reward-related circuits, specifically striatal network. Their ultimate goal is to eliminate bottom-up impulses and emotions for drug intake using withdrawal reducers or to target impaired executive control with cognitive enhancers (George and Koob, 2010; Potenza et al., 2011). These interventions mostly block or mimic the reinforcing effects of drugs, e.g., by affecting the same types of neurotransmitter receptors as those affected by drugs, or target cognitive deficits in addicted individuals. Various cognitive enhancers influence neurotransmitters such as glutamate,

GABA, acetylcholine, and monoamines which modulate cognitive functions such as attention and response inhibition in prefrontal cortex (Potenza et al., 2012). These approaches have been fully discussed in other chapters of this volume (Rezapour et al., 2015).

II. *Noninvasive transcranial electrical and magnetic stimulation techniques*, which can change the release of different neurotransmitters in various brain regions, modify cortical excitability, and modulate the neuronal activity in a region or a network of the brain (Ekhtiari and Bashir, 2010; Parkin et al., 2015). Therefore, these techniques might be able to ameliorate the neuroadaptation induced by chronic drug use and also affect the circuits which mediate addiction-related cognitive processes such as decision-making and inhibitory control (see Bellamoli et al., 2014 for a review). Existing and possible applications of this category of interventions in addiction medicine have been addressed in another chapter of this volume (Yavari et al., 2015).

III. *Cognitive-based (psychological) interventions*, such as cognitive-behavioral therapy and motivational enhancement/motivational interviewing, target both reward and executive control processes. These interventions have various pathogenic or healthy cognitive targets and strengthen frontal inhibitory and executive control, and motivate patients for change (Konova et al., 2013).

Here, we divided main cognitive-based interventions in addiction medicine and introduced their corresponding brain network targets within eight categories, which are shown in Table 2 and explained in the following.

6.1 ENVIRONMENT ENGINEERING

The environment encompasses a wide range of components such as family and peer relationships, exposure to different drug-associated, emotional- or stress-inducing cues, and socioeconomic conditions. These environmental factors can be negative, such as poor relationship with family members and friends, poverty, difficulties in workplace, stressful conditions, and exposure to drug stimuli, or positive, such as good family and peer relations and acceptable socioeconomic status (Sinha, 2001; Solinas et al., 2010). Different types of drug-related cues (visual, auditory, smell, taste, tactile, imaginary processes), negative emotional states (anger, anxiety, depression, emotional distress, etc.), and different stressors (fatigue, hunger, etc.) can induce drug craving by activating striatal-limbic network and thereby increasing subjects' vulnerability to drug taking and relapse (Sinha, 2001). Modifying life conditions of the patients is a crucial part of their treatment. The first step in craving prevention during abstinence is to reduce cue exposure and provides a clean and cue-free environment. One step further would be to provide positive life experiences. These are the main goals pursued by environment engineering interventions. The DAN, which has been shown to be overactive during drug cue exposure (Luijten et al., 2011), and striatal-limbic network which is activated by drug-related cues (Jentsch and Taylor, 1999) are affected by this category of interventions.

Table 2 Existing Cognitive Interventions Potentially Efficient on Drug Craving with Their Target Brain Networks

No.	Intervention			Target Brain Networks
1	Environment engineering (cue abstinence)			L, A
2	Goal setting and motivational enhancement			S, EC
3	Behavioral activation (natural reward replacement)			L, S
4	Attention training techniques	Explicit attention training		A, EC
		Implicit attention training		
		Attentional bias modification		
		Meditation training		
5	Mindfulness training	Psychoeducation for metacognition		A, D
		Interoceptive mindfulness		
		Mindful craving surfing/ acceptance		
6	Reappraisal training	Stimulus reappraisal	Internal stimuli	S, EC
			External stimuli	
		Action and outcome reappraisal		
7	Memory reconsolidation	Activating drug-related memories, therefore making them vulnerable		A, S, M
		Modifying episodic drug-related memories		
		Cue devaluation (recurrent cue exposure)		
		Overwriting drug-related memories		
		Modifying compulsive behaviors (procedural memories)		
8	Effortful active suppression	Shifting attention to another subject (thought suppression)		EC, A
		Using motor control system for effortful and voluntary generation of opposing actions.		

Abbreviations: L, striatal-limbic network; A, dorsal attention (top-down) network; S, salience network; ECN, executive control network; D, default mode network; M, memory networks.

6.2 GOAL SETTING AND MOTIVATIONAL ENHANCEMENT

Motivational interviews and motivational enhancement interventions provide a wide range of client-centered therapies in which patients set their own goals and manage their motivation toward these goals. Counselors evoke patients' intrinsic motivation to change, help them to explore and resolve ambivalence, and consolidate a personal decision and plan for change (Miller and Rollnick, 1991; Smedslund et al., 1996). These techniques develop a focus in a patient's life other than their addiction and improve their commitment to behavioral changes. Motivational interventions help subjects to control their craving and maintain their abstinence by augmenting self-referential processing, salience attribution to abstinence, and inhibitory control. Positive effects of these interventions on craving could be attributed to salience network and ECN (Ewing et al., 2011).

6.3 BEHAVIORAL ACTIVATION

Behavioral activation (BA) strategies have been widely employed as a major component of treatment for depression and other mood disorders. Their goal is to increase environmental reinforcement and improve patients' mood and positive affect by engaging them in valued and pleasant activities, such as physical exercising, focusing on actions related to an important goal, learning new skills, and improving relationship with other people. Modifying sleep and eating habits of patients is also important in BA therapy (Hopko et al., 2003; Mazzucchelli et al., 2009). Using BA, patients are asked to act "outside-in" rather than "inside-out" according to a defined schedule and independent from their mood (Martell et al., 2013). BA, which targets striatal-limbic and saliency networks, can be applied as a preventive and promising adjunct therapeutic approach for drug craving and addiction (MacPherson et al., 2010).

6.4 ATTENTION TRAINING TECHNIQUES

Drug-related attentional bias is an important characteristic of people with drug use disorder and is associated with addiction severity, craving, treatment outcome, and relapse (Hekmat et al., 2011; Marhe et al., 2013; Schoenmakers et al., 2010).

Cognitive remediation strategies in general and attention training techniques in particular usually employ computerized exercises to strengthen different aspects of attention and reduce attentional bias toward drug cues. Attention training techniques, which train subjects to disengage their attention from drug-related stimuli, have been successfully employed in different studies and resulted in drug consumption decrease (Fadardi and Cox, 2009; Schoenmakers et al., 2010; Wiers et al., 2011). The main target network in this category of interventions could be considered attention network and ECN.

6.5 MINDFULNESS TRAINING

Mindfulness training and mindfulness-based therapies are systematic training of attention and self-regulation (Holzel et al., 2011). Mindfulness-based therapies have been shown to reduce negative mood states such as stress and anxiety, increase positive emotion, and improve attention, self-control, self-regulation, and metacognitive awareness. Therefore, they might normalize negative reinforcement processes in substance-using individuals (e.g., stress-induced craving). Also, they might have positive effects on some addiction symptoms such as bingeing (Garland et al., 2014; Potenza et al., 2012; Tang et al., 2007, 2010, 2013). Mindfulness-based therapies modify activation in different brain areas and networks (Table 2). For example, it has been shown that they induce changes in ACC, insula, temporo-parietal junction, and fronto-limbic circuits (Holzel et al., 2011). They also have been shown to induce changes in white matter integrity in the tract which connects ACC to other brain structures (Tang et al., 2010); these finding show that mindfulness-based therapies impact the attention network and DMN (Holzel et al., 2011; Tang et al., 2013; Westbrook et al., 2013). Considering the absence of prefrontal activation during mindful attention (Westbrook et al., 2013), these interventions might act through reducing the reactivity to drug cues rather than cognitive control enhancement.

6.6 REAPPRAISAL TRAINING

Reappraisal is a commonly used form of cognitive regulation for emotion/appetites including drug craving. Reappraisal training provides the client with deliberate strategies for reinterpreting the meaning of a stimulus, situation, action, or outcome, usually in order to decrease its emotional/appetitive impact. Various explicit cognitive coping strategies are employed in reappraisal to reduce the emotional/appetitive impact of the craving-related situations. During reappraisal, patients try to use propositional thinking to replace the self-defeating, craving-related thoughts with more helpful ones (Beadman et al., 2015; Gross, 2002). Reappraisal can act at different levels, i.e., subjects can consciously devaluate the internal drug-related stimuli (e.g., hedonic or dysphoric experiences), devaluate external drug-related stimuli (e.g., drugs, or people, places or things associated with drug use), use possible harmful outcomes of the drug taking for reappraisal, or use cognitive reappraisal. Cognitive reappraisal is a type of cognitive change that is used for allocating new valuation processing. Target brain networks in reappraisal training could include the salience network and ECN (Staudinger et al., 2009).

6.7 MEMORY RECONSOLIDATION

Memory is one of the most important cognitive functions affected by addiction. Pathological associative learning causes compulsive/impulsive drug use despite its negative consequences. Therefore, memory can be considered an important intervention target. Interventions in this category aim to modify the powerfully remembered associations between conditioned drug-related cues and drug experience in the brain. These conditioned drug memories can be interfered with at different levels; recurrent cue exposure without the associated drug-taking response (unconditioned stimulus) reduces conditioned responding (extinction). Imagination, written scripts, and live exposure techniques can be used for exposing the patients to the stimuli associated with their addictive behavior. This ultimately leads to behavioral desensitization

(Kaplan et al., 2011; Volkow et al., 2004). What happens in extinction learning is not just forgetting a conditional behavior, but is an active learning process which diminishes the frequency or intensity of conditioned responses to drug cues and actively reduces the value or salience of drug-related cues and their emotional responses (Kaplan et al., 2011). One step further is to not only extinguish but also overwrite drug-related memories with other emotional memories, i.e., to develop new patterns of healthy behaviors to replace conditioned drug-induced ones (Bouton, 2004).

The major neural circuits affected by these interventions are considered to be attention, memory, and salience networks.

6.8 EFFORTFUL ACTIVE SUPPRESSION

Two ends of craving management strategies can be considered "reappraisal" and "effortful, active suppression." The former acts on the initial phases of craving generation process with the main goal of decreasing appetitive experience, while the latter acts on already elicited impulses with the main goal of decreasing behavioral expression (Gross, 2002). In suppression, individuals purposefully try to prevent drug and craving-related thoughts coming to mind. Using the suppression strategy in dealing with drug craving is very common, yet has some disadvantages such as lower levels of confidence and more depressive symptoms in patients who use these strategy (Rogojanski et al., 2011). It might even increase responsivity to drug cues (Beadman et al., 2015). The ECN is the key network in effortful active suppression.

7 CLINICAL IMPLICATIONS FOR AN INTEGRATIVE NEUROCOGNITIVE MODEL OF CRAVING

Various pharmacological, cognitive, and NIBS interventions have been proposed to manage craving and prevent relapse. In spite of their efforts to target some aspects of drug craving, no existing treatment is completely effective (Dutra et al., 2008; Martin and Rehm, 2012). Furthermore, there is little evidence regarding interactions of these interventions and the possibilities for designing multiapproach interventions. A neurocognitive model of craving may help address this gap in different ways:

1. *Individualized treatment planning and monitoring*: Drug-taking behaviors are often considered the final output of drug craving, but the level of correspondence of different underlying cognitive processes are different in each patient. Since neural activity during cue exposure has been shown to be a predictor of relapse following treatment (Janes et al., 2010; Jia et al., 2011), an integrative model helps us to better understand the functional connectivity between several networks and their cognitive processes involved in drug craving and substance use disorders, and also to determine the neurocognitive target of treatments more specifically. Furthermore, it provides us with an opportunity to use appropriate

assessment tools for evaluating changes during and posttreatment and monitoring the course of treatments in an individualized manner.

2. *Integrated cognitive therapies*: As described above, various cognitive therapies target different processes and underlying networks. Identifying complimentary combination of cognitive interventions seems a promising strategy for improving treatment outcomes (Barrowclough et al., 2001; Cavallo et al., 2007). A neurocognitive model can suggest novel ways to intelligently combine different cognitive therapies such as cognitive remediation strategies, attentional bias modifications, and reappraisal trainings with each other in a comprehensive program to have a more effective psychotherapy package.
3. *Multidimensional treatment interventions*: A strategy to enhance the effectiveness of different types of therapies is to combine them with one or more adjuvant treatments. A neurocognitive model of craving would be able to suggest novel ways to design multiapproach interventions. These multiapproach interventions such as a combination of different pharmacological and cognitive therapies can address weaknesses of each therapy and cover a range of craving components. NIBS techniques including transcranial electrical and magnetic stimulation (tES and TMS) are new emerging tools that have shown promise for the treatment of different neuropsychiatric disorders (Lefaucheur et al., 2014; Tortella et al., 2015). Several studies have employed a combination of cognitive therapies and transcranial direct current stimulation (tDCS) (Brunoni et al., 2014a; Segrave et al., 2014) and pharmacotherapy with tDCS (Brunoni et al., 2014b) for the treatment of depression, in some cases yielding significant therapeutic responses. Hence, NIBS techniques can be used combined with pharmacological and cognitive therapies based on the integrative neurocognitive mode, augmenting the positive impact of the primary interventions.

8 CONCLUSIONS AND FUTURE DIRECTIONS

It has been long assumed that the isolated operation of single brain areas generates various functions in the brain (simplified structure–function mapping viewpoint). Some of these isolated brain regions are connected through direct physical pathways and form small networks in the brain. An example is the structural connectivity between dorsolateral prefrontal cortex, as a reflective top-down system, and the amygdala, as a reflexive bottom-up system, which specifically becomes imbalanced in addiction. More recent evidence emphasizes the importance of interacting and distributed brain areas or large-scale brain networks for complex brain functions, such as memory and language. Nodes of these networks might correspond to neurons, populations of neurons, or anatomically isolated brain regions. These nodes can be linked via structural (anatomical), functional (correlational), and/or effective (causal) connections (Shafi et al., 2012). Development of neuroimaging techniques and also powerful network modeling tools from graph theory and dynamical systems have made it feasible to analyze the structure and function of brain networks (Sporns,

2013). Finally, it has been suggested that various dynamic cognitive processes in the brain, such as drug craving, emerge from the interaction and different levels of contribution of these specialized large-scale networks; i.e., specific levels of contribution and interaction of different networks, such as DMN, SN, and ECN enable the brain to generate different states and switch between them (Shafi et al., 2012). Each of these four different viewpoints, regional, small-scale networks, large-scale networks, and brain states, which are schematically shown in Fig. 2, provides insights into the brain function and can be employed for specific applications.

All these advances in cognitive neuroscience have enriched our knowledge of brain function and raised serious hopes for an evolution in the understanding and treatment of neuropsychological disorders and specifically substance use disorder. Currently, addiction medicine is mostly focused on behavioral outputs, such as drug abstinence and reducing high-risk behaviors in different therapeutic processes; but research in recent years suggests that this approach is not able to cover all potentially important aspects of addiction treatment. Attention to the cognitive roots of behavior suggests an extended spectrum of interventions for addiction medicine. In light of accumulated knowledge about addiction neurobiology, developing an integrative model of drug craving can help us to reach more successful results in relapse prevention and craving management. More future studies are needed to determine the interventions based on this model and also suggest tools for evaluation and monitoring of treatment outcomes.

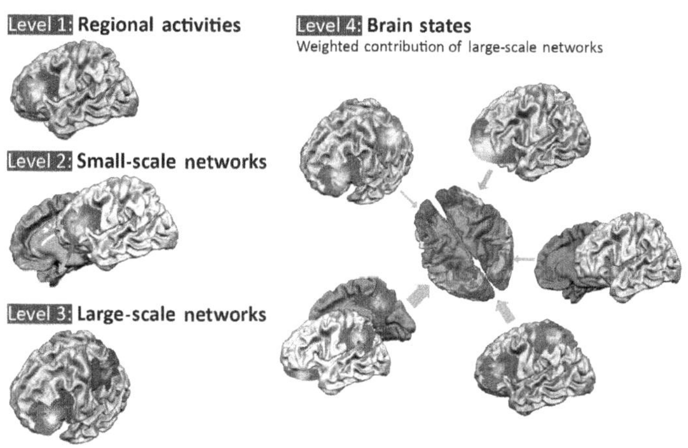

FIGURE 2

Four levels of approach to the neural correspondence of drug craving. (1) Regional models, (2) small-scale network models, (3) large-scale network models, and (4) brain state models with weighted contribution of the large-scale networks. Brain activities related to different cognitive components of drug craving can be studied in "Static" or "Time Variant and Dynamic" modes in these levels. All these models are wrong in terms of depicting the complete nature of this phenomenon, but they are all useful to make implications in addiction medicine.

REFERENCES

Anton, R.F., 1999. What is craving? Models and implications for treatment. Alcohol Res. Health 23 (3), 165–173.

Barrowclough, C., et al., 2001. Randomized controlled trial of motivational interviewing, cognitive behavior therapy, and family intervention for patients with comorbid schizophrenia and substance use disorders. Am. J. Psychiatry 158 (10), 1706–1713.

Beadman, M., et al., 2015. A comparison of emotion regulation strategies in response to craving cognitions: effects on smoking behaviour, craving and affect in dependent smokers. Behav. Res. Ther. 69, 29–39. Retrieved, http://linkinghub.elsevier.com/retrieve/pii/S0005796715000522.

Bechara, A., 2005. Decision making, impulse control and loss of willpower to resist drugs: a neurocognitive perspective. Nat. Neurosci. 8 (11), 1458–1463.

Bellamoli, E., et al., 2014. RTMS in the treatment of drug addiction: an update about human studies. Behav. Neurol. 2014, 815215. Retrieved, http://www.pubmedcentral.nih.gov/articlerender.fcgi?artid=4006612&tool=pmcentrez&rendertype=abstract.

Bonson, K.R., et al., 2002. Neural systems and cue-induced cocaine craving. Neuropsychopharmacology 26 (3), 376–386.

Bouton, M.E., 2004. Context and behavioral processes in extinction. Learn. Mem. 11 (5), 485–494.

Brody, A.L., et al., 2002. Brain metabolic changes during cigarette craving. Arch. Gen. Psychiatry 59 (12), 1162–1172.

Broyd, S.J., et al., 2009. Default-mode brain dysfunction in mental disorders: a systematic review. Neurosci. Biobehav. Rev. 33 (3), 279–296.

Brunoni, A.R., et al., 2014a. Cognitive control therapy and transcranial direct current stimulation for depression: a randomized, double-blinded, controlled trial. J. Affect. Disord. 162, 43–49.

Brunoni, A.R., et al., 2014b. Differential improvement in depressive symptoms for tDCS alone and combined with pharmacotherapy: an exploratory analysis from the sertraline vs. electrical current therapy for treating depression clinical study. Int. J. Neuropsychopharmacol. 17 (1), 53–61.

Cavallo, D.A., et al., 2007. Combining cognitive behavioral therapy with contingency management for smoking cessation in adolescent smokers: a preliminary comparison of two different CBT formats. Am. J. Addict. 16 (6), 468–474.

Charboneau, E.J., et al., 2013. Cannabis cue-induced brain activation correlates with drug craving in limbic and visual salience regions: preliminary results. Psychiatry Res. 214 (2), 122–131.

Chase, H.W., Eickhoff, S.B., Laird, A.R., Lee, H., 2011. The neural basis of drug stimulus processing and craving: an activation likelihood estimation meta-analysis. Biol. Psychiatry 70 (8), 785–793.

Cloutman, L.L., Binney, R.J., Drakesmith, M., Parker, G.J.M., Lambon Ralph, M.A., 2012. The variation of function across the human insula mirrors its patterns of structural connectivity: evidence from in vivo probabilistic tractography. NeuroImage 59 (4), 3514–3521.

Cole, D.M., et al., 2010. Nicotine replacement in abstinent smokers improves cognitive withdrawal symptoms with modulation of resting brain network dynamics. NeuroImage 52 (2), 590–599.

Corbetta, M., Shulman, G.L., 2002. Control of goal-directed and stimulus-driven attention in the brain. Nat. Rev. Neurosci. 3 (3), 201–215.

Courtney, K.E., Ghahremani, D.G., Ray, L.A., 2015. The effect of alcohol priming on neural markers of alcohol cue-reactivity. Am. J. Drug Alcohol Abuse 41, 300–308.

Cox, W.M., Fadardi, J.S., Intriligator, J.M., Klinger, E., 2014. Attentional bias modification for addictive behaviors: clinical implications. CNS Spectr. 19 (3), 215–224.

Dackis, C.A., Kampman, K.M., Lynch, K.G., Pettinati, H.M., O'Brien, C.P., 2005. A double-blind, placebo-controlled trial of modafinil for cocaine dependence. Neuropsychopharmacology 30 (1), 205–211.

David, S.P., et al., 2007. Effects of acute nicotine abstinence on cue-elicited ventral striatum/nucleus accumbens activation in female cigarette smokers: a functional magnetic resonance imaging study. Brain Imaging Behav. 1 (3–4), 43–57.

DeWitt, S.J., Ketcherside, A., McQueeny, T.M., Dunlop, J.P., Filbey, F.M., 2015. The hyper-sentient addict: an exteroception model of addiction. Am. J. Drug Alcohol Abuse 41 (5), 374–381.

Dosenbach, N.U.F., et al., 2007. Distinct brain networks for adaptive and stable task control in humans. Proc. Natl. Acad. Sci. U. S. A. 104 (26), 11073–11078.

Drummond, D.C., 2001. Theories of drug craving, ancient and modern. Addiction 96 (1), 33–46.

Due, D.L., Huettel, S.A., Hall, W.G., Rubin, D.C., 2002. Activation in mesolimbic and visuospatial neural circuits elicited by smoking cues: evidence from functional magnetic resonance imaging. Am. J. Psychiatr. 159 (6), 954–960.

Dutra, L., et al., 2008. A meta-analytic review of psychosocial interventions for substance use disorders. Am. J. Psychiatry 165 (2), 179–187.

Eichele, T., et al., 2008. Prediction of human errors by maladaptive changes in event-related brain networks. Proc. Natl. Acad. Sci. U.S.A. 105 (16), 6173–6178.

Ekhtiari, H., 2010. Addiction studies in Iran; neuroscientists need to do more. Basic Clin. Neurosci. 1 (3), 3–4.

Ekhtiari, H., Bashir, S., 2010. Brain stimulation technology in addiction medicine; main problems waiting for solutions. Basic Clin. Neurosci. 1 (4), 3–4.

Ekhtiari, H., et al., 2008. Functional neuroimaging study of brain activation due to craving in heroin intravenous users. Iran. J. Psychiatry Clin. Psychol. 14, 269–280.

Engelmann, J.M., et al., 2012. Neural substrates of smoking cue reactivity: a meta-analysis of fMRI studies. NeuroImage 60 (1), 252–262.

Everitt, B.J., Robbins, T.W., 2005. Neural systems of reinforcement for drug addiction: from actions to habits to compulsion. Nat. Neurosci. 8 (11), 1481–1489.

Ewing, F., Sarah, W., Filbey, F.M., Hendershot, C.S., McEachern, A.D., Hutchison, K.E., 2011. Proposed model of the neurobiological mechanisms underlying psychosocial alcohol interventions: the example of motivational interviewing. J. Stud. Alcohol Drugs 72, 903–916.

Fadardi, J.S., Cox, W.M., 2009. Reversing the sequence: reducing alcohol consumption by overcoming alcohol attentional bias. Drug Alcohol Depend. 101, 137–145.

Field, M., Cox, W.M., 2008. Attentional bias in addictive behaviors: a review of its development, causes, and consequences. Drug Alcohol Depend. 97 (1–2), 1–20.

Field, M., Marhe, R., Franken, I.H.A., 2014. The clinical relevance of attentional bias in substance use disorders. CNS Spectr. 19 (3), 225–230.

Floresco, S.B., Blaha, C.D., Yang, C.R., Phillips, A.G., 2001. Modulation of hippocampal and amygdalar-evoked activity of nucleus accumbens neurons by dopamine: cellular mechanisms of input selection. J. Neurosci. 21 (8), 2851–2860.

Fox, M.D., Buckner, R.L., White, M.P., Greicius, M.D., Pascual-Leone, A., 2012. Efficacy of transcranial magnetic stimulation targets for depression is related to intrinsic functional connectivity with the subgenual cingulate. Biol. Psychiatry 72 (7), 595–603.

Franken, I.H.A., 2003. Drug craving and addiction: integrating psychological and neuropsychopharmacological approaches. Prog. Neuropsychopharmacol. Biol. Psychiatry 27 (4), 563–579.

Garavan, H., et al., 2000. Cue-induced cocaine craving: neuroanatomical specificity for drug users and drug stimuli. Am. J. Psychiatry 157 (11), 1789–1798.

Garland, E.L., Froeliger, B., Howard, M.O., 2014. Mindfulness training targets neurocognitive mechanisms of addiction at the attention-appraisal-emotion interface. Front. Psychiatry 4, 1–16.

George, O., Koob, G.F., 2010. Individual differences in prefrontal cortex function and the transition from drug use to drug dependence. Neurosci. Biobehav. Rev. 35 (2), 232–247. Retrieved July 11, 2015. http://www.pubmedcentral.nih.gov/articlerender.fcgi?artid=2955797&tool=pmcentrez&rendertype=abstract.

Goldstein, R.Z., et al., 2007. Role of the anterior cingulate and medial orbitofrontal cortex in processing drug cues in cocaine addiction. Neuroscience 144 (4), 1153–1159.

Gorelick, D.A., Zangen, A., George, M.S., 2014. Transcranial magnetic stimulation in the treatment of substance addiction. Ann. N. Y. Acad. Sci. 1327, 79–93.

Goudriaan, A.E., de Ruiter, M.B., van den Brink, W., Oosterlaan, J., Veltman, D.J., 2010. Brain activation patterns associated with cue reactivity and craving in abstinent problem gamblers, heavy smokers and healthy controls: an fMRI study. Addict. Biol. 15 (4), 491–503.

Grant, S., et al., 1996. Activation of memory circuits during cue-elicited cocaine craving. Proc. Natl. Acad. Sci. U.S.A. 93 (21), 12040–12045.

Gross, J.J., 2002. Emotion regulation: affective, cognitive, and social consequences. Psychophysiology 39 (3), 281–291.

Hekmat, S., et al., 2011. Cognitive flexibility, attention and speed of mental processing in opioid and methamphetamine addicts in comparison with non-addicts. Basic Clin. Neurosci. 2, 12–19. Retrieved, http://bcn.iums.ac.ir/browse.php?a_code=A-10-2-15&slc_lang=en&sid=1.

Hoelscher, T.J., Klinger, E., Barta, S.G., 1981. Incorporation of concern- and nonconcern-related verbal stimuli into dream content. J. Abnorm. Psychol. 90 (1), 88–91.

Holzel, B.K., et al., 2011. How does mindfulness meditation work? Proposing mechanisms of action from a conceptual and neural perspective. Perspect. Psychol. Sci. 6, 537–559.

Hopko, D.R., Lejuez, C.W., Ruggiero, K.J., Eifert, G.H., 2003. Contemporary behavioral activation treatments for depression: procedures, principles, and progress. Clin. Psychol. Rev. 23 (5), 699–717. Retrieved June 22, 2015 http://www.ncbi.nlm.nih.gov/pubmed/12971906.

Hyman, S.E., Malenka, R.C., Nestler, E.J., 2006. Neural mechanisms of addiction: the role of reward-related learning and memory. Annu. Rev. Neurosci. 29, 565–598.

Janes, A.C., et al., 2010. Brain reactivity to smoking cues prior to smoking cessation predicts ability to maintain tobacco abstinence. Biol. Psychiatry 67 (8), 722–729.

Janes, A.C., Farmer, S., Peechatka, A.L., Frederick, B. de B., Lukas, S.E., 2015. Insula-dorsal anterior cingulate cortex coupling is associated with enhanced brain reactivity to smoking cues. Neuropsychopharmacology 40 (7), 1561–1568.

Jasinska, A.J., Stein, E.A., Kaiser, J., Naumer, M.J., Yalachkov, Y., 2014. Factors modulating neural reactivity to drug cues in addiction: a survey of human neuroimaging studies. Neurosci. Biobehav. Rev. 38 (1), 1–16.

Jentsch, J.D., Taylor, J.R., 1999. Impulsivity resulting from frontostriatal dysfunction in drug abuse: implications for the control of behavior by reward-related stimuli. Psychopharmacology (Berl.) 146 (4), 373–390. Retrieved, http://www.ncbi.nlm.nih.gov/pubmed/10550488.

Jia, Z., et al., 2011. An initial study of neural responses to monetary incentives as related to treatment outcome in cocaine dependence. Biol. Psychiatry 70 (6), 553–560.

Johansson, B.A., Berglund, M., Lindgren, A., 2006. Efficacy of maintenance treatment with naltrexone for opioid dependence: a meta-analytical review. Addiction 101 (4), 491–503.

Kaplan, G.B., Heinrichs, S.C., Carey, R.J., 2011. Treatment of addiction and anxiety using extinction approaches: neural mechanisms and their treatment implications. Pharmacol. Biochem. Behav. 97 (3), 619–625. Retrieved http://dx.doi.org/10.1016/j.pbb.2010.08.004.

Karoly, H.C., Harlaar, N., Hutchison, K.E., 2013. Substance use disorders: a theory-driven approach to the integration of genetics and neuroimaging. Ann. N. Y. Acad. Sci. 1282 (1), 71–91.

Kassel, J.D., Shiffman, S., 1992. What can hunger teach us about drug craving? A comparative analysis of the two constructs. Adv. Behav. Res. Ther. 14 (3), 141–167.

Kilts, C.D., et al., 2001. Neural activity related to drug craving in cocaine addiction. Arch. Gen. Psychiatry 58 (4), 334.

Konova, A.B., Moeller, S.J., Goldstein, R.Z., 2013. Common and distinct neural targets of treatment: changing brain function in substance addiction. Neurosci. Biobehav. Rev. 37 (10 Pt. 2), 2806–2817. Retrieved August 14, 2015. http://www.sciencedirect.com/science/article/pii/S0149763413002261.

Krmpotich, T.D., et al., 2013. Resting-state activity in the left executive control network is associated with behavioral approach and is increased in substance dependence. Drug Alcohol Depend. 129 (1–2), 1–7.

Larimer, M.E., Palmer, R.S., Marlatt, G.A., 1999. Relapse prevention. An overview of Marlatt's cognitive-behavioral model. Alcohol Res. Health 23 (2), 151–160.

Lefaucheur, J.-P., et al., 2014. Evidence-based guidelines on the therapeutic use of repetitive transcranial magnetic stimulation (rTMS). Clin. Neurophysiol. 125 (11), 2150–2206.

Lerman, C., et al., 2014. Large-scale brain network coupling predicts acute nicotine abstinence effects on craving and cognitive function. JAMA Psychiatry 71 (5), 523–530.

Lou, M., Wang, E., Shen, Y., Wang, J., 2012. Cue-elicited craving in heroin addicts at different abstinent time: an fMRI pilot study. Subst. Use Misuse 47 (6), 631–639.

Lubman, D.I., Yücel, M., Pantelis, C., 2004. Addiction, a condition of compulsive behaviour? Neuroimaging and neuropsychological evidence of inhibitory dysregulation. Addiction 99 (12), 1491–1502.

Ludwig, A.M., Wikler, A., Stark, L.H., 1974. The first drink: psychobiological aspects of craving. Arch. Gen. Psychiatry 30 (4), 539–547.

Luijten, M., et al., 2011. Neurobiological substrate of smoking-related attentional bias. NeuroImage 54 (3), 2374–2381.

Maarefvand, M., Ghiasvand, H.R., Ekhtiari, H., 2013. Drug craving terminology among opiate dependents; a mixed method study. Iran. J. Psychiatry 8, 97–103.

MacPherson, L., et al., 2010. Randomized controlled trial of behavioral activation smoking cessation treatment for smokers with elevated depressive symptoms. J. Consult. Clin. Psychol. 78, 55–61.

Marhe, R., Luijten, M., van de Wetering, B.J., Smits, M., Franken, I.H., 2013. Individual differences in anterior cingulate activation associated with attentional bias predict cocaine use after treatment. Neuropsychopharmacology 38 (6), 1085–1093. Retrieved, http://dx.doi.org/10.1038/npp.2013.7.

Martell, C.R., Dimidjian, S., Herman-Dunn, R., 2013. Behavioral Activation for Depression: A Clinician's Guide. Guilford Press, New York.

Martin, G.W., Rehm, J., 2012. The effectiveness of psychosocial modalities in the treatment of alcohol problems in adults: a review of the evidence. Can. J. Psychiatry 57 (6), 350–358.

Mazzucchelli, T., Kane, R., Rees, C., 2009. Behavioral Activation Treatments for Depression in Adults: A Meta-Analysis and Review. Retrieved June 22, 2015. http://www.ncbi.nlm.nih.gov/books/PMH0027573/.

McBride, D., Barrett, S.P., Kelly, J.T., Aw, A., Dagher, A., 2006. Effects of expectancy and abstinence on the neural response to smoking cues in cigarette smokers: an fMRI study. Neuropsychopharmacology 31 (12), 2728–2738.

McClernon, F.J., Kozink, R.V., Lutz, A.M., Rose, J.E., 2009. 24-H smoking abstinence potentiates fMRI-BOLD activation to smoking cues in cerebral cortex and dorsal striatum. Psychopharmacology 204 (1), 25–35.

Mel'nikov, M.E., et al., 2014. Dynamic mapping of the brain in substance-dependent individuals: functional magnetic resonance imaging. Bull. Exp. Biol. Med. 158 (2), 260–263.

Menon, V., 2011. Large-scale brain networks and psychopathology: a unifying triple network model. Trends Cogn. Sci. 15 (10), 483–506.

Miller, W.R., Rollnick, S., 1991. Motivational Interviewing: Preparing People to Change Addictive Behavior. Guilford Press, New York.

Miquel, M., Toledo, R., García, L.I., Coria-Avila, G.A., Manzo, J., 2009. Why should we keep the cerebellum in mind when thinking about addiction? Curr. Drug Abuse Rev. 2 (1), 26–40.

Moeller, S.J., et al., 2014. Functional, structural, and emotional correlates of impaired insight in cocaine addiction. JAMA Psychiatry 71 (1), 61–70.

Müller-Oehring, E.M., et al., 2013. Midbrain-driven emotion and reward processing in alcoholism. Neuropsychopharmacology 38 (10), 1844–1853.

Park, M.-S., et al., 2007. Brain substrates of craving to alcohol cues in subjects with alcohol use disorder. Alcohol Alcohol. 42 (5), 417–422.

Parkin, B.L., Ekhtiari, H., Walsh, V.F., 2015. Non-invasive human brain stimulation in cognitive neuroscience: a primer. Neuron 87 (5), 932–945. Retrieved September 3, 2015. http://www.sciencedirect.com/science/article/pii/S0896627315006741.

Potenza, M.N., Sofuoglu, M., Carroll, K.M., Rounsaville, B.J., 2011. Neuroscience of behavioral and pharmacological treatments for addictions. Neuron 69 (4), 695–712. Retrieved July 11, 2015. http://www.pubmedcentral.nih.gov/articlerender.fcgi?artid=3063555&tool=pmcentrez&rendertype=abstract.

Potenza, M.N., Sofuoglu, M., Carroll, K.M., Rounsaville, B.J., 2012. Neuroscience of behavioral and pharmacological treatments for addictions. Neuron 69 (4), 695–712.

Potvin, S., Tikàsz, A., Dinh-Williams, L.L.-A., Bourque, J., Mendrek, A., 2015. Cigarette cravings, impulsivity, and the brain. Front. Psychiatry 6, 125.

Prisciandaro, J.J., et al., 2014. The relationship between years of cocaine use and brain activation to cocaine and response inhibition cues. Addiction 109 (12), 2062–2070.

Rezapour, T., DeVito, E.E., Sofuoglu, M., Ekhtiari, H., 2015. Perspectives on neurocognitive rehabilitation as an adjunct treatment for addictive disorders: from cognitive improvement to relapse prevention. In: Ekhtiari, H., Paulus, M. (Eds.), Progress in Brain Research. Elsevier, Netherlands. [Epub ahead of Print].

Robbins, T.W., Ersche, K.D., Everitt, B.J., 2008. Drug addiction and the memory systems of the brain. Ann. N. Y. Acad. Sci. 1141, 1–21.

Robinson, T.E., Berridge, K.C., 1993. The neural basis of drug craving: an incentive-sensitization theory of addiction. Brain Res. Brain Res. Rev. 18 (3), 247–291.

Robinson, T.E., Berridge, K.C., 2008. The incentive sensitization theory of addiction: some current issues. Philos. Trans. R Soc. Lond. B Biol. Sci. 363 (1507), 3137–3146.

Rogojanski, J., Vettese, L.C., Antony, M.M., 2011. Coping with cigarette cravings: comparison of suppression versus mindfulness-based strategies. Mindfulness 2 (1), 14–26.

Sara, S.J., Segal, M., 1991. Plasticity of sensory responses of locus coeruleus neurons in the behaving rat: implications for cognition. Prog. Brain Res. 88, 571–585.

Sayette, M.A., et al., 2000. The measurement of drug craving. Addiction 95 (Suppl. 2), S189–S210.

Schoenmakers, T.M., et al., 2010. Clinical effectiveness of attentional bias modification training in abstinent alcoholic patients. Drug Alcohol Depend. 109 (1–3), 30–36. Retrieved, http://dx.doi.org/10.1016/j.drugalcdep.2009.11.022.

Schultz, W., 2007. Behavioral dopamine signals. Trends Neurosci. 30 (5), 203–210.

Seeley, W.W., et al., 2007. Dissociable intrinsic connectivity networks for salience processing and executive control. J. Neurosci. 27 (9), 2349–2356.

Segrave, R.A., Arnold, S., Hoy, K., Fitzgerald, P.B., 2014. Concurrent cognitive control training augments the antidepressant efficacy of tDCS: a pilot study. Brain Stimul. 7 (2), 325–331.

Selden, N.R., Everitt, B.J., Jarrard, L.E., Robbins, T.W., 1991. Complementary roles for the amygdala and hippocampus in aversive conditioning to explicit and contextual cues. Neuroscience 42 (2), 335–350.

Seo, D., Sinha, R., 2014. The neurobiology of alcohol craving and relapse. Handb. Clin. Neurol. 125, 355–368.

Shackman, A.J., et al., 2011. The integration of negative affect, pain and cognitive control in the cingulate cortex. Nat. Rev. Neurosci. 12 (3), 154–167.

Shafi, M.M., Brandon Westover, M., Fox, M.D., Pascual-Leone, A., 2012. Exploration and modulation of brain network interactions with noninvasive brain stimulation in combination with neuroimaging. Eur. J. Neurosci. 35, 805–825.

Shahbabaie, A., et al., 2014. State dependent effect of transcranial direct current stimulation (tDCS) on methamphetamine craving. Int. J. Neuropsychopharmacol. 17 (10), 1591–1598.

Shariatirad, S., Maarefvand, M., Ekhtiari, H., 2013. Emergence of a methamphetamine crisis in Iran. Drug Alcohol Rev. 32 (2), 223–224.

Siegel, S., 1989. Pharmacological Conditioning and Drug Effects. Humana Press, pp. 115–180.

Singleton, E.G., Gorelick, D.A., 1998. Mechanisms of alcohol craving and their clinical implications. In: Galanter, M. (Ed.), Recent Developments in Alcoholism. Springer US, Boston, MA, pp. 177–195.

Sinha, R., 2001. How does stress increase risk of drug abuse and relapse? Psychopharmacology 158, 343–359.

Smedslund, G., et al., 1996. Cochrane Database of Systematic Reviews. John Wiley & Sons, Ltd., Chichester, UK.

Smolka, M.N., et al., 2006. Severity of nicotine dependence modulates cue-induced brain activity in regions involved in motor preparation and imagery. Psychopharmacology 184 (3–4), 577–588.

Solinas, M., Thiriet, N., Chauvet, C., Jaber, M., 2010. Prevention and treatment of drug addiction by environmental enrichment. Prog. Neurobiol. 92 (4), 572–592. Retrieved September 13, 2015. http://www.sciencedirect.com/science/article/pii/S0301008210001450.

Sporns, O., 2013. Structure and function of complex brain networks. Dialogues Clin. Neurosci. 15 (3), 247–262.

Staudinger, M.R., Erk, S., Abler, B., Walter, H., 2009. Cognitive reappraisal modulates expected value and prediction error encoding in the ventral striatum. NeuroImage 47 (2), 713–721. Retrieved September 13, 2015. http://www.sciencedirect.com/science/article/pii/S1053811909004844.

Stewart, J., de Wit, H., Eikelboom, R., 1984. Role of unconditioned and conditioned drug effects in the self-administration of opiates and stimulants. Psychol. Rev. 91 (2), 251–268.

Sutherland, M.T., McHugh, M.J., Pariyadath, V., Stein, E.A., 2012. Resting state functional connectivity in addiction: lessons learned and a road ahead. NeuroImage 62 (4), 2281–2295.

Tabatabaei-Jafari, H., et al., 2014. Patterns of brain activation during craving in heroin dependents successfully treated by methadone maintenance and abstinence-based treatments. J. Addict. Med. 8, 123–129.

Tang, Y.-Y., et al., 2007. Short-term meditation training improves attention and self-regulation. Proc. Natl. Acad. Sci. U.S.A. 104 (43), 17152–17156.

Tang, Y.-Y., et al., 2010. Short-term meditation induces white matter changes in the anterior cingulate. Proc. Natl. Acad. Sci. U.S.A. 107 (35), 15649–15652.

Tang, Y.-Y., Tang, R., Posner, M.I., 2013. Brief meditation training induces smoking reduction. Proc. Natl. Acad. Sci. U.S.A. 110 (34), 13971–13975. Retrieved, http://www.pubmedcentral.nih.gov/articlerender.fcgi?artid=3752264&tool=pmcentrez&rendertype=abstract.

Tiffany, S.T., 1990. A cognitive model of drug urges and drug-use behavior: role of automatic and nonautomatic processes. Psychol. Rev. 97 (2), 147–168.

Tortella, G., et al., 2015. Transcranial direct current stimulation in psychiatric disorders. World J. Psychiatry 5 (1), 88–102.

Volkow, N.D., Fowler, J.S., Wang, G.J., 2004. The addicted human brain viewed in the light of imaging studies: brain circuits and treatment strategies. Neuropharmacology 47 (Suppl. 1), 3–13.

Volkow, N.D., et al., 2006. Cocaine cues and dopamine in dorsal striatum: mechanism of craving in cocaine addiction. J. Neurosci. 26 (24), 6583–6588.

Volkow, N.D., Baler, R.D., Goldstein, R.Z., 2011. Addiction: pulling at the neural threads of social behaviors. Neuron 69 (4), 599–602.

Waters, A.J., et al., 2003. Attentional bias predicts outcome in smoking cessation. Health Psychol. 22 (4), 378–387.

Wang, G.J., et al., 1999. Regional brain metabolic activation during craving elicited by recall of previous drug experiences. Life Sci. 64 (9), 775–784.

Wang, Y., et al., 2013. Altered fronto-striatal and fronto-cerebellar circuits in heroin-dependent individuals: a resting-state FMRI study. PLoS One 8 (3), e58098.

Westbrook, C., et al., 2013. Mindful attention reduces neural and self-reported cue-induced craving in smokers. Soc. Cogn. Affect. Neurosci. 8, 73–84.

Westerberg, C.E., Miller, B.B., Reber, P.J., Cohen, N.J., Paller, K.A., 2011. Neural correlates of contextual cueing are modulated by explicit learning. Neuropsychologia 49 (12), 3439–3447.

Wiers, R.W., Eberl, C., Rinck, M., Becker, E.S., Lindenmeyer, J., 2011. Retraining automatic action tendencies changes alcoholic patients' approach bias for alcohol and improves treatment outcome. Psychol. Sci. 22, 490–497.

Witkiewitz, K., Marlatt, G.A., Walker, D., 2005. Mindfulness-based relapse prevention for alcohol and substance use disorders. J. Cogn. Psychother. 19 (3), 211–228.

Yalachkov, Y., Kaiser, J., Naumer, M.J., 2009. Brain regions related to tool use and action knowledge reflect nicotine dependence. J. Neurosci. 29 (15), 4922–4929.

Yalachkov, Y., Kaiser, J., Naumer, M.J., 2010. Sensory and motor aspects of addiction. Behav. Brain Res. 207 (2), 215–222.

Yavari, F., et al., 2015. Noninvasive brain stimulation for addiction medicine: from monitoring to modulation. In: Ekhtiari, H., Paulus, M. (Eds.), Progress in Brain Research. Elsevier, Netherlands.

CHAPTER

Response inhibition and addiction medicine: from use to abstinence

8

Philip A. Spechler[*,†,1], Bader Chaarani[*], Kelsey E. Hudson[†], Alexandra Potter[*,†], John J. Foxe[‡,§], Hugh Garavan[*,†]

[*]*Department of Psychiatry, Vermont Center on Behavior and Health, University of Vermont, Burlington, VT, USA*

[†]*Department of Psychological Science, University of Vermont, Burlington, VT, USA*

[‡]*Department of Pediatrics, Albert Einstein College of Medicine, Bronx, NY, USA*

[§]*Department of Neuroscience, Albert Einstein College of Medicine, Bronx, NY, USA*

[1]*Corresponding author: Tel.: +1-802-656-3774, e-mail address: philip.spechler@uvm.edu*

Abstract

Historically, neuroscientific research into addiction has emphasized affective and reinforcement mechanisms as the essential elements underlying the pursuit of drugs, their abuse, and difficulties associated with abstinence. However, research over the last decade or so has shown that cognitive control systems, associated largely but not exclusively with the frontal lobes, are also important contributors to drug use behaviors. Here, we focus on inhibitory control and its contribution to both current use and abstinence. A body of evidence points to impaired inhibitory abilities across a range of drugs of abuse. Typically, studies suggest that substance-abusing individuals are characterized by relative hypoactivity in brain systems underlying inhibitory control. In contrast, abstinent users tend to show either normal or supernormal levels of activity in the same systems attesting to the importance of inhibitory control in suppressing the drug use urges that plague attempts at abstinence. In this chapter, the brain and behavioral basis of response inhibition will be reviewed, with a focus on neuroimaging studies of response inhibition in current and abstinent drug abusers.

Keywords

Response inhibition, Addiction, Abstinence, Neuroimaging

1 INTRODUCTION

A defining characteristic of addiction is the loss of control over one's behavior. It is central to the diagnosis of a substance use disorder, it is characteristic of the all-too-common relapses of abstinent users attempting to stay clean, and it is apparent when

Progress in Brain Research, Volume 223, ISSN 0079-6123, http://dx.doi.org/10.1016/bs.pbr.2015.07.024

initial intentions to have just one drink escalate into a binge drinking session. Although hedonic processes such as liking and craving may form the core motivation to consume drugs, certain cognitive processes, such as attention and memory, likely contribute to these drives whereas others, such as response inhibition, likely contribute to the individual's efforts to resist these drives. For instance, Bechara's cognitive theory of addiction posits that the augmented bottom-up signal of appetitive salience for drugs, in part, attenuates top-down inhibitory control (Bechara, 2005).

Cognitive control processes, also commonly referred to as executive functions, are attentionally demanding, and consciously available, volitional processes that initiate a certain action or interrupt ongoing actions (Atkinson and Shiffrin, 1968; Schneider and Shiffrin, 1977; Shiffrin and Schneider, 1977). Cognitive control takes on many forms, including, but not limited to, attentional control and inhibitory control. Attentional control involves the interaction between perceiving environmental cues and the allocation of perceptual processing resources (Norman and Shallice, 1985) whereas inhibitory control broadly refers to counteracting behaviors preceding, accompanying, or resulting from cues. With regard to addiction, initiation of drug cravings may involve mechanisms by which stimuli associated with previous drug use are detected and processed (Grant et al., 1996; Hester et al., 2006), while the inhibition of behavior may involve mechanisms related to monitoring and regulation (Forman et al., 2004; Kaufman et al., 2003). This chapter will focus on inhibitory control, largely operationalized as response inhibition, and its contribution to substance abuse. Specifically, response inhibition will be considered as a means to characterize substance use, abstinence, and recovery in substance-dependent individuals.

2 RESPONSE INHIBITION TASKS

Inhibitory control is broadly conceptualized as the ability to suppress or countermand a thought, action, or feeling. Many investigators study inhibitory control using carefully designed tasks like the stop-signal task, or the go/no-go task, that measure an individual's ability to suppress a prepotent motor response. During the stop-signal task, subjects perform a primary task such as identifying with button-press responses if a visually presented arrow (the target stimulus) points to the left or the right. On a minority of trials, often one quarter of trials, a unique auditory or visual stimulus (the stop-signal) follows the target and instructs the subject to countermand their response. Task difficulty is manipulated by varying the delay between the target and stop stimulus, such that the longer the delay the more difficult it is to inhibit the response. By calculating how fast subjects respond on trials without a stop-signal and the average stop-signal delay on trials in which they successful inhibit 50% of the time, one can estimate the speed of the response inhibition process known as the stop-signal reaction time (SSRT) (Logan and Cowan, 1984). During go/no-go tasks, subjects are presented with a continuous stream of stimuli, the majority of which require a button-press response (go trial), and a minority requiring no response (no-go trial) with the inhibitory demand being induced through the prepotency to respond even on no-go trials. While both tasks are arguably very rudimentary examples of inhibitory control, there is evidence, outlined below, that the neural circuitry subserving response inhibition is also involved in other types of cognitive and emotional inhibition, thereby indicating that they may serve as reasonable probes for more complex inhibitory demands, including those related to resisting drugs. As the neural circuitry of response inhibition is relatively well understood and yields reliable and sensitive behavioral measures of inhibitory ability, it has generated a significant number of studies focused on the role of response inhibition in addiction (Luijten et al., 2014; Smith et al., 2014).

3 THE NEUROBIOLOGY OF CONTROL

Neuroimaging research has identified the dorsolateral prefrontal cortex (dlPFC) as a brain region critical for cognitive control. Evidence suggests the dlPFC is implicated during dual-task coordination (D'Esposito et al., 1995; Mansouri et al., 2009), task switching (Badre and Wagner, 2006; Dove et al., 2000; Sohn et al., 2000), memory updating (Edin et al., 2009; Salmon et al., 1996), and response sequencing, monitoring, and manipulation (Kim et al., 2013; Owen et al., 1996). This is consistent with the human lesion literature implicating the frontal lobes in organizing, regulating, and producing coherent behavior (Luria and Pribram, 1973; Stuss and Frank Benson, 1987). Frontal lobe-damaged patients appear to lose important aspects of autonomous cognitive control as evidenced by the loss of behavioral control to environmental contingencies (e.g., capture errors and utilization behaviors; Lhermitte, 1986). Although the focus of this review will be on prefrontal systems mediating control, these systems operate in conjunction with extensive parietal, premotor, cingulate, subcortical, and cerebellar networks. Further, despite the evidence implicating the frontal lobes in cognitive control, the assignment of specific frontal loci to specific functions is far from resolved due, perhaps, to one of the defining characteristics of the frontal lobes being their ability to flexibly adapt to task demands. Dosenbach and colleagues suggest that different brain networks are involved in distinct aspects of control with the frontoparietal cortex implicated in initiating and adapting behavior, while sustained stable task performance is associated with the anterior cingulate cortex (ACC), anterior insula, frontal operculum, and anterior prefrontal cortex (Dosenbach et al., 2007).

With a specific focus on inhibitory control, a body of research (lesion, transcranial magnetic stimulation (TMS), and fMRI methodologies) implicates the right inferior frontal cortex (rIFC) in motor response inhibition (Aron et al., 2003; Chambers et al., 2006; Garavan et al., 1999, 2006). More broadly, the rIFC is one node of a motor inhibition network which also includes the pre-supplementary motor area (pre-SMA), and subthalamic nucleus (STN) (Aron et al., 2014). It is unclear about the exact causal pathways of these regions (Duann et al., 2009; Neubert et al., 2010; Swann et al., 2012), but research proposes that the STN receives input from both the rIFC and pre-SMA, and the STN inhibits motor activity at the basal ganglia (Aron and Poldrack, 2006; Schmidt et al., 2013). Figure 1 shows a number of the main cortical areas activated during response inhibition from the largest neuroimaging study of the STOP task (Whelan et al., 2012). Human lesion studies provide converging evidence that lesions in the right pre-SMA (Floden and Stuss, 2006; Nachev et al., 2007) and the right inferior frontal gyrus (IFG) subregion pars opercularis impair response inhibition (Aron et al., 2003, 2004, 2014). The first study using TMS found that temporary deactivation of the right IFG pars opercularis selectively impaired the ability to stop an already initiated action, whereas the deactivation of the same region did not affect physiological arousal or the ability to execute responses, confirming the important role of the IFG in the regulation of response inhibition (Chambers et al., 2006). In addition, Cai and colleagues showed that stimulation of the right pre-SMA slowed the implementation of stopping (measured via SSRT) but had no influence on modulation of response tendencies and suggested that this region impairs stopping behavior through a specific disruption of response inhibition (Cai et al., 2012). These studies are supported by the temporal and spatial precision afforded by electrocorticography studies, which have found the rIFC responds prior to successful inhibition (Swann et al., 2009, 2012). Recent studies suggest that this may reflect a broader role for this region in detecting attentionally salient events (Hampshire et al., 2010), although it may be the case that in order to evoke right IFG activity, the salience of these events must be relevant to response control (Dodds et al., 2011).

Although typically not activated in imaging studies of motor response inhibition, there is considerable evidence of a role for the orbitofrontal cortex (OFC) in impulse control. For example, OFC damage in a rodent model increases SSRT (Eagle et al., 2008), while patients with lesion damage to the OFC show increased self-report and cognitive measures of impulsivity and altered time perception relative to healthy controls and non-OFC lesioned patients (Berlin et al., 2004). That said, many

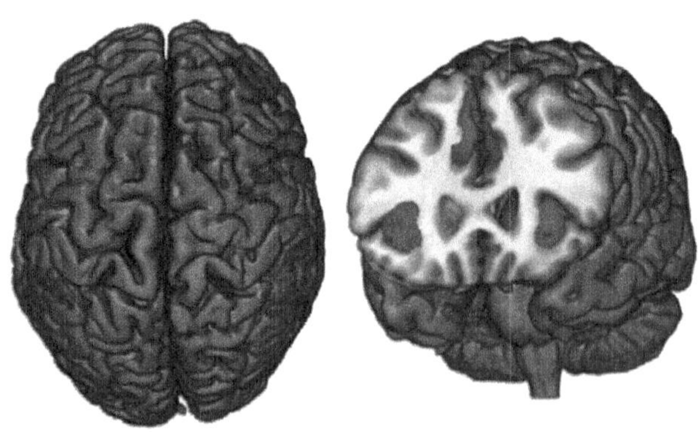

FIGURE 1

Response inhibition on the STOP task produces robust activation in parietal and frontal cortex, including bilateral inferior frontal gyrus.

behaviors that appear impulsive might not be driven by a deficit in impulsivity *per se*. For example, Torregrossa and colleagues argue that the most robust deficit in OFC-damaged animals is in reversal learning. Seemingly impulsive behaviors, such as perseverative responding, and failure to alter responding when rewards for a learned behavior are devalued, may in fact reflect impairment in the ability to update the value of an outcome, especially under changing circumstances (Torregrossa et al., 2008).

There is evidence that regions implicated in motor inhibition and, in particular, right frontal cortex, are involved in aspects of inhibitory control beyond response inhibition. This includes the suppression of drug cravings elicited by a cocaine video: brain activation in the rIFC was increased when inhibiting a craving response and was negatively coupled with activation levels in the right nucleus accumbens (Volkow et al., 2010). In a think/no-think paradigm, in which paired associates are actively suppressed, activation in rIFC was associated with suppressing the sensory components of memories (Depue et al., 2007). de Fockert and colleagues showed that increasing working memory load increased activity levels in bilateral inferior and middle frontal gyri while simultaneously increasing the distraction caused by (and sensory processing of) irrelevant faces (de Fockert et al., 2001). Hester and colleagues modified this paradigm to show that irrelevant drug stimuli produced heightened activity in visual cortex in cocaine users relative to drug–naïve controls (Hester and Garavan, 2009). Critically, those users with the greatest levels of activity in right prefrontal cortex showed the smallest behavioral interference caused by the distracting drug stimuli. In a similar manner, a study of the ability to ignore ecstasy-related stimuli produced greater occipital activation but reduced right prefrontal activation in ecstasy users relative to controls (Roberts and Garavan, 2013). Tabibnia and colleagues identified the rIFC in a number of inhibitory deficits of methamphetamine-dependent subjects (Tabibnia et al., 2011). Results indicated lower gray matter in the rIFC in dependent subjects relative to controls, and gray matter in this region was correlated with drug craving, response inhibition performance, and a test of affect regulation. Finally, Behan and colleagues have recently shown that the rIFC is more active when subjects suppress reward anticipation (Behan et al., 2015). Here, a novel task required subjects to prepare for either a target to which they must respond as fast as possible to receive a reward, or, a stop-signal indicating they should make no response. A psychophysiological interaction analysis suggested the possibility of having to inhibit, rather than respond quickly, produced activity increases in the rIFC, which were correlated with activity decreases in the ventral striatum. Further, the rIFC activity was adjacent to a distinct rIFC region associated with motor inhibition. Combined, this brief review suggests that the rIFC may have a broad role in inhibitory processes that extend beyond motor inhibition. That said, there remains a lack of a comprehensive theory relating the similarities and differences between the various types of inhibitory control to their neurobiological and psychological overlap. Further research probing the multiple types of inhibitory control in the same sample may be a valuable advance.

4 RESPONSE INHIBITION AND DRUGS OF ABUSE

Substance using populations are characterized by deficits in response inhibition. A recent meta-analysis (Smith et al., 2014) of 97 studies found evidence for impaired response inhibition among those dependent on alcohol, cocaine, methamphetamine, tobacco, and MDMA.

4.1 NICOTINE

Although findings in the literature are mixed, a recent meta-analysis found a small but significant effect relating cigarette smoking to response inhibition deficits (Smith et al., 2014). Results from neuroimaging investigations have generally found alterations in the neural circuitry associated with response inhibition in smokers compared to nonsmoking controls (de Ruiter et al., 2012; Luijten et al., 2013; Nestor et al., 2011; but see Galván et al., 2011). For example, Nestor et al. (2011) found that smokers showed reduced activation compared to nonsmokers in a widely distributed network including the ACC, left IFG, bilateral inferior parietal lobules, and bilateral insula. This is similar to the findings of de Ruiter et al. (2012) who found reduced activation of the rostral ACC during inhibition in smokers.

One interesting line of research has examined the relationship between neural activity during successful response inhibition and craving for cigarettes. Berkman et al. (2011) demonstrated that subjects with greater task-related neural activity in nodes of the response inhibition network (bilateral inferior frontal gyrus, SMA, putamen, and left caudate) smoked less in response to subsequent, naturally occurring occasions of cigarette craving. These results suggest that functioning in the circuitry underlying motor inhibition translated to greater behavioral control in response to craving. Further, these investigators found an inverse relationship between amygdala activation during response inhibition and behavior, such that subjects with greater amygdala activation had a stronger positive relationship between craving and smoking behavior. These findings link altered patterns of neural activation with behavioral constructs known to be critical in addiction. Further, as studies have reported hypoactivation in the neural circuitry for response inhibition without differences in task performance, this study underscores the potential utility of neuroimaging as a sensitive measure of neurobiological alterations related to impulsive behavior. Finally, there is considerable value in studies that link lab-based measures of neurobiological function with assessments of inhibitory control in the real world. Real-world behaviors as assessed, for example, by mobile technologies, open up valuable opportunities to relate the neurobiology of inhibitory control to avoid drug use in the natural environment, which in many cases is laden with cues to use.

4.2 ALCOHOL

Alcohol abusers have increased commission error rates compared to nondrinkers or social drinkers on go/no-go tasks (Kamarajan et al., 2005; Murphy and Garavan, 2011), and longer SSRTs on the stop-signal task compared to controls (e.g., Goudriaan et al., 2011; Lawrence et al., 2009; Rubio et al., 2007). However, mixed results have been reported with a number of studies showing no difference in response inhibition related to alcohol consumption (Li et al., 2009; Papachristou et al., 2013; Schmaal et al., 2013). It has been suggested (Smith and Mattick, 2013) that this may relate to sex differences, based on evidence that heavy drinking may be preferentially associated with impaired response inhibition in females (Nederkoorn et al., 2009; Smith and Mattick, 2013; Townshend and Duka, 2005). That said, few studies have been sufficiently powered to specifically examine sex differences in response inhibition related to alcohol consumption. Nonetheless, Smith and colleagues reported an overall impairment in response inhibition in their meta-analysis and suggested that a dose response relationship may exist between impaired response inhibition and drinking patterns (Smith et al., 2014). However, there have been no systematic studies addressing this possibility.

Studies using functional neuroimaging to examine response inhibition in problem drinkers are limited. Li and colleagues found no performance differences on SSRT but lower activation in left dlPFC in alcohol-dependent patients (Li et al., 2009). However, these subjects were all successfully abstinent in alcohol treatment at the time of scanning, making it difficult to determine if activation patterns were related to alcohol withdrawal or early recovery from alcohol dependence. Recent findings have shown that alcohol-use disorders are associated with lower activation in the IFG, insula, inferior parietal lobule, and ACC compared to controls (Claus et al., 2013). When comparing heavy to light alcohol consumption in college drinkers, the heavy drinkers showed impaired performance and altered patterns of neural activity during response inhibition in areas including the ACC, portions of the frontal lobe, hippocampus, and thalamus (Ahmadi et al., 2013). Structural neuroimaging experiments have suggested that chronic alcohol abuse is associated with global volume reduction, cortical and subcortical gray matter reductions, and enlargement of the ventricles. The volume loss in frontal, cerebellar, and subcortical regions are believed to play a critical role in individual differences related to task performance (Chanraud et al., 2007; Scheurich, 2005; Sullivan, 2003). Therefore, as the neural architecture supporting response inhibition deteriorates, behavioral inhibition capacity is likely to suffer.

4.3 CANNABIS

Studies in both adolescent and adult cannabis users have found little evidence for disrupted cognitive performance (Grant et al., 2012; Jager et al., 2010; Schweinsburg et al., 2010; Tapert et al., 2007); however, see Moreno et al. (2012). Interestingly, several studies have demonstrated that while there are inconsistent effects of cannabis use on inhibitory performance, there are neural differences that can be detected via fMRI (Behan et al., 2014; Hester et al., 2009; Roberts and Garavan, 2010; Schweinsburg et al., 2008; Tapert et al., 2007). For example, Roberts and Garavan investigated neural activity using fMRI during response inhibition in adolescent cannabis users and nondrug using controls. While users had equal performance to control subjects, the users had increased activation in frontal and parietal regions during successful inhibitions. This pattern of activation was interpreted to indicate increased neural resources required of the users to achieve performance levels comparable to controls (Roberts and Garavan, 2010). Similar results were found in a study of college students (cannabis users compared to nondrug users) where there was equal task performance but increased activation in the right inferior parietal lobule, the right putamen, and the supplementary motor area in the users (Hester et al., 2009).

It is notable that the pattern of effects in cannabis users (comparable performance but greater activation relative to controls) differs from the hypoactivity associated with other drugs of abuse. Some evidence suggests that heavier use, earlier onset, and greater cumulative cannabis consumption is associated with smaller increases in activation in frontal and parietal regions compared to lighter users or those who begin using later (Schweinsburg et al., 2008, 2010). Such findings indicate that there may be an interaction of brain development and cannabis exposure on brain function and may additionally suggest a compensatory mechanism in heavy cannabis users (Jacobus et al., 2009). Another possibility is that the increased activation of cannabis users may compensate for altered functional connectivity between regions. Recently, Orr and colleagues showed increased intrahemispheric and decreased interhemispheric resting-state connectivity in adolescent heavy cannabis users (Orr et al., 2013). The same sample of adolescent users, when performing a go/no-go task showed impaired performance but no regional activation differences relative to controls. Instead, the users showed increased connectivity during the task between bilateral parietal lobes and left cerebellum, and these same regions showed increased resting-state connectivity (Behan et al., 2014). Although these results may suggest that atypical patterns of activation in cannabis users may be related to differences in inter- and intrahemispheric connectivity, the full set of results fails to offer a straightforward message. As cannabis is the most commonly used illicit drug and the onset of use is common during the sensitive adolescent neurodevelopmental period, it is important that the effects of cannabis on neurocognitive function vis-à-vis inhibitory control be the subject of further inquiry.

4.4 COCAINE

There is strong evidence that cocaine users have poorer response inhibition than nonusers. This is observed in studies using the stop-signal task (Colzato et al., 2007; Fillmore and Craig, 2002; Li et al., 2006; Morie et al., 2014; but see Vonmoos et al., 2013) and in go/no-go tasks (Fernández-Serrano et al., 2011; Hester and Garavan, 2004; Hester et al., 2007; Kaufman et al., 2003; Lane et al., 2007). A review by Spronk and colleagues calculated pooled effect sizes for both SSRT on the stop-signal task and errors of commission on go/no-go tasks and found a moderate pooled effect size (0.50) of cocaine user status on the stop-signal task and a moderate to large (0.64) pooled effect size for errors of commission on the go/no-go task (Spronk et al., 2013). fMRI studies have generally shown reduced neural activity in the PFC including rostral ACC and SMA (Hester and Garavan, 2004; Kaufman et al., 2003; Li et al., 2007).

Using independent component analysis on a stop-signal task, Elton and colleagues discriminated cocaine users from nonusers based on activity patterns decomposed into 11 components. Two of these components were specifically related to stop-signal success, and cocaine users exhibited decreased activation in these networks compared to controls. One network comprised the bilateral IFG, angular gyri, middle temporal, and posterior parietal gyri, and the other network comprised the dlPFC, ventrolateral PFC, dorsomedial PFC, anterior insula, and middle temporal gyrus (Elton et al., 2014).

4.5 MDMA/ECSTASY

A meta-analysis found that overall there is a small effect size on inhibitory errors in heavy MDMA users compared to controls (Smith et al., 2014). Among individual studies, there are several that reported no behavioral performance differences (von Geusau et al. 2004; Quednow et al., 2006; Roberts et al., 2013; Roberts and Garavan, 2010). However, two of these studies used neuroimaging and found altered neural processing in MDMA users. For example, Roberts and colleagues found that ecstasy/polydrug users showed altered EEG patterns suggestive of attentional or inhibitory deficits (Roberts et al., 2013). Similarly, Roberts and Garavan (2010) found intact performance but increased activation in the response inhibition network (right DLPFC, inferior frontal gyrus, and parietal lobule) in recreational ecstasy users. Other studies of current MDMA users have reported moderately impaired behavioral performance in response inhibition (Hoshi et al., 2007). Taken together, the available literature suggests a small impairment in response inhibition associated with MDMA use and altered neural processing in users with intact behavioral performance.

5 RESPONSE INHIBITION AND ABSTINENCE

Relapse is, in many regards, a defining characteristic of drug dependence. Successful abstinence might be viewed within a framework whereby prefrontal cognitive systems seek to control biased attention and pathological behaviors. Hence, successful abstinence may rest on the outcome of the antagonism between drug-wanting systems driven, for example, by ventral striatally mediated salience attribution systems (Robinson and Berridge, 2003), and drug-denying systems governed by the prefrontal cortex (Goldstein and Volkow, 2002; Fig. 2).

There is, however, relatively little empirical data on the neurobiology of successful abstinence despite its potential value for informing therapeutic interventions. The extant literature has typically investigated short-term abstinence and has revealed many persistent deficits, which, for example, for cocaine users, are more pronounced in heavy users in lateral and medial prefrontal regions associated with cognitive control (Bolla et al., 2003, 2004). Abstinent cannabis users show a similar pattern

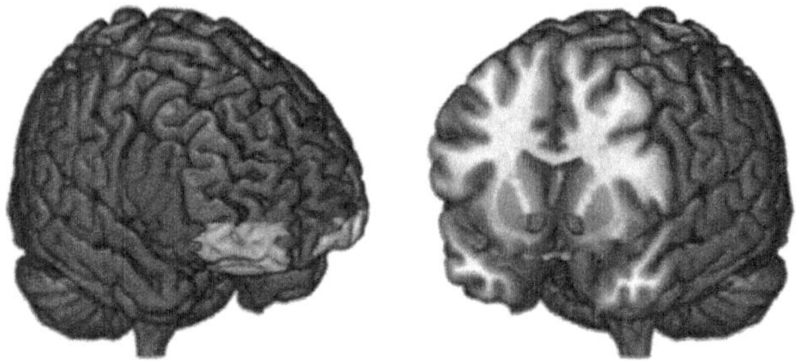

FIGURE 2

We hypothesize that abstinence relies upon recovery of prefrontal systems involved in inhibitory control (regions such as the right IFG and OFC shown on the left). Vulnerability to relapse may be reflected in reinforcement or salience systems (involving regions such as the ventral striatum shown on the right). We hypothesize that relapse may arise from lapses in the prefrontal regulatory systems.

of lateral and medial hypoactivity but have also been reported to show bilateral hippocampal hyperactivity (Eldreth et al., 2004). There is, however, evidence to suggest that prolonged abstinence will correct the general pattern of prefrontal hypoactivity in users (see below) with, for example, cocaine abstinence reducing high-risk responses on a gambling task (Bartzokis et al., 2000). Structural MRI studies have found reduced gray matter volume in prefrontal, orbitofrontal, and cingulate regions in cocaine abstinent individuals (Fein et al., 2002; Matochik et al., 2003), which, some argue, can last even with prolonged abstinence (Tanabe et al., 2009). Interestingly, Connolly and colleagues found in a cross-sectional analysis that cocaine abstinent individuals reached control-like levels of gray matter volumes in the cingulate, insula, and dlPFC by 35 weeks of abstinence (Connolly et al., 2013).

During abstinence, impulse control might be important for suppressing drug-seeking behaviors and drug cravings. Although subjective reports of craving often prove to be poor predictors of subsequent abstinence, cognitive and neuroimaging measures can sometimes do better (Grüsser et al., 2004; Kosten et al., 2005). For example, higher scores on a self-report measure of impulsivity (the Barratt Impulsiveness Scale) have been shown to predict poorer treatment outcome (Moeller et al., 2001; Patkar et al., 2004). With regard to brain predictors, unfortunately, the neuroimaging literature on predicting relapse is small and has employed a variety of tasks that were not necessarily designed to induce a craving response or to assess the user's ability to exercise inhibitory control over that response. Nonetheless, the existing results do identify prefrontal systems, among other regions, as effective predictors of treatment outcome. For example, using a two-button prediction task, Paulus and colleagues showed activation levels in prefrontal, temporal, and posterior cingulate regions early in abstinence to predict subsequent relapse for methamphetamine users (Paulus et al., 2005). Grüsser et al. (2004) found that activity in response to alcohol-related stimuli in the putamen, ACC, and medial prefrontal cortex predicted relapse. In cocaine treatment-seeking individuals, fMRI error-related processing (stop-error vs. stop-success) revealed blunted activity in the dorsal ACC predicted relapse in both sexes, while females exhibited reduced thalamic activity, and males exhibited reduced insular activity (Luo et al., 2013). Although it does not follow from these findings that behavioral measures of impulse control should also predict abstinence, the predictive value of prefrontal cortex suggests that regulatory processes may be involved.

Given the important role that cognitive processes may play in avoiding relapse in drug users and gamblers (Cox et al., 2002; Goudriaan et al., 2008; Passetti et al., 2008; Waters et al., 2003), it may be the case that the best predictors of treatment outcome are those that reflect cognitive control over drug urges rather than the drug urges themselves. This is supported by a study by Brewer et al. (2008) who identified cognitive control prefrontal regions, in addition to other subcortical and posterior cingulate regions, as being the best predictors of treatment outcome in a treatment-receiving sample of cocaine users. Further evidence for the assertion that impulse control might contribute to successful abstinence arises from cross-sectional research of abstinent former users using a go/no-go task. These studies show an apparent reversal in activation patterns, such that prefrontal hypoactivity in current users is paired with prefrontal hyperactivity in abstinent users. For example, Connolly et al. (2012) showed that both short-term abstinent cocaine users (1–5 weeks) and long-term abstinent users (4–24 months) present with fMRI hyperactivity in cognitive control regions relative to drug–naïve controls. That is, the brain regions involved in impulse control (e.g., right middle and rIFC), which are consistently shown to be hypoactive in current users, show elevated activity in former users compared to drug–naïve controls. Subsequent studies have shown former users to be either comparable in performance, fMRI activation levels, and motor-inhibition-related ERP components to controls (Bell et al., 2014; Morie et al., 2014) or to show elevated activation associated with successful inhibitions (Hester et al., 2013). The latter study also revealed blunted activation in response to errors and punishments in the former users suggesting some deficits may persist longer into abstinence.

Evidence for enhanced cognitive control contributing to successful abstinence is also observed in former cigarette smokers (abstinent for 2 years). Using a go/no-go task, current smokers showed reduced activity relative to controls in the dlPFC and the ACC while the former smokers revealed greater inhibition- and error-related activation in the ACC relative to the current smokers (Nestor et al., 2011). A recent study in cigarette smokers highlighted behavioral effects of practicing self-control (i.e., small acts of impulse control such as avoiding sweets were practiced over 2 weeks before quitting) which significantly improved abstinence rates; 27% in the self-control group, relative to 12% in a control condition, were still abstinent 1 month after quitting (Muraven, 2010).

TMS and tDCS have shown some efficacy in enhancing cognitive control. Jacobson and colleagues demonstrated faster SSRTs while stimulating the right inferior frontal gyrus (Jacobson et al., 2011). Applying this technique to substance-abusing individuals may prove fruitful. One study in alcohol detoxification found a single session of TMS over the rIFC facilitated cognitive control performance a week later (Herremans et al., 2013). Similarly, pharmacological interventions targeting cognitive enhancement in cigarette smokers have provided some support for the facilitation of abstinence. Focusing on studies directly assessing response inhibition performance, galantamine, a cholinesterase inhibitor, has reduced subjective craving for cigarettes, while improving performance on a go/no-go task (Sofuoglu et al., 2012). Another study suggests that the use of an NMDA partial agonist, D-cylcoserine, attenuates subjective ratings of cigarette "stimulation" and "relaxation," while improving performance on a go/no-go task (Nesic et al., 2011). Lastly, in a combined fMRI-pharmacological study of guanfacine, a noradrenergic agonist, smokers exhibited reduced cigarette consumption. While no effect was found on task performance, the fMRI results indicate guanfacine attenuated dlPFC responses. The authors interpret this finding as a possible guanfacine-related facilitation of cognitive efficiency.

In summary, the extant literature suggests compromised inhibitory control in active users and normalized or enhanced control in abstinent users. If inhibitory control is shown to be an important contributor to abstinence then this raises exciting possibilities for pharmacological or behavioral interventions. In time, neuroimaging measures may enable us to predict who is most likely to abstain (e.g., related to the integrity of the circuitry underlying inhibitory control) and, by tracking recovery in this circuitry, give guidance on who is most at risk for subsequent relapse.

6 CONCLUSION AND FUTURE DIRECTIONS

The preceding review suggests that deficits in inhibitory control characterize substance dependence. There are, however, drug-specific effects that require further elaboration (e.g., the mixed findings in cannabis users). The integration of functional activation, functional connectivity, and brain structural data is important, but so too is a much richer phenotypic characterization of the users including their drug use histories (age of onset, polydrug use), mental health comorbidities, family and environmental influences, and so on. In reviewing the literature, there persists a lack of a comprehensive understanding on how the various types of inhibitory control relate to one another, psychologically and neurobiologically. More assessments of drug use and other types of inhibitory control (e.g., delaying gratification) or inhibitory control in reward-related contexts may yield new insights. It is a conundrum that although different aspects of inhibitory control appear to be uncorrelated with one another (e.g., self-report personality measures, impulsive choice, and impulsive responding; Reynolds et al., 2006), drug users score highly impulsive on all. Combining this with the evidence that inhibitory control is related to reward processes such as drug-induced euphoria and drug self-administration (Cervantes et al., 2013; Weafer and de Wit, 2013), suggests that more conceptual work is required to integrate these constructs. Finally, as noted above, relating lab-based measures of inhibitory control to drug urges and craving in the natural environment is an important extension of the existing research.

With regard to abstinence, there are two questions of primary importance, and for both, inhibitory control appears to be a central construct. First, to what extent does inhibitory control predicts abstinence? This is important clinically (i.e., identifying who is most likely to relapse can help in allocating interventions and additional services) and also theoretically (i.e., the predictors of relapse give good guidance on the mechanisms that may contribute in a causal manner to abstinence; Garavan et al., 2013). Second, what is the time-course of recovery of inhibitory control and other processes pertinent to addiction? One speculation is that certain processes (e.g., the incentive salience attributed to drugs and drug cues mediated by structures such as the ventral striatum) may persist long after the cessation of use and may underlie relapse risk. It may be the case that inhibitory control recovers to normal (or greater than normal) levels relatively early in abstinence, and while inhibitory control exercised over drug cravings and behaviors is essential to abstinence, relapse is highly likely when this regulatory function becomes disrupted as happens, for example, under stressful situations. Large sample, longitudinal studies of abstainers that assess multiple functions at multiple time-points are required to fully elaborate the role that inhibitory control contributes to avoiding relapse.

REFERENCES

Ahmadi, A., Pearlson, G.D., Meda, S.A., Dager, A., Potenza, M.N., Rosen, R., Austad, C.S., et al., 2013. Influence of alcohol use on neural response to go/no-go task in college drinkers. Neuropsychopharmacology 38 (11), 2197–2208. http://dx.doi.org/10.1038/npp.2013.119.

Aron, A.R., Poldrack, R.A., 2006. Cortical and subcortical contributions to stop signal response inhibition: role of the subthalamic nucleus. J. Neurosci. 26 (9), 2424–2433. http://dx.doi.org/10.1523/JNEUROSCI.4682-05.2006.

Aron, A.R., Fletcher, P.C., Bullmore, E.T., Sahakian, B.J., Robbins, T.W., 2003. Stop-signal inhibition disrupted by damage to right inferior frontal gyrus in humans. Nat. Neurosci. 6 (2), 115–116. http://dx.doi.org/10.1038/nn1003.

Aron, A.R., Robbins, T.W., Poldrack, R.A., 2004. Inhibition and the right inferior frontal cortex. Trends Cogn. Sci. 8 (4), 170–177. http://dx.doi.org/10.1016/j.tics.2004.02.010.

Aron, A.R., Robbins, T.W., Poldrack, R.A., 2014. Inhibition and the right inferior frontal cortex: one decade on. Trends Cogn. Sci. 18 (4), 177–185. http://dx.doi.org/10.1016/j.tics.2013.12.003.

Atkinson, R.C., Shiffrin, R.M., 1968. Human Memory: A Proposed System and Its Control Processes. Academic Press, Oxford, England. xi, 249, http://search.proquest.com/psycinfo/docview/615905557/24D0F579AE654572PQ/1?accountid=14679.

Badre, D., Wagner, A.D., 2006. Computational and neurobiological mechanisms underlying cognitive flexibility. Proc. Natl. Acad. Sci. U. S. A. 103 (18), 7186–7191. http://dx.doi.org/10.1073/pnas.0509550103.

Bartzokis, G., Lu, P.H., Beckson, M., Rapoport, R., Grant, S., Wiseman, E.J., London, E.D., 2000. Abstinence from cocaine reduces high-risk responses on a gambling task.

Neuropsychopharmacology 22 (1), 102–103. http://dx.doi.org/10.1016/S0893-133X(99)00077-9.

Bechara, A., 2005. Decision making, impulse control and loss of willpower to resist drugs: a neurocognitive perspective. Nat. Neurosci. 8 (11), 1458–1463. http://dx.doi.org/10.1038/nn1584.

Behan, B., Connolly, C.G., Datwani, S., Doucet, M., Ivanovic, J., Morioka, R., Stone, A., Watts, R., Smyth, B., Garavan, H., 2014. Response inhibition and elevated parietal-cerebellar correlations in chronic adolescent cannabis users. Neuropharmacology 84, 131–137. http://dx.doi.org/10.1016/j.neuropharm.2013.05.027.

Behan, B., Stone, A., Garavan, H., 2015. Right prefrontal and ventral striatum interactions underlying impulsive choice and impulsive responding. Hum. Brain Mapp. 36 (1), 187–198. http://dx.doi.org/10.1002/hbm.22621.

Bell, R.P., Foxe, J.J., Ross, L.A., Garavan, H., 2014. Intact inhibitory control processes in abstinent drug abusers (I): a functional neuroimaging study in former cocaine addicts. Neuropharmacology 82, 143–150. http://dx.doi.org/10.1016/j.neuropharm.2013.02.018.

Berkman, E.T., Falk, E.B., Lieberman, M.D., 2011. In the trenches of real-world self-control neural correlates of breaking the link between craving and smoking. Psychol. Sci. 22 (4), 498–506. http://dx.doi.org/10.1177/0956797611400918.

Berlin, H.A., Rolls, E.T., Kischka, U., 2004. Impulsivity, time perception, emotion and reinforcement sensitivity in patients with orbitofrontal cortex lesions. Brain 127 (5), 1108–1126. http://dx.doi.org/10.1093/brain/awh135.

Bolla, K.I., Eldreth, D.A., London, E.D., Kiehl, K.A., Mouratidis, M., Contoreggi, C., Matochik, J.A., et al., 2003. Orbitofrontal cortex dysfunction in abstinent cocaine abusers performing a decision-making task. Neuroimage 19 (3), 1085–1094.

Bolla, K.I., Ernst, M., Kiehl, K., Mouratidis, M., Eldreth, D., Contoreggi, C., Matochik, J., et al., 2004. Prefrontal cortical dysfunction in abstinent cocaine abusers. J. Neuropsychiatry Clin. Neurosci. 16 (4), 456–464. http://dx.doi.org/10.1176/appi.neuropsych.16.4.456.

Brewer, J.A., Worhunsky, P.D., Carroll, K.M., Rounsaville, B.J., Potenza, M.N., 2008. Pretreatment brain activation during stroop task is associated with outcomes in cocaine-dependent patients. Biol. Psychiatry 64 (11), 998–1004. http://dx.doi.org/10.1016/j.biopsych.2008.05.024.

Cai, W., George, J.S., Verbruggen, F., Chambers, C.D., Aron, A.R., 2012. The role of the right presupplementary motor area in stopping action: two studies with event-related transcranial magnetic stimulation. J. Neurophysiol. 108 (2), 380–389. http://dx.doi.org/10.1152/jn.00132.2012.

Cervantes, M.C., Laughlin, R.E., David Jentsch, J., 2013. Cocaine self-administration behavior in inbred mouse lines segregating different capacities for inhibitory control. Psychopharmacology (Berl) 229 (3), 515–525. http://dx.doi.org/10.1007/s00213-013-3135-4.

Chambers, C.D., Bellgrove, M.A., Stokes, M.G., Henderson, T.R., Garavan, H., Robertson, I.H., Morris, A.P., Mattingley, J.B., 2006. Executive 'brake failure' following deactivation of human frontal lobe. J. Cogn. Neurosci. 18 (3), 444–455. http://dx.doi.org/10.1162/089892906775990606.

Chanraud, S., Martelli, C., Delain, F., Kostogianni, N., Douaud, G., Aubin, H.-J., Reynaud, M., Martinot, J.-L., 2007. Brain morphometry and cognitive performance in detoxified alcohol-dependents with preserved psychosocial functioning. Neuropsychopharmacology 32 (2), 429–438. http://dx.doi.org/10.1038/sj.npp.1301219.

Claus, E.D., Feldstein Ewing, S.W., Filbey, F.M., Hutchison, K.E., 2013. Behavioral control in alcohol use disorders: relationships with severity. J. Stud. Alcohol Drugs 74 (1), 141–151.

Colzato, L.S., van den Wildenberg, W.P., Hommel, B., 2007. Impaired inhibitory control in recreational cocaine users. PLoS One 2 (11), e1143. http://dx.doi.org/10.1371/journal.pone.0001143.

Connolly, C.G., Foxe, J.J., Nierenberg, J., Shpaner, M., Garavan, H., 2012. The neurobiology of cognitive control in successful cocaine abstinence. Drug Alcohol Depend. 121 (1–2), 45–53. http://dx.doi.org/10.1016/j.drugalcdep.2011.08.007.

Connolly, C.G., Bell, R.P., Foxe, J.J., Garavan, H., 2013. Dissociated grey matter changes with prolonged addiction and extended abstinence in cocaine users. PLoS One 8 (3), e59645. http://dx.doi.org/10.1371/journal.pone.0059645.

Cox, W.M., Hogan, L.M., Kristian, M.R., Race, J.H., 2002. Alcohol attentional bias as a predictor of alcohol abusers' treatment outcome. Drug Alcohol Depend. 68 (3), 237–243. http://dx.doi.org/10.1016/S0376-8716(02)00219-3.

D'Esposito, M., Detre, J.A., Alsop, D.C., Shin, R.K., Atlas, S., Grossman, M., 1995. The neural basis of the central executive system of working memory. Nature 378 (6554), 279–281. http://dx.doi.org/10.1038/378279a0.

de Fockert, J.W., Rees, G., Frith, C.D., Lavie, N., 2001. The role of working memory in visual selective attention. Science 291 (5509), 1803–1806. http://dx.doi.org/10.1126/science.1056496.

Depue, B.E., Curran, T., Banich, M.T., 2007. Prefrontal regions orchestrate suppression of emotional memories via a two-phase process. Science 317 (5835), 215–219. http://dx.doi.org/10.1126/science.1139560.

Dodds, C.M., Morein-Zamir, S., Robbins, T.W., 2011. Dissociating inhibition, attention, and response control in the frontoparietal network using functional magnetic resonance imaging. Cereb. Cortex 21 (5), 1155–1165. http://dx.doi.org/10.1093/cercor/bhq187.

Dosenbach, N.U., Fair, D.A., Miezin, F.M., Cohen, A.L., Wenger, K.K., Dosenbach, R.A., Fox, M.D., et al., 2007. Distinct brain networks for adaptive and stable task control in humans. Proc. Natl. Acad. Sci. 104 (26), 11073–11078. http://dx.doi.org/10.1073/pnas.0704320104.

Dove, A., Pollmann, S., Schubert, T., Wiggins, C.J., von Cramon, D.Y., 2000. Prefrontal cortex activation in task switching: an event-related fMRI study. Cogn. Brain Res. 9 (1), 103–109. http://dx.doi.org/10.1016/S0926-6410(99)00029 4.

Duann, J.-R., Ide, J.S., Luo, X., Li, C.-s.R., 2009. Functional connectivity delineates distinct roles of the inferior frontal cortex and presupplementary motor area in stop signal inhibition. J. Neurosci. 29 (32), 10171–10179. http://dx.doi.org/10.1523/JNEUROSCI.1300-09.2009.

Eagle, D.M., Baunez, C., Hutcheson, D.M., Lehmann, O., Shah, A.P., Robbins, T.W., 2008. Stop-signal reaction-time task performance: role of prefrontal cortex and subthalamic nucleus. Cereb. Cortex 18 (1), 178–188. http://dx.doi.org/10.1093/cercor/bhm044.

Edin, F., Klingberg, T., Johansson, P., McNab, F., Tegnér, J., Compte, A., 2009. Mechanism for top-down control of working memory capacity. Proc. Natl. Acad. Sci. 106 (16), 6802–6807. http://dx.doi.org/10.1073/pnas.0901894106.

Eldreth, D.A., Matochik, J.A., Cadet, J.L., Bolla, K.I., 2004. Abnormal brain activity in prefrontal brain regions in abstinent marijuana users. Neuroimage 23 (3), 914–920. http://dx.doi.org/10.1016/j.neuroimage.2004.07.032.

Elton, A., Young, J., Smitherman, S., Gross, R.E., Mletzko, T., Kilts, C.D., 2014. Neural network activation during a stop-signal task discriminates cocaine-dependent from non-drug-abusing men. Addict. Biol. 19 (3), 427–438. http://dx.doi.org/10.1111/adb.12011.

Fein, G., Di Sclafani, V., Meyerhoff, D.J., 2002. Prefrontal cortical volume reduction associated with frontal cortex function deficit in 6-week abstinent crack-cocaine dependent men. Drug Alcohol Depend. 68 (1), 87–93.

Fernández-Serrano, M.J., Perales, J.C., Moreno-López, L., Pérez-García, M., Verdejo-García, A., 2011. Neuropsychological profiling of impulsivity and compulsivity in cocaine dependent individuals. Psychopharmacology (Berl) 219 (2), 673–683. http://dx.doi.org/10.1007/s00213-011-2485-z.

Fillmore, M.T., Craig, R.R., 2002. Impaired inhibitory control of behavior in chronic cocaine users. Drug Alcohol Depend. 66 (3), 265–273. http://dx.doi.org/10.1016/S0376-8716(01)00206-X.

Floden, D., Stuss, D.T., 2006. Inhibitory control is slowed in patients with right superior medial frontal damage. J. Cogn. Neurosci. 18 (11), 1843–1849. http://dx.doi.org/10.1162/jocn.2006.18.11.1843.

Forman, S.D., Dougherty, G.G., Casey, B.J., Siegle, G.J., Braver, T.S., Barch, D.M., Stenger, V.A., Wick-Hull, C., Pisarov, L.A., Lorensen, E., 2004. Opiate addicts lack error-dependent activation of rostral anterior cingulate. Biol. Psychiatry 55 (5), 531–537. http://dx.doi.org/10.1016/j.biopsych.2003.09.011.

Galván, A., Poldrack, R.A., Baker, C.M., McGlennen, K.M., London, E.D., 2011. Neural correlates of response inhibition and cigarette smoking in late adolescence. Neuropsychopharmacology 36 (5), 970–978. http://dx.doi.org/10.1038/npp.2010.235.

Garavan, H., Ross, T.J., Stein, E.A., 1999. Right hemispheric dominance of inhibitory control: an event-related functional MRI study. Proc. Natl. Acad. Sci. 96 (14), 8301–8306. http://dx.doi.org/10.1073/pnas.96.14.8301.

Garavan, H., Hester, R., Murphy, K., Fassbender, C., Kelly, C., 2006. Individual differences in the functional neuroanatomy of inhibitory control. Brain Res. 1105 (1), 130–142. http://dx.doi.org/10.1016/j.brainres.2006.03.029.

Garavan, H., Brennan, K.L., Hester, R., Whelan, R., 2013. The neurobiology of successful abstinence. Curr. Opin. Neurobiol. 23 (4), 668–674. http://dx.doi.org/10.1016/j.conb.2013.01.029.

Goldstein, R.Z., Volkow, N.D., 2002. Drug addiction and its underlying neurobiological basis: neuroimaging evidence for the involvement of the frontal cortex. Am. J. Psychiatr. 159 (10), 1642–1652. http://dx.doi.org/10.1176/appi.ajp.159.10.1642.

Goudriaan, A.E., Oosterlaan, J., De Beurs, E., Van Den Brink, W., 2008. The role of self-reported impulsivity and reward sensitivity versus neurocognitive measures of disinhibition and decision-making in the prediction of relapse in pathological gamblers. Psychol. Med. 38 (01), 41–50. http://dx.doi.org/10.1017/S0033291707000694.

Goudriaan, A.E., Grekin, E.R., Sher, K.J., 2011. Decision making and response inhibition as predictors of heavy alcohol use: a prospective study. Alcohol. Clin. Exp. Res. 35 (6), 1050–1057. http://dx.doi.org/10.1111/j.1530-0277.2011.01437.x.

Grant, S., London, E.D., Newlin, D.B., Villemagne, V.L., Liu, X., Contoreggi, C., Phillips, R.L., Kimes, A.S., Margolin, A., 1996. Activation of memory circuits during cue-elicited cocaine craving. Proc. Natl. Acad. Sci. 93 (21), 12040–12045.

Grant, J.E., Chamberlain, S.R., Schreiber, L., Odlaug, B.L., 2012. Neuropsychological deficits associated with cannabis use in young adults. Drug Alcohol Depend. 121 (1–2), 159–162. http://dx.doi.org/10.1016/j.drugalcdep.2011.08.015.

Grüsser, S.M., Wrase, J., Klein, S., Hermann, D., Smolka, M.N., Ruf, M., Weber-Fahr, W., et al., 2004. Cue-induced activation of the striatum and medial prefrontal cortex is associated with subsequent relapse in abstinent alcoholics. Psychopharmacology (Berl) 175 (3), 296–302. http://dx.doi.org/10.1007/s00213-004-1828-4.

Hampshire, A., Chamberlain, S.R., Monti, M.M., Duncan, J., Owen, A.M., 2010. The role of the right inferior frontal gyrus: inhibition and attentional control. Neuroimage 50 (3), 1313–1319. http://dx.doi.org/10.1016/j.neuroimage.2009.12.109.

Herremans, S.C., Vanderhasselt, M.-A., De Raedt, R., Baeken, C., 2013. Reduced intra-individual reaction time variability during a Go–NoGo task in detoxified alcohol-dependent patients after one right-sided dorsolateral prefrontal HF-rTMS session. Alcohol Alcohol. 48 (5), 552–557. http://dx.doi.org/10.1093/alcalc/agt054.

Hester, R., Garavan, H., 2004. Executive dysfunction in cocaine addiction: evidence for discordant frontal, cingulate, and cerebellar activity. J. Neurosci. 24 (49), 11017–11022. http://dx.doi.org/10.1523/JNEUROSCI.3321-04.2004.

Hester, R., Garavan, H., 2009. Neural mechanisms underlying drug-related cue distraction in active cocaine users. Pharmacol. Biochem. Behav. 93 (3), 270–277. http://dx.doi.org/10.1016/j.pbb.2008.12.009.

Hester, R., Dixon, V., Garavan, H., 2006. A consistent attentional bias for drug-related material in active cocaine users across word and picture versions of the emotional Stroop task. Drug Alcohol Depend. 81 (3), 251–257. http://dx.doi.org/10.1016/j.drugalcdep.2005.07.002.

Hester, R., Simões-Franklin, C., Garavan, H., 2007. Post-error behavior in active cocaine users: poor awareness of errors in the presence of intact performance adjustments. Neuropsychopharmacology 32 (9), 1974–1984. http://dx.doi.org/10.1038/sj.npp.1301326.

Hester, R., Nestor, L., Garavan, H., 2009. Impaired error awareness and anterior cingulate cortex hypoactivity in chronic cannabis users. Neuropsychopharmacology 34 (11), 2450–2458. http://dx.doi.org/10.1038/npp.2009.67.

Hester, R., Bell, R.P., Foxe, J.J., Garavan, H., 2013. The influence of monetary punishment on cognitive control in abstinent cocaine-users. Drug Alcohol Depend. 133 (1), 86–93. http://dx.doi.org/10.1016/j.drugalcdep.2013.05.027.

Hoshi, R., Mullins, K., Boundy, C., Brignell, C., Piccini, P., Curran, H.V., 2007. Neurocognitive function in current and ex-users of ecstasy in comparison to both matched polydrug-using controls and drug-naive controls. Psychopharmacology 194 (3), 371–379.

Jacobson, L., Javitt, D.C., Lavidor, M., 2011. Activation of inhibition: diminishing impulsive behavior by direct current stimulation over the inferior frontal gyrus. J. Cogn. Neurosci. 23 (11), 3380–3387. http://dx.doi.org/10.1162/jocn_a_00020.

Jacobus, J., Bava, S., Cohen-Zion, M., Mahmood, O., Tapert, S.F., 2009. Functional consequences of marijuana use in adolescents. Pharmacol. Biochem. Behav. 92 (4), 559–565. http://dx.doi.org/10.1016/j.pbb.2009.04.001.

Jager, G., Block, R.I., Luijten, M., Ramsey, N.F., 2010. Cannabis use and memory brain function in adolescent boys: a cross-sectional multicenter functional magnetic resonance imaging study. J. Am. Acad. Child Adolesc. Psychiatry 49 (6). http://dx.doi.org/10.1016/j.jaac.2010.02.001. 561–72.e3.

Kamarajan, C., Porjesz, B., Jones, K.A., Choi, K., Chorlian, D.B., Padmanabhapillai, A., Rangaswamy, M., Stimus, A.T., Begleiter, H., 2005. Alcoholism is a disinhibitory disorder: neurophysiological evidence from a Go/No-Go task. Biol. Psychol. 69 (3), 353–373. http://dx.doi.org/10.1016/j.biopsycho.2004.08.004.

Kaufman, J.N., Ross, T.J., Stein, E.A., Garavan, H., 2003. Cingulate hypoactivity in cocaine users during a GO-NOGO task as revealed by event-related functional magnetic resonance imaging. J. Neurosci. 23 (21), 7839–7843.

Kim, C., Chung, C., Kim, J., 2013. Task-dependent response conflict monitoring and cognitive control in anterior cingulate and dorsolateral prefrontal cortices. Brain Res. 1537, 216–223. http://dx.doi.org/10.1016/j.brainres.2013.08.055.

Kosten, T.R., Scanley, B.E., Tucker, K.A., Oliveto, A., Prince, C., Sinha, R., Potenza, M.N., Skudlarski, P., Wexler, B.E., 2005. Cue-induced brain activity changes and relapse in cocaine-dependent patients. Neuropsychopharmacology 31 (3), 644–650. http://dx.doi.org/10.1038/sj.npp.1300851.

Lane, S.D., Gerard Moeller, F., Steinberg, J.L., Buzby, M., Kosten, T.R., 2007. Performance of cocaine dependent individuals and controls on a response inhibition task with varying levels of difficulty. Am. J. Drug Alcohol Abuse 33 (5), 717–726. http://dx.doi.org/10.1080/00952990701522724.

Lawrence, A.J., Luty, J., Bogdan, N.A., Sahakian, B.J., Clark, L., 2009. Impulsivity and response inhibition in alcohol dependence and problem gambling. Psychopharmacology (Berl) 207 (1), 163–172. http://dx.doi.org/10.1007/s00213-009-1645-x.

Lhermitte, F., 1986. Human autonomy and the frontal lobes. Part II: patient behavior in complex and social situations: the 'environmental dependency syndrome'. Ann. Neurol. 19 (4), 335–343. http://dx.doi.org/10.1002/ana.410190405.

Li, C.-s.R., Milivojevic, V., Kemp, K., Hong, K., Sinha, R., 2006. Performance monitoring and stop signal inhibition in abstinent patients with cocaine dependence. Drug Alcohol Depend. 85 (3), 205–212. http://dx.doi.org/10.1016/j.drugalcdep.2006.04.008.

Li, C.-s.R., Huang, C., Yan, P., Bhagwagar, Z., Milivojevic, V., Sinha, R., 2007. Neural correlates of impulse control during stop signal inhibition in cocaine-dependent Men. Neuropsychopharmacology 33 (8), 1798–1806. http://dx.doi.org/10.1038/sj.npp.1301568.

Li, C.-s.R., Luo, X., Yan, P., Bergquist, K., Sinha, R., 2009. Altered impulse control in alcohol dependence: neural measures of stop signal performance. Alcohol. Clin. Exp. Res. 33 (4), 740–750. http://dx.doi.org/10.1111/j.1530-0277.2008.00891.x.

Logan, G.D., Cowan, W.B., 1984. On the ability to inhibit thought and action: a theory of an act of control. Psychol. Rev. 91 (3), 295–327. http://dx.doi.org/10.1037/0033-295X.91.3.295.

Luijten, M., O'Connor, D.A., Rossiter, S., Franken, I.H., Hester, R., 2013. Effects of reward and punishment on brain activations associated with inhibitory control in cigarette smokers. Addiction 108 (11), 1969–1978. http://dx.doi.org/10.1111/add.12276.

Luijten, M., Machielsen, M.W.J., Veltman, D.J., Hester, R., de Haan, L., Franken, I.H.A., 2014. Systematic review of ERP and fMRI studies investigating inhibitory control and error processing in people with substance dependence and behavioural addictions. J. Psychiatry Neurosci. 39 (3), 149–169. http://dx.doi.org/10.1503/jpn.130052.

Luo, X., Zhang, S., Hu, S., Bednarski, S.R., Erdman, E., Farr, O.M., Hong, K.-I., Sinha, R., Mazure, C.M., Chiang-Shan, R.L., 2013. Error processing and gender-shared and -specific neural predictors of relapse in cocaine dependence. Brain 136 (Pt. 4), 1231–1244. http://dx.doi.org/10.1093/brain/awt040.

Luria, A.R., Pribram, K.H., 1973. The frontal lobes and the regulation of behavior. Academic Press, Oxford, England. xii, 332, http://search.proquest.com/psycinfo/docview/615922321/809F37D77359444DPQ/1?accountid=14679.

Mansouri, F.A., Tanaka, K., Buckley, M.J., 2009. Conflict-induced behavioural adjustment: a clue to the executive functions of the prefrontal cortex. Nat. Rev. Neurosci. 10 (2), 141–152. http://dx.doi.org/10.1038/nrn2538.

Matochik, J.A., London, E.D., Eldreth, D.A., Cadet, J.-L., Bolla, K.I., 2003. Frontal cortical tissue composition in abstinent cocaine abusers: a magnetic resonance imaging study. Neuroimage 19 (3), 1095–1102. http://dx.doi.org/10.1016/S1053-8119(03)00244-1.

Moeller, F.G., Dougherty, D.M., Barratt, E.S., Schmitz, J.M., Swann, A.C., Grabowski, J., 2001. The impact of impulsivity on cocaine use and retention in treatment. J. Subst. Abuse Treat. 21 (4), 193–198. http://dx.doi.org/10.1016/S0740-5472(01)00202-1.

Moreno, M., Estevez, A.F., Zaldivar, F., Montes, J.M.G., Gutiérrez-Ferre, V.E., Esteban, L., Sánchez-Santed, F., Flores, P., 2012. Impulsivity differences in recreational cannabis users and binge drinkers in a university population. Drug Alcohol Depend. 124 (3), 355–362. http://dx.doi.org/10.1016/j.drugalcdep.2012.02.011.

Morie, K.P., De Sanctis, P., Garavan, H., Foxe, J.J., 2014. Executive dysfunction and reward dysregulation: a high-density electrical mapping study in cocaine abusers. Neuropharmacology 85, 397–407. http://dx.doi.org/10.1016/j.neuropharm.2014.05.016.

Muraven, M., 2010. Practicing self-control lowers the risk of smoking lapse. Psychol. Addict. Behav. 24 (3), 446–452. http://dx.doi.org/10.1037/a0018545.

Murphy, P., Garavan, H., 2011. Cognitive predictors of problem drinking and AUDIT scores among college students. Drug Alcohol Depend. 115 (1–2), 94–100. http://dx.doi.org/10.1016/j.drugalcdep.2010.10.011.

Nachev, P., Wydell, H., O'Neill, K., Husain, M., Kennard, C., 2007. The role of the pre-supplementary motor area in the control of action. Neuroimage 36, T155–T163. http://dx.doi.org/10.1016/j.neuroimage.2007.03.034.

Nederkoorn, C., Baltus, M., Guerrieri, R., Wiers, R.W., 2009. Heavy drinking is associated with deficient response inhibition in women but not in men. Pharmacol. Biochem. Behav. 93 (3), 331–336. http://dx.doi.org/10.1016/j.pbb.2009.04.015.

Nesic, J., Duka, T., Rusted, J.M., Jackson, A., 2011. A role for glutamate in subjective response to smoking and its action on inhibitory control. Psychopharmacology (Berl) 216 (1), 29–42. http://dx.doi.org/10.1007/s00213-011-2189-4.

Nestor, L., McCabe, E., Jones, J., Clancy, L., Garavan, H., 2011. Differences in 'bottom-up' and 'top-down' neural activity in current and former cigarette smokers: evidence for neural substrates which May promote nicotine abstinence through increased cognitive control. Neuroimage 56 (4), 2258–2275. http://dx.doi.org/10.1016/j.neuroimage.2011.03.054.

Neubert, F.-X., Mars, R.B., Buch, E.R., Olivier, E., Rushworth, M.F., 2010. Cortical and subcortical interactions during action reprogramming and their related white matter pathways. Proc. Natl. Acad. Sci. 107 (30), 13240–13245. http://dx.doi.org/10.1073/pnas.1000674107.

Norman, D.A., Shallice, T., 1985. Consciousness and self-regulation. In: Davidson, R.J., Schwarts, G.E., Shapiro, D. (Eds.), Advances in Research and Theory, 4. Plenum Press, New York, pp. 2–18.

Orr, C., Morioka, R., Behan, B., Datwani, S., Doucet, M., Ivanovic, J., Kelly, C., et al., 2013. Altered resting-state connectivity in adolescent cannabis users. Am. J. Drug Alcohol Abuse 39 (6), 372–381. http://dx.doi.org/10.3109/00952990.2013.848213.

Owen, A.M., Evans, A.C., Petrides, M., 1996. Evidence for a two-stage model of spatial working memory processing within the lateral frontal cortex: a positron emission tomography study. Cereb. Cortex 6 (1), 31–38. http://dx.doi.org/10.1093/cercor/6.1.31.

Papachristou, H., Nederkoorn, C., Havermans, R., Bongers, P., Beunen, S., Jansen, A., 2013. Higher levels of trait impulsiveness and a less effective response inhibition are linked to more intense cue-elicited craving for alcohol in alcohol-dependent patients. Psychopharmacology (Berl) 228 (4), 641–649. http://dx.doi.org/10.1007/s00213-013-3063-3.

Passetti, F., Clark, L., Mehta, M.A., Joyce, E., King, M., 2008. Neuropsychological predictors of clinical outcome in opiate addiction. Drug Alcohol Depend. 94 (1–3), 82–91. http://dx.doi.org/10.1016/j.drugalcdep.2007.10.008.

Patkar, A.A., Murray, H.W., Mannelli, P., Gottheil, E., Weinstein, S.P., Vergare, M.J., 2004. Pre-treatment measures of impulsivity, aggression and sensation seeking are associated with treatment outcome for African-American cocaine-dependent patients. J. Addict. Dis. 23 (2), 109–122.

Paulus, M.P., Tapert, S.F., Schuckit, M.A., 2005. Neural activation patterns of methamphetamine-dependent subjects during decision making predict relapse. Arch. Gen. Psychiatry 62 (7), 761–768. http://dx.doi.org/10.1001/archpsyc.62.7.761.

Quednow, B.B., Kühn, K.-U., Hoppe, C., Westheide, J., Maier, W., Daum, I., Wagner, M., 2006. Elevated impulsivity and impaired decision-making cognition in heavy users of MDMA ('ecstasy'). Psychopharmacology (Berl) 189 (4), 517–530. http://dx.doi.org/10.1007/s00213-005-0256-4.

Reynolds, B., Ortengren, A., Richards, J.B., de Wit, H., 2006. Dimensions of impulsive behavior: personality and behavioral measures. Personal. Individ. Differ. 40 (2), 305–315. http://dx.doi.org/10.1016/j.paid.2005.03.024.

Roberts, G.M.P., Garavan, H., 2010. Evidence of increased activation underlying cognitive control in ecstasy and cannabis users. Neuroimage 52 (2), 429–435. http://dx.doi.org/10.1016/j.neuroimage.2010.04.192.

Roberts, G.M.P., Garavan, H., 2013. Neural mechanisms underlying ecstasy-related attentional bias. Psychiatry Res. Neuroimaging 213 (2), 122–132. http://dx.doi.org/10.1016/j.pscychresns.2013.03.011.

Roberts, C.A., Fairclough, S., Fisk, J.E., Tames, F.T., Montgomery, C., 2013. Electrophysiological indices of response inhibition in human polydrug users. J. Psychopharmacol. 27 (9), 779–789. http://dx.doi.org/10.1177/0269881113492899.

Robinson, T.E., Berridge, K.C., 2003. Addiction. Annu. Rev. Psychol. 54, 25–53. http://dx.doi.org/10.1146/annurev.psych.54.101601.145237.

Rubio, G., Jiménez, M., Rodríguez-Jiménez, R., Martínez, I., Iribarren, M.M., Jiménez-Arriero, M.A., Ponce, G., Avila, C., 2007. Varieties of impulsivity in males with alcohol dependence: the role of cluster-B personality disorder. Alcohol. Clin. Exp. Res. 31 (11), 1826–1832. http://dx.doi.org/10.1111/j.1530-0277.2007.00506.x.

Ruiter, D., Michiel, B., Oosterlaan, J., Veltman, D.J., van den Brink, W., Goudriaan, A.E., 2012. Similar hyporesponsiveness of the dorsomedial prefrontal cortex in problem gamblers and heavy smokers during an inhibitory control task. Drug Alcohol Depend. 121 (1–2), 81–89. http://dx.doi.org/10.1016/j.drugalcdep.2011.08.010.

Salmon, E., Van der Linden, M., Collette, F., Delfiore, G., Maquet, P., Degueldre, C., Luxen, A., Franck, G., 1996. Regional brain activity during working memory tasks. Brain 119 (5), 1617–1625. http://dx.doi.org/10.1093/brain/119.5.1617.

Scheurich, A., 2005. Neuropsychological functioning and alcohol dependence. Curr. Opin. Psychiatry 18 (3), 319–323. http://dx.doi.org/10.1097/01.yco.0000165602.36671.de.

Schmaal, L., Joos, L., Koeleman, M., Veltman, D.J., van den Brink, W., Goudriaan, A.E., 2013. Effects of modafinil on neural correlates of response inhibition in alcohol-dependent patients. Biol. Psychiatry 73 (3), 211–218. http://dx.doi.org/10.1016/j.biopsych.2012.06.032.

Schmidt, R., Leventhal, D.K., Mallet, N., Chen, F., Berke, J.D., 2013. Canceling actions involves a race between basal ganglia pathways. Nat. Neurosci. 16 (8), 1118–1124. http://dx.doi.org/10.1038/nn.3456.

Schneider, W., Shiffrin, R.M., 1977. Controlled and automatic human information processing: I. Detection, search, and attention. Psychol. Rev. 84 (1), 1–66. http://dx.doi.org/10.1037/0033-295X.84.1.1.

Schweinsburg, A.D., Nagel, B.J., Schweinsburg, B.C., Park, A., Theilmann, R.J., Tapert, S.F., 2008. Abstinent adolescent marijuana users show altered fMRI response during spatial working memory. Psychiatry Res. Neuroimaging 163 (1), 40–51. http://dx.doi.org/10.1016/j.pscychresns.2007.04.018.

Schweinsburg, A.D., Schweinsburg, B.C., Medina, K.L., McQueeny, T., Brown, S.A., Tapert, S.F., 2010. The influence of recency of use on fMRI response during spatial working memory in adolescent marijuana users. J. Psychoactive Drugs 42 (3), 401–412. http://dx.doi.org/10.1080/02791072.2010.10400703.

Shiffrin, R.M., Schneider, W., 1977. Controlled and automatic human information processing: II. Perceptual learning, automatic attending and a general theory. Psychol. Rev. 84 (2), 127–190. http://dx.doi.org/10.1037/0033-295X.84.2.127.

Smith, J.L., Mattick, R.P., 2013. Evidence of deficits in behavioural inhibition and performance monitoring in young female heavy drinkers. Drug Alcohol Depend. 133 (2), 398–404. http://dx.doi.org/10.1016/j.drugalcdep.2013.06.020.

Smith, J.L., Mattick, R.P., Jamadar, S.D., Iredale, J.M., 2014. Deficits in behavioural inhibition in substance abuse and addiction: a meta-analysis. Drug Alcohol Depend. 145, 1–33. http://dx.doi.org/10.1016/j.drugalcdep.2014.08.009.

Sofuoglu, M., Herman, A.I., Li, Y., Waters, A.J., 2012. Galantamine attenuates some of the subjective effects of intravenous nicotine and improves performance on a go no-go task in abstinent cigarette smokers: a preliminary report. Psychopharmacology (Berl) 224 (3), 413–420. http://dx.doi.org/10.1007/s00213-012-2763-4.

Sohn, M.-H., Ursu, S., Anderson, J.R., Andrew Stenger, V., Carter, C.S., 2000. The role of prefrontal cortex and posterior parietal cortex in task switching. Proc. Natl. Acad. Sci. 97 (24), 13448–13453. http://dx.doi.org/10.1073/pnas.240460497.

Spronk, D.B., Van Wel, J.H., Ramaekers, J.G., Verkes, R.J., 2013. Characterizing the cognitive effects of cocaine: a comprehensive review. Neurosci. Biobehav. Rev. 37 (8), 1838–1859. http://dx.doi.org/10.1016/j.neubiorev.2013.07.003.

Stuss, D.T., Frank Benson, D., 1987. The Frontal Lobes and Control of Cognition and Memory. In: The IRBN Press, New York, pp. 141–158. http://search.proquest.com/psycinfo/docview/617353807/941B8AE17B7E49A2PQ/1?accountid=14679.

Sullivan, E.V., 2003. Compromised pontocerebellar and cerebellothalamocortical systems: speculations on their contributions to cognitive and motor impairment in nonamnesic alcoholism. Alcohol. Clin. Exp. Res. 27 (9), 1409–1419. http://dx.doi.org/10.1097/01.ALC.0000085586.91726.46.

Swann, N.C., Tandon, N., Canolty, R., Ellmore, T.M., McEvoy, L.K., Dreyer, S., DiSano, M., Aron, A.R., 2009. Intracranial EEG reveals a time- and frequency-specific role for the right inferior frontal gyrus and primary motor cortex in stopping initiated responses. J. Neurosci. 29 (40), 12675–12685. http://dx.doi.org/10.1523/JNEUROSCI.3359-09.2009.

Swann, N.C., Cai, W., Conner, C.R., Pieters, T.A., Claffey, M.P., George, J.S., Aron, A.R., Tandon, N., 2012. Roles for the pre-supplementary motor area and the right inferior frontal gyrus in stopping action: electrophysiological responses and functional and structural connectivity. Neuroimage 59 (3), 2860–2870. http://dx.doi.org/10.1016/j.neuroimage.2011.09.049.

Tabibnia, G., Monterosso, J.R., Baicy, K., Aron, A.R., Poldrack, R.A., Chakrapani, S., Lee, B., London, E.D., 2011. Different forms of self-control share a neurocognitive substrate. J. Neurosci. 31 (13), 4805–4810. http://dx.doi.org/10.1523/JNEUROSCI.2859-10.2011.

Tanabe, J., Tregellas, J.R., Dalwani, M., Thompson, L., Owens, E., Crowley, T., Banich, M., 2009. Medial orbitofrontal cortex gray matter is reduced in abstinent substance-dependent individuals. Biol. Psychiatry 65 (2), 160–164. http://dx.doi.org/10.1016/j.biopsych.2008.07.030.

Tapert, S.F., Schweinsburg, A.D., Drummond, S.P., Paulus, M.P., Brown, S.A., Yang, T.T., Frank, L.R., 2007. Functional MRI of inhibitory processing in abstinent adolescent marijuana users. Psychopharmacology (Berl) 194 (2), 173–183. http://dx.doi.org/10.1007/s00213-007-0823-y.

Torregrossa, M.M., Quinn, J.J., Taylor, J.R., 2008. Impulsivity, compulsivity, and habit: the role of orbitofrontal cortex revisited. Biol. Psychiatry 63 (3), 253–255. http://dx.doi.org/10.1016/j.biopsych.2007.11.014.

Townshend, J.M., Duka, T., 2005. Binge drinking, cognitive performance and mood in a population of young social drinkers. Alcohol. Clin. Exp. Res. 29 (3), 317–325. http://dx.doi.org/10.1097/01.ALC.0000156453.05028.F5.

Volkow, N.D., Fowler, J.S., Wang, G.-J., Telang, F., Logan, J., Jayne, M., Ma, Y., Pradhan, K., Wong, C., Swanson, J.M., 2010. Cognitive control of drug craving inhibits brain reward regions in cocaine abusers. Neuroimage 49 (3), 2536–2543. http://dx.doi.org/10.1016/j.neuroimage.2009.10.088.

von Geusau, N.A., Stalenhoef, P., Huizinga, M., Snel, J., Richard Ridderinkhof, K., 2004. Impaired executive function in male MDMA ('ecstasy') users. Psychopharmacology (Berl) 175 (3), 331–341. http://dx.doi.org/10.1007/s00213-004-1832-8.

Vonmoos, M., Hulka, L.M., Preller, K.H., Jenni, D., Schulz, C., Baumgartner, M.R., Quednow, B.B., 2013. Differences in self-reported and behavioral measures of impulsivity in recreational and dependent cocaine users. Drug Alcohol Depend. 133 (1), 61–70. http://dx.doi.org/10.1016/j.drugalcdep.2013.05.032.

Waters, A.J., Shiffman, S., Sayette, M.A., Paty, J.A., Gwaltney, C.J., Balabanis, M.H., 2003. Attentional bias predicts outcome in smoking cessation. Health Psychol. 22 (4), 378–387. http://dx.doi.org/10.1037/0278-6133.22.4.378.

Weafer, J., de Wit, H., 2013. Inattention, impulsive action, and subjective response to D-amphetamine. Drug Alcohol Depend. 133 (1), 127–133. http://dx.doi.org/10.1016/j.drugalcdep.2013.05.021.

Whelan, R., Conrod, P., Poline, J.B., Banaschewski, T., Barker, G.J., Bellgrove, M.A., Büchel, C., Byrne, M., Cummins, T., Fauth-Bühler, M., Flor, H., Gallinat, J., Heinz, A., Ittermann, B., Lourdusamy, A., Mann, K., Martinot, J.-L., Lalor, E.C., Lathrop, M., Loth, E., Paus, T., Rietschel, M., Smolka, M.N., Spanagel, R., Stephens, D., Struve, M., Thyreau, B., Vollstaedt-Klein, S., Robbins, T.W., Schumann, G., Garavan, H., the IMAGEN consortium, 2012. Adolescent impulsivity phenotypes characterized by distinct brain networks. Nat. Neurosci. 15, 920–925.

CHAPTER

Neuroscience of inhibition for addiction medicine: from prediction of initiation to prediction of relapse

9

Scott J. Moeller[*,†,1], **Lucia Bederson**[‡], **Nelly Alia-Klein**[*,†], **Rita Z. Goldstein**[*,†,1]

[*]*Department of Psychiatry, Icahn School of Medicine at Mount Sinai, New York, NY, USA*

[†]*Department of Neuroscience, Icahn School of Medicine at Mount Sinai, New York, NY, USA*

[‡]*Department of Psychology, New York University, New York, NY, USA*

[1]*Corresponding authors: Tel.: +1-212-824-8973; Fax: +1-212-803-6743. Tel.: +1-212-824-9312; Fax: +1-212-996-8931, e-mail address: scott.moeller@mssm.edu; rita.goldstein@mssm.edu*

Abstract

A core deficit in drug addiction is the inability to inhibit maladaptive drug-seeking behavior. Consistent with this deficit, drug-addicted individuals show reliable cross-sectional differences from healthy nonaddicted controls during tasks of response inhibition accompanied by brain activation abnormalities as revealed by functional neuroimaging. However, it is less clear whether inhibition-related deficits predate the transition to problematic use, and, in turn, whether these deficits predict the transition out of problematic substance use. Here, we review longitudinal studies of response inhibition in children/adolescents with little substance experience and longitudinal studies of already addicted individuals attempting to sustain abstinence. Results show that response inhibition and its underlying neural correlates predict both substance use outcomes (onset and abstinence). Neurally, key roles were observed for multiple regions of the frontal cortex (e.g., inferior frontal gyrus, dorsal anterior cingulate cortex, and dorsolateral prefrontal cortex). In general, less activation of these regions during response inhibition predicted not only the onset of substance use, but interestingly also better abstinence-related outcomes among individuals already addicted. The role of subcortical areas, although potentially important, is less clear because of inconsistent results and because these regions are less classically reported in studies of healthy response inhibition. Overall, this review indicates that response inhibition is not simply a manifestation of current drug addiction, but rather a core neurocognitive dimension that predicts key substance use outcomes. Early intervention in inhibitory deficits could have high clinical and public health relevance.

Progress in Brain Research, Volume 223, ISSN 0079-6123, http://dx.doi.org/10.1016/bs.pbr.2015.07.007

Keywords

Response inhibition, Inhibitory control, Drug addiction, Developmental trajectories, Clinical outcome, fMRI, Longitudinal designs

1 INTRODUCTION

Drug addiction is a chronically relapsing disorder marked by dysregulated inhibitory control, which may contribute to or exacerbate the addicted individual's ability to restrain drug-taking (Goldstein and Volkow, 2011; Kalivas and Volkow, 2005). Neuroimaging studies utilizing functional magnetic resonance imaging (fMRI) have consistently identified abnormalities in brain function during response inhibition in currently addicted individuals across multiple drugs of abuse (Luijten et al., 2014; Smith et al., 2014). Nevertheless, an enduring problem of such cross-sectional studies is the inability to infer the direction of association. Longitudinal studies offer an exciting opportunity to test whether core drug-relevant neurocognitive deficits (e.g., in response inhibition) predate the transition into and out of problematic drug use. In this way, one can evaluate whether such deficits in drug addiction represent an epiphenomenon or an actual predisposing factor.

Accordingly, the goal of the current review is to examine the extent to which performance- and/or neural-related decrements during tasks of inhibitory control precede the transition to drug use/addiction, and then whether such decrements predict clinical outcomes when already addicted individuals seek treatment or attempt to abstain. In particular, we seek to evaluate the hypothesis that impaired response inhibition is not simply a concurrent symptom of drug addiction, but instead a core neurocognitive dimension that predicts key substance use outcomes. We concentrate on longitudinal studies, largely those reported within the last 10 years, which have examined prospective associations between inhibitory control and the dependent variable of interest (drug use initiation or escalation, dependence, relapse, or abstinence). Most of the fMRI studies reviewed here report the results of task-induced activations (e.g., activity that occurs during a condition of response inhibition contrasted with activity during a condition of prepotent response). Other studies used task-related functional connectivity (i.e., the covariation between the fMRI time courses of a given voxel and other voxels in the brain), which offers a promising complement to task-based activation studies. The main literature review itself is organized into two parts. Part 1 discusses adolescent longitudinal studies that use tasks of inhibitory control to predict future drug use or transition into drug dependence. Part 2 discusses adult longitudinal studies that use tasks of inhibitory control to predict clinical and treatment outcomes in already addicted individuals. We conclude with a summary of findings and a discussion of future research directions.

We exclude from this review studies that involved passive exposure to drug-related stimuli, studies that used tasks associated with the receipt of reward, or studies that reported addiction-related abnormalities in brain structural integrity. Reviews that address these important topics can be found elsewhere (e.g., Garavan et al., 2013; Heitzeg et al., 2015; Jasinska et al., 2014; note some overlap in currently included studies with those from Heitzeg et al. (2015). This review also excludes behavioral addictions (e.g., gambling, food, sex, or video games) and studies that use event-related potentials, as more longitudinal studies in these fields are needed before firm conclusions about prospective relationships can be drawn. Studies that focus on family history (or other risk factors) as the main grouping variable are also excluded (e.g., Hardee et al., 2014). Finally, for brevity and focus, we also exclude tasks of inhibition that measure related constructs (e.g., error awareness; Hester et al., 2009), or studies that incorporate pharmacological (Moeller et al., 2014; Schmaal et al., 2013) or genetic (Filbey et al., 2012) modulation.

2 COMMONLY USED RESPONSE INHIBITION TASKS IN DRUG ADDICTION

Three of the most commonly used inhibitory control tasks, in order from simplest to most cognitively complex, include go/no-go tasks (Chambers et al., 2009), stop-signal tasks (Aron et al., 2014; Verbruggen and Logan, 2008), and Stroop tasks (MacLeod, 1991; Smith and Ersche, 2014). These tasks collectively measure a person's ability to modify or stop a behavior, particularly when the behavior may not be optimal or advantageous, or is perceived as incorrect. In go/no-go tasks, participants respond as quickly as possible to frequent go stimuli and inhibit responses to infrequent no-go stimuli. Correct nonresponses on no-go trials reflect the ability to exert inhibitory control over behavior. In stop-signal tasks, the goal is to successfully inhibit (stop) an action that has already begun. Participants respond to an ongoing sequence of stimuli; on some (stop) trials, however, a signal is presented (e.g., a tone, a change in stimulus display) after the stimulus onset that instructs participants to halt their response on that trial. The paradigm is typically configured to find the inflection point in which 50% of stop trials are unsuccessful relative to the mean reaction time; the longer this stop-signal reaction time (SSRT), the worse the inhibitory control. In Stroop tasks, participants must override a more automatic response tendency (reading a word) and instead respond with a task-specific demand (responding to the ink color of the word). Stroop tasks can be purely cognitive: in the classical color-word Stroop, participants respond to the ink color of color words (e.g., "blue") presented in either the congruent font (blue font) or an interfering incongruent font (e.g., red font). Stroop tasks can also be emotional: interference can be introduced by attentional bias or current concerns of the individual. In the case of drug addiction, individuals can be instructed to ignore the semantic content of drug-related words (e.g., "pipe") and instead respond to their font color; typically, the reaction time to drug words is longer than for neutral words (e.g., "vase"), indicating impaired response inhibition (Cox et al., 2006). An important caveat is that these tasks, while tapping into inhibitory control, also depend on other executive, attentional, or emotional processing functions. For example, some have argued that Stroop tasks tap into different higher-order executive functions than go/no-go and stop-signal tasks, such as compulsivity and impulsivity, respectively (Fineberg et al., 2014).

All three of these tasks have reliably yielded activations in regions of interest (ROIs)/networks known to be engaged during inhibitory control. These include the inferior frontal gyrus (IFG), anterior cingulate cortex (ACC) (especially its dorsal/motor subregion), middle frontal and superior frontal gyri (MFG/SFG) [which includes the dorsolateral prefrontal cortex (DLPFC)], parietal lobe, and pre-supplementary motor area (pre-SMA) (Bari and Robbins, 2013; Cieslik et al., 2015; Fig. 1). Importantly, some of these same regions are consistently identified as being disrupted in currently addicted individuals performing the tasks (for recent, comprehensive reviews on this topic, see Luijten et al., 2014; Smith et al., 2014). These studies and reviews in current drug dependence suggest pertinent regions/networks to spotlight for longitudinal prediction (Fig. 1), which is the focus of the remainder of this review.

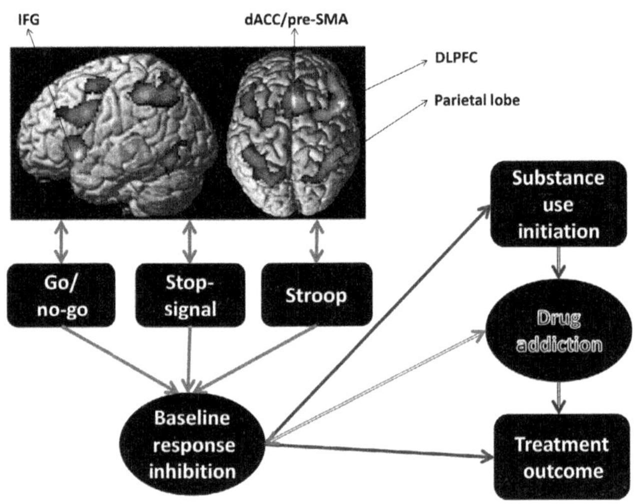

FIGURE 1

Schematic of the current review. Response inhibition is associated with performance and neural correlates of response inhibition (regions/networks include IFG, dACC/pre-SMA, DLPFC, and parietal lobe), which together prospectively predict substance use initiation and clinical/treatment outcomes. Blue arrows (gray in the print version) reflect concurrent associations; red arrows (dark gray in the print version) reflect longitudinal predictions; skinny black arrows are descriptive. Rectangles reflect measured variables; circles reflect latent variables (i.e., variables defined by other measured variables, whether explicitly included in the schematic or not). The broken text and arrows of the addiction circle signify implied relationships (i.e., not the focus of the current review). IFG = inferior frontal gyrus, dACC = dorsal anterior cingulate cortex, pre-SMA = pre-supplementary motor area, DLPFC = dorsolateral prefrontal cortex.

Brain activation maps are adapted from a previous meta-analysis of response inhibition in health (Cieslik et al., 2015) (with permission from Elsevier).

3 PART 1: PROGRESSION INTO ADDICTION/PROBLEMATIC SUBSTANCE USE (TABLE 1)

3.1 GO/NO-GO STUDIES

A moderately large sample of adolescents performed a go/no-go task during fMRI at baseline and then again 18 months later. More left angular/supramarginal gyrus activation and less ventromedial prefrontal cortex (vmPFC) activation to the no-go versus go trials at baseline predicted an increase in drug use occasions at follow-up (i.e., accounting for baseline drug use)—particularly in those who were already heavier users (Mahmood et al., 2013). Another longitudinal fMRI study tested for changes over the first year of college during an emotionally salient go/no-go task that instructed participants to respond to alcohol cues compared with nonalcohol cues; here, the dependent variable was task-related functional connectivity. Young adults were scanned three times (summer, first semester, second semester). At the second assessment (during which respondents reported an increase relative to the first assessment in the negative consequences of alcohol use, such as losing consciousness during drinking or performing poorly on an exam because of drinking), functional connectivity was increased among a network of regions implicated in response inhibition and cognitive control (e.g., bilateral DLPFC, rostral ACC, dorsal ACC) (Beltz et al., 2013).

Even more illuminating, however, are studies that begin tracking youth before they have begun experimenting with addictive substances. In one study, an fMRI go/no-go task was used at two study sessions to compare adolescents who were initially nondrinkers but later transitioned into heavy drinking against adolescents who remained nondrinkers during both assessments. Adolescents who later transitioned into heavy drinking showed less fMRI response to no-go versus go trials in the MFG, parietal cortex, putamen, and cerebellum. Interestingly, these effects were reversed at the second scanning session such that the adolescent heavy drinkers showed increased fMRI activation in these regions (except in the putamen, where no group differences were observed in the second session) (Wetherill et al., 2013). Results were interpreted to indicate that the reduced fMRI activation before drinking could reflect vulnerability, whereas the increased fMRI activation after drinking could reflect compensation. Another fMRI study investigated adolescents again with initially very limited substance use experience, classifying them at follow-up into those who transitioned to heavy use of alcohol versus those who remained nonusers. Similar results were reported, whereby youth who later transitioned into heavy alcohol use had less activation in multiple brain regions encompassing the IFG, DLPFC, putamen, middle temporal gyri, and inferior parietal lobules (Norman et al., 2011). More recently, preteens (9–12 years) performed an fMRI go/no-go task at baseline; 4 years later, participants completed assessments of substance use, which were used to create matched groups of substance users and nonsubstance users. In contrast to the other studies, there were no significant fMRI differences between the groups during successful no-go inhibition at baseline. Instead, nonusers showed increased

Table 1 Prediction of Drug Use Initiation

Study	Participants (Retention Rate)	Mean Age (SD)	Sex: M/F	Race	Disorder Status	RI Task	Time Between Assessments	Main Results, Behavioral	Main Results, Imaging
Go/no-go tasks									
Beltz et al. (2013)	*N* = 11 (91.01%)	18–19	5/6	NR	NR	Go/no-go fMRI	T1–T2: 2 months T2–T3: 3 months	NS	Task functional interconnectivity: ↑ B DLPFC, ACC over time
Heitzeg et al. (2014)	*N* = 45			100% C	*Externalizing*: CBCL & YSR	Go/no-go fMRI	T1–T2: ~4 years	RT: PU ↓ NU	*Contrast: failed RI > correct RI*; NU ↑ PU: L MFG; L MFG neg corr externalizing
	HC, *n* = 19	10.9 (1.1)	15/4						
	NU, *n* = 13	10.9 (0.9)	10/3						
	PU, *n* = 13	11.0 (1.0)	10/3						
Norman et al. (2011)	*N* = 38			~80% C	*Substance use*: CDDR & DSM-IV *Externalizing*: CBCL	Go/no-go fMRI	T1–T2: 4.2 years	NS	*Contrast: HU no-go < HC no-go*: ↑ L DLPFC, L superior MFG, R IFG, B medial FG, B paracentral lobules, pre-SMA, L ACC, L Put, L MTG, R MTG, B IPL; ↑ R IFG, L ACC, R MTG, L IPL neg corr T2 CBCL attention problems
	HU: *n* = 21	13.9 (0.9)	11/10						
	HC: *n* = 17	13.4 (0.7)	8/9						
Mahmood et al. (2013)	*N* = 80				CDDR	Go/no-go fMRI	T1–T2: 18 months	NS	*Contrast: no-go > go*: HF users: ↓ vmPFC pos corr T2 Drug & Alc Sx
	HF, *n* = 39	17.4 (0.9)	28/11	76% C					
	LF, *n* = 41	17.6 (1.0)	30/11	79% C					
Wetherill et al. (2013)	*N* = 40			55% C, 20% H, 15% multi, 5% As, 5% A	*Substance use*: CDDR and DSM-IV *Psychopathology/ externalizing:* CBCL, YSR, & ASR	Go/no-go fMRI	T1–T2: 3+ years	NS group difference; RI ↑ with age	*Contrast: HU no-go < HC no-go*: T1: ↑ B MFG, IPL, L Put, L cerebellum; T2: ↓ B MFG, R inferior parietal lobule, L cerebellum; HU: ↑ R MFG pos corr T2 drinking
	HU, *n* = 20	14.7 (1.1)	11/9						
	HC, *n* = 20	14.1 (1.2)	11/9						

Stop-signal tasks									
Fernie et al. (2013)	$N=287$ (94.4%)	12–13	NR	NR	NR	Stop-signal behavior	T1–T2: 6 months T2–T3: 6 months T3–T4: 6 months T4–T5: 6 months	↑ SSRT pos corr T2 Drinking	NA
Nigg et al. (2006)	$N=498$	12–14	362/136	100% C	*Alc/drug use*: DDHQ	Stop-signal behavior	T1–T2: ~3 years	↑ SSRT pos corr T2 drinking & drug problems	NA
Wong et al. (2010)	$N=386$	15–17	292/94	100% C	DDHQ-Y & DISC	Stop-signal behavior	T1–T2: ~3 years T2–T3: ~3 years T3–T4: ~3 years T4–T5: ~3 years	↑ SSRT pos corr T2 Alc Sx	NA
Whelan et al. (2014)	$N=2650$ BD, $n=115$ FBD, $n=121$ HC, $n=150$	 14.62 (0.39) 14.45 (0.40) 14.53 (0.43)	 4/66 69/52 70/80	NR	ESPAD	Stop-signal fMRI	T1–T2: ~2 years	NR	*Contrast: failed RI*: ↑ R precentral gyrus pos corr with T2 binge drinking
Stroop tasks									
Peeters et al. (2013)	$N=347$ (72.0%)	13.6 (0.9)	330/44	~75% Dutch ~25 non-Dutch	*Alc quantity*: 14-item scale *Severity*: CRAFFT	cwStroop behavior; approach-avoidance behavior	T1–T2: 6 months	Poor RI: Alc approach pos corr T2 Alc use	NA

Notes: *As, Asian; A, African American; Alc, alcohol; ASR, Adult Self-Report; B, bilateral; C, Caucasian; CBCL, Child Behavioral Checklist; CDDR, Customary Drinking and Drug Use Record; cwStroop, Stroop with classical color-word stimuli; DDHQ, Drinking and Other Drug Use History Questionnaire; DDHQ-Y Drinking and Other Drug Use History Questionnaire—Youth Version; DISC, Diagnostic Interview Schedule for Children; DLPFC, dorsolateral prefrontal cortex; ESPAD, European School Survey Project on Alcohol and Drugs; FBD, future binge drinker; FH+, family history AUD positive; Fas, false alarms; FH−, family history AUD negative; H, Hispanic; HC, healthy control; HF, high frequency; HU, heavy user; IFG, inferior frontal gyrus; IPL, inferior parietal lobe; L, left; LF, low frequency; MFG, middle frontal gyrus; MTG, middle temporal gyrus; neg corr, negative correlation; NA, not applicable; NR, not reported; NS, not significant; NU, nonuser; PFC, prefrontal cortex; pos corr, positive correlation; PU, problem-user; Put, putamen; pre-SMA, pre-supplementary motor area; SSRT, stop-signal reaction time; R, right; RI, response inhibition; Sx, symptoms; T, time; TRF, Teacher's Report Form; YAAPST, Young Adult Problems Screening Test; vmPFC, ventromedial prefrontal cortex; YSR, Youth Self-Report.*

activation relative to users during unsuccessful inhibition versus successful inhibition in the left MFG (DLPFC); DLPFC activation predicted outcome (group membership) over and above the effects of externalizing behavior (Heitzeg et al., 2014). This different pattern of effects could be due to the different fMRI contrast (error-related processing).

3.2 STOP-SIGNAL STUDIES

Four hundred ninety-eight children from 275 families from a high-risk, prospectively followed cohort completed executive function measures in early and late adolescence, with the goal of predicting lifetime drinking and drug-related ratings in late adolescence; multilevel models controlling for various potential confounds showed that poorer response inhibition (i.e., higher SSRT) predicted the onset of future drug and alcohol use (Nigg et al., 2006). Moreover, in the same high-risk sample, poorer response inhibition in late adolescence predicted alcohol-related problems in young adulthood (e.g., driving while intoxicated or experiencing an alcohol-induced blackout) (Wong et al., 2010).

A different research group similarly used a multiwave longitudinal study (five assessments over 2 years) to test associations between the behavioral stop-signal task and alcohol use in adolescents (Fernie et al., 2013). Data were analyzed using sophisticated cross-lagged analyses, which enable investigation of the relationships between response inhibition at time 1 and alcohol use at time 2 while controlling for cross-sectional associations between these variables at both time points and for their stability over time. Results showed that stop-signal performance prospectively predicted alcohol involvement, whereas the reverse association (alcohol involvement predicting response inhibition) did not reach significance. These analyses, which approximate causal relationships between variables in a nonexperimental design, suggest that response inhibition in adolescence confers vulnerability toward future substance use.

In an elegant, recent fMRI study, machine learning techniques were used to integrate multimodal self-report, structural and functional imaging, and genetics data in service of predicting concurrent and future binge drinking in a large sample of adolescents. In the longitudinal arm, fMRI activation in the precentral gyrus to response inhibition failures predicted future binge drinking (Whelan et al., 2014).

3.3 STROOP STUDIES

Few studies have used Stroop tasks to predict emerging substance use problems. One behavior-only study used a Stroop task to stratify adolescents into those with stronger or weaker response inhibition (weaker response inhibition was defined as higher incongruent > neutral response reaction time). The task itself was an approach-avoidance paradigm that used stimuli depicting alcohol or soda, and participants were instructed to either pull (approach) or push (avoid) a lever in response to the stimuli. Results showed that greater alcohol approach tendencies (i.e., faster reaction time to pull the lever toward than push the lever away) predicted alcohol use at 6-month follow-up only in the adolescents with weaker Stroop-assessed inhibition (Peeters et al., 2013).

3.4 PART 1 SUMMARY

These studies suggest that performance on tasks of response inhibition in adolescence/young adulthood predicts future initiation into substance use. Despite some exceptions (and although additional studies utilizing the Stroop task are needed), the general pattern of results suggests underactivations during the response inhibition trials in key inhibition-related regions in the individuals who would later become substance users; an opposite (hyperactivation) pattern seemed to occur when examining response failure. The most consistent neural correlate of response inhibition was the DLPFC, which is a core region in response inhibition but also in the implementation of cognitive control more generally (Egner et al., 2008; Kerns et al., 2004). Other regions identified in multiple studies included the parietal cortex and the putamen. The precentral gyrus also deserves mention, given this region's emergence in a well-powered and well-controlled study (Whelan et al., 2014). These neural underactivations during response inhibition in the individuals who would later develop problematic drug use were typically observed in the absence of behavioral differences between the groups (Table 1). Lack of group differences on task performance suggests that these fMRI differences are potentially marking abnormal neural activity (rather than, for example, an inability to perform the task). Taken together, inhibition problems and associated aberrant brain response during the exertion of inhibitory control appear to predate substance use.

4 PART 2: PREDICTION OF CLINICAL OUTCOME IN ALREADY ADDICTED INDIVIDUALS (TABLE 2)

4.1 GO/NO-GO STUDIES

In an interesting study of smokers motivated to quit, fMRI during a go/no-go task (successful no-go versus go events) was used to predict outcome via an experience sampling method (a unique contribution to this literature); as part of these assessments, participants responded eight times per day for 3 weeks about their craving and cigarettes smoked (Berkman et al., 2011). The IFG, pre-SMA, and basal ganglia were selected as ROIs. Results revealed a positive correlation between craving at one time point and smoking at the next time point. Interestingly, this relationship was moderated by all three ROIs such that individuals who had higher fMRI activations in these regions to the no-go stimuli had a blunted correlation between craving and smoking. This finding could suggest that enhanced neural response during response inhibition reflects a greater capacity to exert top-down control over impulses (e.g., craving) [although it should be noted that other studies have interpreted

Table 2 Prediction of Clinical Outcome

Study	Participants (Retention Rate)	Mean Age (SD)	Sex: M/F	Race	Abst Length	Disorder Status	Treatment Status	RI Task	Time Between Assessments	Main Results, Behavioral	Main Results, Imaging
Go/no-go tasks											
Berkman et al. (2011)	$N=31$	46 (9.7)	16/15	52% C, 26% H 19% A 3% O	NR	>10 cig/day	Cessation program	Go/no-go fMRI	T1–T2: 4 weeks	NS	*Contrast: no-go > go*: ↑ IFG, BG, & pre-SMA pos corr with attenuation of craving-smoking link
Prisciandaro et al. (2013)	CD: $N=30$ (ReL: $n=6$; no-ReL:	41.2 (8.3)	5/1	83% A	≥72 h	DSM-IV for CD	Outpatient	Drug no/ no-go fMRI	T1–T2: 1 week	NS	*Contrast: no-go > go:* ↑ B postcentral gyri pos corr with urine+
	$n=24$)	48.4 (8.7)	20/4	83% A							
Stop-signal tasks											
Jakubczyk et al. (2013)	AD: $N=254$	44.2 (10.2)	189/65		NR	DSM-IV for AD	Inpatient	Stop-signal behavior	T1–T2: 12 months	NS	NA
Luo et al. (2013)	CD: $N=97$			29% C	2–4 weeks	DSM-IV for CD	Inpatient	Stop-signal fMRI	T1–T2: 14 days T1–T3:	NS	*Contrast: stop error > stop correct*: in female CD, ↓ thalamus & dACC pos
	ReL: $n=80$	39.1 (7.5)	27/53	67% A							
	No-ReL: $n=17$	43.0 (7.3)	7/10	4% O					30 days T1–T4: 60 days T1–T5: 90 days		corr relapse; in male CD, ↓ L insula & dACC pos corr relapse
Stroop tasks											
Brewer et al. (2008)	CD: $N=20$	38.6 (9.3)	12/8	30% C 50% A 20% H	28 days	DSM-IV for CD	Outpatient	cwStroop fMRI	NR	RT: ↑ incong > cong pos corr Tx retention	*Contrast: incong < cong;* ↑ R Put pos corr urine− & longest abst; R putamen, L vmPFC & left PCC pos corr longest abst; ↓ DLPFC pos corr Tx retention
Carpenter et al. (2012)	CD: $N=25$	37 (7.1)	22/3	36% C 28% A	14 days	DSM-IV for CD	Outpatient	dStroop behavior	24 weeks	RT: ↑ drug > neutral	NA

				24% H 3% O						pos corr Phase II Tx and urine−	
Carpenter et al. (2006)	$N=80$				NR	DSM-IV for CD/MJ	Outpatient	(Substance-matched) dStroop behavior	NR	RT: In CD, ↑ cocaine > neutral pos corr urine+ and shorter Tx; in MJ and HD, NS	NA
	CD: $n=45$	38.6 (8.1)	33/12	31% C							
	MJ: $n=25$	32.4 (8.9)	20/5	24% A							
	HD: $n=10$	32.4 (6.6)	9/1	33% H							
Cox et al. (2002)	$N=30$			NR	NR	DSM-IV for AD	Inpatient	dStroop behavior	AD: ~24 days HC: ~28 days	RT: in AD ReL but not non-ReL, ↑ alcohol > neutral	NA
	AD: $n=14$ (ReL: $n=9$; No-ReL: $n=5$)	41.9 (10.6)	11/3								
	HC: $n=16$	37.3 (10.3)	4/12								
Devito et al. (2012)	SUD: $n=12$	37.2 (9.5)	7/5	NR	NR	DSM-IV for SUD	RCT	cwStroop fMRI	T1–T2: 8 weeks	RT: SUD < HC, but ↓ incong RT in SUD at T2	*Contrast: incong > cong T2 < T1* *SUD < HC*: ↓ STN/VTA, GP, thal & hypothal
	HC: $n=12$	31.0 (8.6)	5/7								
Marhe et al. (2013)	CD: $N=26$	38.7 (9.2)	22/4	NR	NR	DSM-IV for CD	Inpatient	dStroop fMRI	T1–T2: 3 months	RT: cocaine > neutral	*Contrast: cocaine > neutral*: ↑ R dACC ROI pos corr cocaine use days
Mitchell et al. (2013)	CD: $N=15$	39.0 (10.4)	6/9	CD: 40% C, 66.7% A	NR	DSM-IV for CD	Outpatient	cwStroop fMRI	T1–T2: 8–12 weeks	NS	*Measure: intrinsic connectivity*: CD < HC: R caudate, B OFC, IFG, insula, thal, SN and VS; ↑ connectivity in B thalamus, VS, & SN neg corr abst during Tx & pos corr urine+
	HC: $N=15$	40.0 (7.4)	7/8	HC: 53% C, 47% A							
Moeller et al. (2012b)	CD: $N=15$ (Tx-seeking)	41.4 (9.1)	11/4	53% A 33% C 13% O	≥3 weeks	DSM-IV for CD	Mix inpatient/outpatient	dStroop fMRI	T1–T2: 6.4 (±1) months	RT: T2 < T1 neutral	*Contrast: T2 > T1*; ↑ B midbrain (VTA/SN) & R thal, neg corr simulated drug-seeking
	CD: $N=13$ (active users)	42.5 (5.9)	12/1	77% A 8% C 15% O							

Continued

Table 2 Prediction of Clinical Outcome—cont'd

Study	Participants (Retention Rate)	Mean Age (SD)	Sex: M/F	Race	Abst Length	Disorder Status	Treatment Status	RI Task	Time Between Assessments	Main Results, Behavioral	Main Results, Imaging
	HC: $N=13$	39.6 (4.9)	11/2	69% A 23% C 8% O							
Verdejo-Garcia et al. (2012)	CD: $N=131$	33.9 (21.6)	120/11	NR	≥15 days	DSM-IV for CD	Therapeutic community	cwStroop behavior	T1–T2: 15–30 days	RT incong > cong: cocaine + heroin comorbid < cocaine only; RT trend pos corr Tx retention	NA
Worhunsky et al. (2013)	CD: $N=20$	38.6 (9.3)	12/8	CD: 30% C, 50% A, 20% O	NR	DSM-IV for CD	Outpatient	cwStroop fMRI	8 weeks treatment	RT: NS Errors: CD > HC	*Measure: ICA*: ↑ subcortical- & ventral frontostriatal networks pos corr urine −; ↓ frontocingular network pos corr Tx wks
	HC: $N=20$	36.8 (8.9)	12/8	HC: 70% C, 30% A							

Notes: *As, Asian; A, African American; Abst, abstinence; ACC, anterior cingulate cortex; AD, alcohol use disorder; ASR, Adult Self-Report; B, bilateral; BG, basal ganglia; C, Caucasian; CBCL, Child Behavior Checklist; CD, cocaine dependence; Cong, congruent; CDDR, Customary Drinking and Drug Use Record; cwStroop, Stroop with classical color-word stimuli; DDHQ-Y, Drinking and Other Drug Use History Questionnaire-Youth Version; DISC, Diagnostic Interview Schedule for Children; DLPFC, dorsolateral prefrontal cortex; dStroop, Stroop with drug-associated stimuli; GP, globus pallidus; H, hispanic; HC, healthy control; HD, heroin dependence; ICA, independent components analysis; IFG, inferior frontal gyrus; Incong, incongruent; ITC, inferior temporal cortex; IRAP, Implicit Relational Assessment Procedure; L, left; Lent Nucl, lentiform nucleus; MFG, middle frontal gyrus; MJ, Marijuana Dependence; MTG, middle temporal gyrus; neg corr, negative correlation; NA, not applicable; NR, not reported; No-ReL, nonrelapsers; NS, not significant; O, other race; PFC, prefrontal cortex; pre-SMA, pre-supplementary motor area; Put, putamen; R, right; ReL, relapsers; RCT, randomized clinical trial; RI, response inhibition; ROI, region of interest; R-SAT, Regulation—Revised Strategy Application Test; SFG, superior frontal gyrus; STG, superior temporal gyrus; SN, substantia nigra; STN, subthalamic nucleus; SUD, substance use disorder; T, time; Thal, thalamus; Tx, treatment; urine+, positive urine result; urine−, negative urine result; vmPFC, ventromedial prefrontal cortex; VS, ventral striatum; VTA, ventral tegmental area; YSR, Youth Self-Report.*

such enhanced activation as reflecting compensation (Wetherill et al., 2013)]. In a secondary analysis of this study, the basal ganglia ROI (but not the other two ROIs) predicted reduced smoking across 4 weeks (objectively measured by breath CO) (Berkman et al., 2011).

In another fMRI study (this one using a more standard analysis methodology), increased activation in a different region (the postcentral gyrus, to all no-go events versus all go events) predicted treatment outcome 1 week later (positive cocaine urine screen) (Prisciandaro et al., 2013). Important caveats of this study are that these participants were also included in a treatment trial that administered D-cycloserine, and they also completed a cue-reactivity task during the same scanning session.

4.2 STOP-SIGNAL STUDIES

In an fMRI study, a moderately sized cohort of treatment-seeking cocaine-dependent individuals completed the stop-signal task and was followed over 3 months to predict clinical outcome. Decreased activation in the dorsal ACC during error-related processing (stop error versus stop success) predicted relapse in males and females (note that males and females also exhibited some differential activations that predicted relapse: decreased thalamus activation in females; decreased insula activation in males) (Luo et al., 2013). In contrast, in a behavior-only study of treatment-seeking alcohol-dependent individuals, the stop-signal task administered at baseline did not predict 12-month outcome; instead, 12-month outcome was predicted by genetic variation (type 2A serotonin receptor polymorphism) (Jakubczyk et al., 2013). Notable differences between these studies include the use of fMRI and the length of the follow-up period (3 versus 12 months).

4.3 STROOP STUDIES

In a behavior-only study examining the variables that predict treatment retention in a therapeutic community, participants completed a battery of neuropsychological measures including the color-word Stroop task. Better Stroop task performance (both the standard inhibition measure and a second measure assessing switching) significantly predicted better 3-month outcome, but these results did not survive the authors' correction for multiple comparisons (Verdejo-Garcia et al., 2012). Other behavioral studies used emotional (drug) Stroop tasks. In an early study of alcohol abusers, participants completed an alcohol Stroop task at baseline and then again 4 weeks later. Compared with control participants and alcohol abusers who completed treatment, alcohol abusers who did not complete treatment had alcohol-related interference scores that increased from baseline to follow-up (Cox et al., 2002). A caveat of this study is the small sample sizes in each group ($n=5$ participants who remained abstinent or had a small lapse; $n=9$ participants who relapsed or failed to maintain contact with a counselor). In another earlier study, treatment-seeking drug-addicted individuals performed a drug Stroop task, with the stimuli content matched to the participants' particular substance problem (e.g., cocaine stimuli for individuals addicted to cocaine) (Carpenter et al., 2006). Results showed that cocaine Stroop interference scores predicted more cocaine positive urines and shorter treatment duration in the cocaine participants, but similar substance-specific analyses were not significant in individuals in treatment for marijuana or heroin (but note smaller sample sizes in these latter two groups compared with the cocaine group). In contrast, another study from the same lead author showed that drug Stroop interference scores were *positively* correlated with a greater likelihood of continuing with treatment (entering a Phase II, which included providing negative cocaine urine screens) (Carpenter et al., 2012). This latter result could indicate that the interference scores in this case were tapping into a hypervigilance toward the cocaine cues to sustain commitment to the treatment process (Moeller and Goldstein, 2014). These conflicting findings remain to be reconciled, but could include variability in the characteristics of the participants (e.g., abstinence lengths) and/or the therapeutic context (e.g., presence of a voucher system) (Carpenter et al., 2012).

A growing number of fMRI studies have used Stroop tasks to predict clinical outcome in already addicted individuals. In one of the first studies of its kind, 20 treatment-seeking cocaine-dependent individuals performed an fMRI color-word Stroop task prior to initiating treatment. Interestingly, *higher* behavioral Stroop interference predicted better clinical outcomes (more weeks in treatment). Analysis of the fMRI data showed that during interference trials (incongruent versus congruent), higher activation of the vmPFC, posterior cingulate, and striatum predicted a longer duration of self-reported abstinence (the striatum additionally predicted percent of negative urine screens); and reduced activation of the DLPFC predicted treatment retention (Brewer et al., 2008). In another study, a drug (cocaine) Stroop task was administered to cocaine-dependent patients during their first week in detoxification treatment and was used to predict cocaine use at 3-month follow-up. Dorsal ACC activation to cocaine versus neutral words positively predicted future cocaine use (i.e., relapse) (Marhe et al., 2013). Interestingly, the direction of correlation was opposite to the previous study, perhaps attributable to the task valence (emotionally neutral in the former versus emotionally salient in the latter).

Other fMRI studies instead examined the predictive effects of task-related connectivity during neutral Stroop tasks. In one study, 16 treatment-seeking cocaine-dependent individuals and matched healthy controls completed an fMRI color-word Stroop task (Mitchell et al., 2013). In addition to the behavioral Stroop predicting abstinence [i.e., more interference at pretreatment correlated with better outcome, supporting the study above (Brewer et al., 2008)], less functional connectivity among the ventral striatum, thalamus, substantia nigra, right insula, and left hippocampus predicted better clinical outcome (longer abstinence) (Mitchell et al., 2013). This finding is somewhat difficult to interpret considering that the addicted individuals had less connectivity among these regions overall than healthy controls. Nevertheless, this prospective finding within the addicted group is consistent with other work showing that subcortical pathways in drug addiction are hyperconnected during resting-state in association with a greater severity of dependence (Konova et al., 2013); less hyperconnectivity of these subcortical structures, then, could be driving the better treatment outcomes observed in this study. In contrast, however, in a second connectivity study of treatment-seeking cocaine-addicted individuals, independent component analysis (ICA) was applied to fMRI data during color-word Stroop interference (Worhunsky et al., 2013). Here, better clinical outcome (higher numbers of negative cocaine urine screens) was predicted by greater engagement of a subcortical network (encompassing the thalamus, striatum, amygdala, and hippocampus) and a "ventral frontostriatal" network (encompassing vmPFC, ventral striatum, and subgenual/rostral components of the ACC). In contrast, more weeks in treatment were associated with reduced engagement of a "frontocingular" network (encompassing ACC, medial PFC, and insula). Thus, additional research is needed to reconcile inconsistencies among the studies, especially with respect to the contribution of subcortical structures.

Other studies have used bookend (pre–post) fMRI sessions to examine neural changes as a function of treatment and/or abstinence. In one study, substance-dependent individuals underwent fMRI during a color-word Stroop task at baseline and follow-up, with 8 weeks of computer-assisted cognitive behavioral therapy for substance abuse in between the two scanning sessions; nonsubstance-using control participants also completed the Stroop task following a similar time interval. At follow-up, the treatment-seekers showed decreased interference-related fMRI signal in multiple brain regions including the DLPFC, ACC, IFG, and a subcortical cluster that encompassed the midbrain and subthalamic nucleus (Devito et al., 2012). In another study, treatment-seeking cocaine-addicted individuals completed a drug Stroop task at baseline and then again at a 6-month follow-up. Results showed that midbrain fMRI signal increased during the entire task (to drug and neutral words) from baseline to follow-up, and this enhanced midbrain response correlated with reduced cocaine-related choice on a simulated drug-choice paradigm (Moeller et al., 2012b).

4.4 PART 2 SUMMARY

Here, many of the behavior-only studies used drug Stroop tasks, which yielded somewhat mixed/contradictory results in predicting clinical outcome in treatment-seeking drug-addicted individuals. When inhibition tasks were combined with neuroimaging, prediction was generally improved. Similarly to the initiation literature, these neuroimaging effects generally emerged in the absence of behavioral task effects (particularly for go/no-go and stop-signal tasks) (Table 2). In contrast, the Stroop tasks were often associated with behavioral differences (between groups, assessment time points, etc.) (Table 2), and therefore, one cannot rule out the possibility that differential fMRI activations are attributable to differential ability of individuals to perform the task. Despite this potential uncertainty, however, these imaging studies were fairly consistent in showing that clinical outcome was prospectively predicted by the DLPFC, dorsal ACC, IFG, and regions of the basal ganglia such as the striatum and midbrain. In general, although with multiple exceptions, better clinical outcome was predicted by decreased PFC activation but enhanced subcortical activation. The predictive effect of subcortical activation could be attributable to recovery of dopaminergic integrity with abstinence (Volkow et al., 2001).

Nevertheless, it is important to replicate this subcortical effect in future work, both because of the inconsistent direction of activation in these studies and because subcortical activations are not as reliably reported during inhibitory control tasks in healthy individuals.

Overall, better response inhibition and less activation during the exertion of inhibitory control predicted a better clinical outcome. As there is no *a priori* reason to suspect that individuals with better response inhibition had a less severe addiction, these studies suggest that better response inhibition helps individuals to refrain from drug-taking when they are motivated to do so. An interesting variable to examine in this regard, which was not routinely reported in these studies, is the number of quit attempts during the course of the addiction. One could anticipate that individuals with better response inhibition would have fewer quit attempts.

5 CONCLUSION AND FUTURE DIRECTIONS

5.1 PARADIGM CONSIDERATIONS

An important future direction is to test whether there are unique neural mechanisms underlying the ability to exert inhibitory control in a drug-related context versus a neutral context. Insofar as inhibitory control in drug-addicted individuals is anticipated to be lowest upon being confronted with drugs or drug-associated stimuli (Goldstein and Volkow, 2011), such task designs could potentially explain unique variance in drug use outcomes—particularly since neuropsychological impairments in drug addiction, while pervasive, are generally mild in magnitude and may require more sensitive neuropsychological probes for their detection (Goldstein et al., 2004, 2007; Moeller et al., 2009; Woicik et al., 2009). Although drug Stroop tasks have been deployed to predict clinical outcome in addicted individuals as reviewed above, these studies generally have not concurrently administered a standard color-word Stroop task for direct comparisons (e.g., drug task minus matched neutral task).

Response inhibition paradigms could also benefit from designs that enable the parametric correlation of trial-by-trial behavioral responses with the associated neural signals for each individual. This type of design could help reduce concerns about interpretation of the fMRI effects when there are also behavioral differences between groups or longitudinal assessments. More broadly, another interesting direction would be to directly contrast an inhibitory control task with another demanding cognitive task (e.g., working memory) that engages similar neural circuitry (e.g., the DLPFC). In this way, one could test whether any cognitively demanding task predicts future drug use, or whether there are uniquely predictive aspects of response inhibition.

5.2 EXPAND STUDY INTO MORE ADDICTIONS

Another important direction is to expand the present literature into different drug classes. Alcohol is overrepresented in studies examining the prediction of drug use initiation, and cocaine is overrepresented in studies examining prediction of clinical outcome. For the former (prediction of initiation), the decision to focus on alcohol use is justified, given the focus of these studies on adolescents and young adults. Nevertheless, it will be important to expand this young adult longitudinal literature into the misuse of opioid prescription medication, which has become a paramount public health concern in recent years (Schrager et al., 2014). Moreover, the recent legalization of marijuana in several states (e.g., Colorado and Washington) has increased concerns about underage use and misuse (Monte et al., 2015). For the latter (prediction of outcome), it will be important to increase the number of studies examining how response inhibition impacts clinical outcomes in other addictions that have high public health implications (e.g., nicotine, alcohol, heroin, methamphetamine). Beyond drug addiction, there is scant inhibitory control longitudinal research on behavioral addictions, such as gambling or internet/video game addiction.

5.3 INDIVIDUAL DIFFERENCES

It is imperative to study addicted individuals with psychiatric comorbidities. Individuals with comorbidities represent a majority of addicted individuals and are more likely to have unmet treatment needs (Melchior et al., 2014). Another potential modulatory variable is the presence of comorbid attention deficit/hyperactivity disorder (ADHD), which is associated with both the initiation of substance abuse and impaired response inhibition (Lee et al., 2011); other externalizing symptomatology, such as anger, could also be important to examine (Aharonovich et al., 2001). Finally, sex differences may modulate response inhibition in drug addiction, as indicated by one of the studies reviewed above (Luo et al., 2013). In further support, in a study examining sex by substance dependence interactions on self-reported impulsivity, female drug-addicted individuals exhibited the highest impulsivity of all participant groupings (Perry et al., 2013). Women may also have greater difficulty inhibiting drug use (e.g., smoking) following cue exposure (Doran, 2014).

5.4 UNDERLYING NEUROCHEMISTRY

The neurochemistry of these effects also remains to be uncovered, especially if these results are to aid the development of innovative pharmacotherapies to treat drug addiction. Dopamine is likely to play an important role, given its reported contribution to higher-order cognitive functions that bear on self-regulation/response inhibition inclusive of cognitive flexibility (Kehagia et al., 2010), exertion/sustaining effort (Niv et al., 2007; Satoh et al., 2003), and motivation (Moeller et al., 2012a). Supporting the latter, in a preliminary sample of cocaine-addicted individuals and healthy controls, we showed that dopamine D2 receptor availability, measured by positron emission tomography (PET) with [^{11}C]raclopride, correlated with fMRI midbrain response to errors during the color-word Stroop task when cognitive resources were presumably most depleted (during the final versus the first task repetition) (Moeller et al., 2012a). In addition, studies administering the stop-signal tasks during

PET with [^{18}F]fallypride in healthy individuals revealed correlations between SSRT and D2/D3 receptor availability in the left OFC, right MFG, and right precentral gyrus (Albrecht et al., 2014) and the striatum (Ghahremani et al., 2012). Accordingly, therapeutic agents that act on this system, such as the indirect dopamine agonist methylphenidate, could be used to modulate the neural correlates of response inhibition in drug addiction as indeed previously demonstrated (Goldstein et al., 2010; Li et al., 2010; Moeller et al., 2014; Sofuoglu et al., 2013).

5.5 SUMMARY, LIMITATIONS, AND CLINICAL IMPLICATIONS

We reviewed behavioral and neuroimaging studies of response inhibition aiming to predict longitudinal outcomes in substance abuse. We identified a larger number of studies relevant to the prediction of clinical outcome than to the prediction of transition into substance abuse, underscoring a need for more studies that can detect at-risk individuals before they transition to addiction. In particular, needed are large-scale, comprehensive studies that can integrate and/or disentangle the influences of multiple and multimodal predictors related to response inhibition; the creation of several collaborative imaging consortiums has begun to address this crucial gap (Paus, 2010) (see results in the current review reported by Whelan et al., 2014). These big data initiatives can also help resolve some of the inconsistencies between studies, as small sample sizes are likely to represent a source of increased variation; this concern is accentuated for the relapse prediction studies, which as a whole had smaller sample sizes than the drug use initiation studies. Another concern for these relapse prediction studies is the abstinence length at the time of scanning: abstinence length was variable between the studies (ranging from hours to weeks), and in many studies, this information was not reported (Table 2). This variable could have crucial bearing on the capacity to exert inhibitory control (e.g., if one is studying participants who are experiencing acute withdrawal symptoms and/or intense craving), or alternatively could be evidence of individuals having already exerted inhibitory control (e.g., if one is studying participants who have sustained abstinence for several weeks). Moreover, one needs also to exercise a degree of caution when interpreting the results of studies that retrospectively test for neuroimaging predictors (e.g., using the outcome, such as relapse versus abstinence, as the basis of creating groups for a baseline neuroimaging analysis). This type of analysis can lead to overfitting that can inflate the magnitude of the observed differences, a problem that has been well-articulated elsewhere (Garavan et al., 2013).

Despite these concerns, results generally support the hypothesis that these tasks, and their underlying neural correlates, predict important prospective outcomes. Behaviorally, better response inhibition generally predicted better outcomes. Neurally, the general pattern of results was that frontal regions were less activated during the exertion of inhibitory control in the individuals who would later become problematic substance users. This finding of blunted frontal activation during response inhibition is also consistent with other externalizing psychopathologies, including ADHD (Rubia et al., 2011) and intermittent explosive disorder (Coccaro et al., 2007). Interestingly, however, less activation of similar frontal regions generally predicted better clinical outcomes when the context was sustaining abstinence.

These results have important clinical implications. Although these results cannot illuminate causal relationships between variables, longitudinal prediction constitutes an improvement over cross-sectional studies and can support the important conclusion that response inhibition deficits could be targeted for intervention to improve future outcomes. These could include targeted cognitive–behavioral exercises, possibly in combination with pharmacotherapy and/or individualized neurofeedback. These types of interventions can help address the vital public health goals of identifying the young individuals most likely to progress from recreational to problematic substance use, and identifying the addicted individuals most likely to relapse after beginning treatment or abstinence. Individuals with reduced inhibitory control could be selected for additional therapeutic/interventional resources to produce better drug-related outcomes.

ACKNOWLEDGMENTS

This work was supported by the National Institute on Drug Abuse (to S.J.M.: 1K01DA037452), and by seed grants from the Icahn School of Medicine at Mount Sinai and the Mount Sinai Brain Imaging Center (to S.J.M., N.A.K., and R.Z.G.). We thank Gabriela Gan, Rebecca Preston-Campbell, and Anna B. Konova for helpful suggestions on concepts presented in this chapter.

Disclosure/Conflict of Interest: None declared.

REFERENCES

Aharonovich, E., Nguyen, H.T., Nunes, E.V., 2001. Anger and depressive states among treatment-seeking drug abusers: testing the psychopharmacological specificity hypothesis. Am. J. Addict. 10 (4), 327–334.

Albrecht, D.S., Kareken, D.A., Christian, B.T., Dzemidzic, M., Yoder, K.K., 2014. Cortical dopamine release during a behavioral response inhibition task. Synapse 68 (6), 266–274. doi:http://dx.doi.org/10.1002/syn.21736.

Aron, A.R., Robbins, T.W., Poldrack, R.A., 2014. Inhibition and the right inferior frontal cortex: one decade on. Trends Cogn. Sci. 18 (4), 177–185. doi:http://dx.doi.org/10.1016/j.tics.2013.12.003.

Bari, A., Robbins, T.W., 2013. Inhibition and impulsivity: behavioral and neural basis of response control. Prog. Neurobiol. 108, 44–79. doi:http://dx.doi.org/10.1016/j.pneurobio.2013.06.005.

Beltz, A.M., Gates, K.M., Engels, A.S., Molenaar, P.C., Pulido, C., Turrisi, R., et al., 2013. Changes in alcohol-related brain networks across the first year of college: a prospective pilot study using fMRI effective connectivity mapping. Addict. Behav. 38 (4), 2052–2059. doi:http://dx.doi.org/10.1016/j.addbeh.2012.12.023.

Berkman, E.T., Falk, E.B., Lieberman, M.D., 2011. In the trenches of real-world self-control: neural correlates of breaking the link between craving and smoking. Psychol. Sci. 22 (4), 498–506. doi:http://dx.doi.org/10.1177/0956797611400918.

Brewer, J.A., Worhunsky, P.D., Carroll, K.M., Rounsaville, B.J., Potenza, M.N., 2008. Pretreatment brain activation during Stroop task is associated with outcomes in cocaine-dependent patients. Biol. Psychiatry 64 (11), 998–1004. doi:http://dx.doi.org/10.1016/j.biopsych.2008.05.024.

Carpenter, K.M., Schreiber, E., Church, S., McDowell, D., 2006. Drug Stroop performance: relationships with primary substance of use and treatment outcome in a drug-dependent outpatient sample. Addict. Behav. 31 (1), 174–181. doi:http://dx.doi.org/10.1016/j.addbeh.2005.04.012.

Carpenter, K.M., Martinez, D., Vadhan, N.P., Barnes-Holmes, D., Nunes, E.V., 2012. Measures of attentional bias and relational responding are associated with behavioral treatment outcome for cocaine dependence. Am. J. Drug Alcohol Abuse 38 (2), 146–154. doi:http://dx.doi.org/10.3109/00952990.2011.643986.

Chambers, C.D., Garavan, H., Bellgrove, M.A., 2009. Insights into the neural basis of response inhibition from cognitive and clinical neuroscience. Neurosci. Biobehav. Rev. 33 (5), 631–646. doi:http://dx.doi.org/10.1016/j.neubiorev.2008.08.016.

Cieslik, E.C., Mueller, V.I., Eickhoff, C.R., Langner, R., Eickhoff, S.B., 2015. Three key regions for supervisory attentional control: evidence from neuroimaging meta-analyses. Neurosci. Biobehav. Rev. 48C, 22–34. doi:http://dx.doi.org/10.1016/j.neubiorev.2014.11.003.

Coccaro, E.F., McCloskey, M.S., Fitzgerald, D.A., Phan, K.L., 2007. Amygdala and orbitofrontal reactivity to social threat in individuals with impulsive aggression. Biol. Psychiatry 62 (2), 168–178. doi:http://dx.doi.org/10.1016/j.biopsych.2006.08.024.

Cox, W.M., Hogan, L.M., Kristian, M.R., Race, J.H., 2002. Alcohol attentional bias as a predictor of alcohol abusers' treatment outcome. Drug Alcohol Depend. 68 (3), 237–243.

Cox, W.M., Fadardi, J.S., Pothos, E.M., 2006. The addiction-Stroop test: theoretical considerations and procedural recommendations. Psychol. Bull. 132 (3), 443–476.

Devito, E.E., Worhunsky, P.D., Carroll, K.M., Rounsaville, B.J., Kober, H., Potenza, M.N., 2012. A preliminary study of the neural effects of behavioral therapy for substance use disorders. Drug Alcohol Depend. 122 (3), 228–235. doi:http://dx.doi.org/10.1016/j.drugalcdep.2011.10.002.

Doran, N., 2014. Sex differences in smoking cue reactivity: craving, negative affect, and preference for immediate smoking. Am. J. Addict. 23 (3), 211–217. doi:http://dx.doi.org/10.1111/j.1521-0391.2014.12094.x.

Egner, T., Etkin, A., Gale, S., Hirsch, J., 2008. Dissociable neural systems resolve conflict from emotional versus nonemotional distracters. Cereb. Cortex 18 (6), 1475–1484. doi:http://dx.doi.org/10.1093/cercor/bhm179.

Fernie, G., Peeters, M., Gullo, M.J., Christiansen, P., Cole, J.C., Sumnall, H., Field, M., 2013. Multiple behavioural impulsivity tasks predict prospective alcohol involvement in adolescents. Addiction 108 (11), 1916–1923. doi:http://dx.doi.org/10.1111/add.12283.

Filbey, F.M., Claus, E.D., Morgan, M., Forester, G.R., Hutchison, K., 2012. Dopaminergic genes modulate response inhibition in alcohol abusing adults. Addict. Biol. 17 (6), 1046–1056. doi:http://dx.doi.org/10.1111/j.1369-1600.2011.00328.x.

Fineberg, N.A., Chamberlain, S.R., Goudriaan, A.E., Stein, D.J., Vanderschuren, L.J., Gillan, C.M., et al., 2014. New developments in human neurocognition: clinical, genetic, and brain imaging correlates of impulsivity and compulsivity. CNS Spectr. 19 (1), 69–89. doi:http://dx.doi.org/10.1017/s1092852913000801.

Garavan, H., Brennan, K.L., Hester, R., Whelan, R., 2013. The neurobiology of successful abstinence. Curr. Opin. Neurobiol. 23 (4), 668–674. doi:http://dx.doi.org/10.1016/j.conb.2013.01.029.

Ghahremani, D.G., Lee, B., Robertson, C.L., Tabibnia, G., Morgan, A.T., De Shetler, N., et al., 2012. Striatal dopamine D(2)/D(3) receptors mediate response inhibition and related activity in frontostriatal neural circuitry in humans. J. Neurosci. 32 (21), 7316–7324. doi:http://dx.doi.org/10.1523/jneurosci.4284-11.2012.

Goldstein, R.Z., Volkow, N.D., 2011. Dysfunction of the prefrontal cortex in addiction: neuroimaging findings and clinical implications. Nat. Rev. Neurosci. 12 (11), 652–669. doi:http://dx.doi.org/10.1038/nrn3119.

Goldstein, R.Z., Leskovjan, A.C., Hoff, A.L., Hitzemann, R., Bashan, F., Khalsa, S.S., et al., 2004. Severity of neuropsychological impairment in cocaine and alcohol addiction: association with metabolism in the prefrontal cortex. Neuropsychologia 42 (11), 1447–1458.

Goldstein, R.Z., Woicik, P.A., Lukasik, T., Maloney, T., Volkow, N.D., 2007. Drug fluency: a potential marker for cocaine use disorders. Drug Alcohol Depend. 89 (1), 97–101.

Goldstein, R.Z., Woicik, P.A., Maloney, T., Tomasi, D., Alia-Klein, N., Shan, J., Volkow, N.D., 2010. Oral methylphenidate normalizes cingulate activity in cocaine addiction during a salient cognitive task. Proc. Natl. Acad. Sci. USA 107 (38), 16667–16672. doi:http://dx.doi.org/10.1073/pnas.1011455107.

Hardee, J.E., Weiland, B.J., Nichols, T.E., Welsh, R.C., Soules, M.E., Steinberg, D.B., et al., 2014. Development of impulse control circuitry in children of alcoholics. Biol. Psychiatry 76 (9), 708–716. doi:http://dx.doi.org/10.1016/j.biopsych.2014.03.005.

Heitzeg, M.M., Nigg, J.T., Hardee, J.E., Soules, M., Steinberg, D., Zubieta, J.K., Zucker, R.A., 2014. Left middle frontal gyrus response to inhibitory errors in children prospectively predicts early problem substance use. Drug Alcohol Depend. 141, 51–57. doi:http://dx.doi.org/10.1016/j.drugalcdep.2014.05.002.

Heitzeg, M.M., Cope, L.M., Martz, M.E., Hardee, J.E., 2015. Neuroimaging risk markers for substance abuse: recent findings on inhibitory control and reward system functioning. Curr. Addict. Rep. 2 (2), 91–103.

Hester, R., Nestor, L., Garavan, H., 2009. Impaired error awareness and anterior cingulate cortex hypoactivity in chronic cannabis users. Neuropsychopharmacology 34, 2450–2458.

Jakubczyk, A., Klimkiewicz, A., Kopera, M., Krasowska, A., Wrzosek, M., Matsumoto, H., et al., 2013. The CC genotype in the T102C HTR2A polymorphism predicts relapse in individuals after alcohol treatment. J. Psychiatr. Res. 47 (4), 527–533. doi:http://dx.doi.org/10.1016/j.jpsychires.2012.12.004.

Jasinska, A.J., Stein, E.A., Kaiser, J., Naumer, M.J., Yalachkov, Y., 2014. Factors modulating neural reactivity to drug cues in addiction: a survey of human neuroimaging studies. Neurosci. Biobehav. Rev. 38, 1–16. doi:http://dx.doi.org/10.1016/j.neubiorev.2013.10.013.

Kalivas, P.W., Volkow, N.D., 2005. The neural basis of addiction: a pathology of motivation and choice. Am. J. Psychiatry 162 (8), 1403–1413.

Kehagia, A.A., Murray, G.K., Robbins, T.W., 2010. Learning and cognitive flexibility: frontostriatal function and monoaminergic modulation. Curr. Opin. Neurobiol. 20 (2), 199–204. doi:http://dx.doi.org/10.1016/j.conb.2010.01.007.

Kerns, J.G., Cohen, J.D., MacDonald 3rd, A.W., Cho, R.Y., Stenger, V.A., Carter, C.S., 2004. Anterior cingulate conflict monitoring and adjustments in control. Science 303 (5660), 1023–1026.

Konova, A.B., Moeller, S.J., Tomasi, D., Volkow, N.D., Goldstein, R.Z., 2013. Effects of methylphenidate on resting-state functional connectivity of the mesocorticolimbic dopamine pathways in cocaine addiction. JAMA Psychiatry 70 (8), 857–868. doi:http://dx.doi.org/10.1001/jamapsychiatry.2013.1129.

Lee, S.S., Humphreys, K.L., Flory, K., Liu, R., Glass, K., 2011. Prospective association of childhood attention-deficit/hyperactivity disorder (ADHD) and substance use and abuse/dependence: a meta-analytic review. Clin. Psychol. Rev. 31 (3), 328–341. doi:http://dx.doi.org/10.1016/j.cpr.2011.01.006.

Li, C.S., Morgan, P.T., Matuskey, D., Abdelghany, O., Luo, X., Chang, J.L., Malison, R.T., 2010. Biological markers of the effects of intravenous methylphenidate on improving inhibitory control in cocaine-dependent patients. Proc. Natl. Acad. Sci. USA 107 (32), 14455–14459.

Luijten, M., Machielsen, M.W., Veltman, D.J., Hester, R., de Haan, L., Franken, I.H., 2014. Systematic review of ERP and fMRI studies investigating inhibitory control and error processing in people with substance dependence and behavioural addictions. J. Psychiatry Neurosci. 39 (3), 149–169.

Luo, X., Zhang, S., Hu, S., Bednarski, S.R., Erdman, E., Farr, O.M., et al., 2013. Error processing and gender-shared and -specific neural predictors of relapse in cocaine dependence. Brain 136 (Pt 4), 1231–1244. doi:http://dx.doi.org/10.1093/brain/awt040.

MacLeod, C.M., 1991. Half a century of research on the Stroop effect: an integrative review. Psychol. Bull. 109 (2), 163–203.

Mahmood, O.M., Goldenberg, D., Thayer, R., Migliorini, R., Simmons, A.N., Tapert, S.F., 2013. Adolescents' fMRI activation to a response inhibition task predicts future substance use. Addict. Behav. 38 (1), 1435–1441. doi:http://dx.doi.org/10.1016/j.addbeh.2012.07.012.

Marhe, R., Luijten, M., van de Wetering, B.J., Smits, M., Franken, I.H., 2013. Individual differences in anterior cingulate activation associated with attentional bias predict cocaine use after treatment. Neuropsychopharmacology 38 (6), 1085–1093. doi:http://dx.doi.org/10.1038/npp.2013.7.

Melchior, M., Prokofyeva, E., Younes, N., Surkan, P.J., Martins, S.S., 2014. Treatment for illegal drug use disorders: the role of comorbid mood and anxiety disorders. BMC Psychiatry 14, 89. doi:http://dx.doi.org/10.1186/1471-244x-14-89.

Mitchell, M.R., Balodis, I.M., Devito, E.E., Lacadie, C.M., Yeston, J., Scheinost, D., et al., 2013. A preliminary investigation of Stroop-related intrinsic connectivity in cocaine dependence: associations with treatment outcomes. Am. J. Drug Alcohol Abuse 39 (6), 392–402. doi:http://dx.doi.org/10.3109/00952990.2013.841711.

Moeller, S.J., Goldstein, R.Z., 2014. Impaired self-awareness in human addiction: deficient attribution of personal relevance. Trends Cogn. Sci. 18 (12), 635–641. doi:http://dx.doi.org/10.1016/j.tics.2014.09.003.

Moeller, S.J., Maloney, T., Parvaz, M.A., Dunning, J.P., Alia-Klein, N., Woicik, P.A., et al., 2009. Enhanced choice for viewing cocaine pictures in cocaine addiction. Biol. Psychiatry 66 (2), 169–176. doi:http://dx.doi.org/10.1016/j.biopsych.2009.02.015.

Moeller, S.J., Tomasi, D., Honorio, J., Volkow, N.D., Goldstein, R.Z., 2012a. Dopaminergic involvement during mental fatigue in health and cocaine addiction. Transl. Psychiatry 2, e176. doi:http://dx.doi.org/10.1038/tp.2012.110.

Moeller, S.J., Tomasi, D., Woicik, P.A., Maloney, T., Alia-Klein, N., Honorio, J., et al., 2012b. Enhanced midbrain response at 6-month follow-up in cocaine addiction, association with reduced drug-related choice. Addict. Biol. 17 (6), 1013–1025. doi:http://dx.doi.org/10.1111/j.1369-1600.2012.00440.x.

Moeller, S.J., Honorio, J., Tomasi, D., Parvaz, M.A., Woicik, P.A., Volkow, N.D., Goldstein, R.Z., 2014. Methylphenidate enhances executive function and optimizes prefrontal function in both health and cocaine addiction. Cereb. Cortex 24 (3), 643–653. doi:http://dx.doi.org/10.1093/cercor/bhs345.

Monte, A.A., Zane, R.D., Heard, K.J., 2015. The implications of marijuana legalization in Colorado. JAMA 313 (3), 241–242. doi:http://dx.doi.org/10.1001/jama.2014.17057.

Nigg, J.T., Wong, M.M., Martel, M.M., Jester, J.M., Puttler, L.I., Glass, J.M., et al., 2006. Poor response inhibition as a predictor of problem drinking and illicit drug use in adolescents at risk for alcoholism and other substance use disorders. J. Am. Acad. Child Adolesc. Psychiatry 45 (4), 468–475. doi:http://dx.doi.org/10.1097/01.chi.0000199028.76452.a9.

Niv, Y., Daw, N.D., Joel, D., Dayan, P., 2007. Tonic dopamine: opportunity costs and the control of response vigor. Psychopharmacology (Berl) 191 (3), 507–520.

Norman, A.L., Pulido, C., Squeglia, L.M., Spadoni, A.D., Paulus, M.P., Tapert, S.F., 2011. Neural activation during inhibition predicts initiation of substance use in adolescence. Drug Alcohol Depend. 119 (3), 216–223. doi:http://dx.doi.org/10.1016/j.drugalcdep.2011.06.019.

Paus, T., 2010. Population neuroscience: why and how. Hum. Brain Mapp. 31 (6), 891–903. doi:http://dx.doi.org/10.1002/hbm.21069.

Peeters, M., Monshouwer, K., van de Schoot, R.A., Janssen, T., Vollebergh, W.A., Wiers, R.W., 2013. Automatic processes and the drinking behavior in early adolescence: a prospective study. Alcohol. Clin. Exp. Res. 37 (10), 1737–1744. doi:http://dx.doi.org/10.1111/acer.12156.

Perry, R.I., Krmpotich, T., Thompson, L.L., Mikulich-Gilbertson, S.K., Banich, M.T., Tanabe, J., 2013. Sex modulates approach systems and impulsivity in substance dependence. Drug Alcohol Depend. 133 (1), 222–227. doi:http://dx.doi.org/10.1016/j.drugalcdep.2013.04.032.

Prisciandaro, J.J., Myrick, H., Henderson, S., McRae-Clark, A.L., Brady, K.T., 2013. Prospective associations between brain activation to cocaine and no-go cues and cocaine relapse. Drug Alcohol Depend. 131 (1-2), 44–49. doi:http://dx.doi.org/10.1016/j.drugalcdep.2013.04.008.

Rubia, K., Halari, R., Cubillo, A., Smith, A.B., Mohammad, A.M., Brammer, M., Taylor, E., 2011. Methylphenidate normalizes fronto-striatal underactivation during interference inhibition in medication-naive boys with attention-deficit hyperactivity disorder. Neuropsychopharmacology 36 (8), 1575–1586. doi:http://dx.doi.org/10.1038/npp.2011.30.

Satoh, T., Nakai, S., Sato, T., Kimura, M., 2003. Correlated coding of motivation and outcome of decision by dopamine neurons. J. Neurosci. 23 (30), 9913–9923.

Schmaal, L., Joos, L., Koeleman, M., Veltman, D.J., van den Brink, W., Goudriaan, A.E., 2013. Effects of modafinil on neural correlates of response inhibition in alcohol-dependent patients. Biol. Psychiatry 73 (3), 211–218. doi:http://dx.doi.org/10.1016/j.biopsych.2012.06.032.

Schrager, S.M., Kecojevic, A., Silva, K., Jackson Bloom, J., Iverson, E., Lankenau, S.E., 2014. Correlates and consequences of opioid misuse among high-risk young adults. J. Addict. 2014, 156954. doi:http://dx.doi.org/10.1155/2014/156954.

Smith, D.G., Ersche, K.D., 2014. Using a drug-word Stroop task to differentiate recreational from dependent drug use. CNS Spectr. 19 (3), 247–255. doi:http://dx.doi.org/10.1017/s1092852914000133.

Smith, J.L., Mattick, R.P., Jamadar, S.D., Iredale, J.M., 2014. Deficits in behavioural inhibition in substance abuse and addiction: a meta-analysis. Drug Alcohol Depend. 145, 1–33. doi:http://dx.doi.org/10.1016/j.drugalcdep.2014.08.009.

Sofuoglu, M., Devito, E.E., Waters, A.J., Carroll, K.M., 2013. Cognitive enhancement as a treatment for drug addictions. Neuropharmacology 64 (1), 452–463. doi:http://dx.doi.org/10.1016/j.neuropharm.2012.06.021.

Verbruggen, F., Logan, G.D., 2008. Response inhibition in the stop-signal paradigm. Trends Cogn. Sci. 12 (11), 418–424. doi:http://dx.doi.org/10.1016/j.tics.2008.07.005.

Verdejo-Garcia, A., Betanzos-Espinosa, P., Lozano, O.M., Vergara-Moragues, E., Gonzalez-Saiz, F., Fernandez-Calderon, F., et al., 2012. Self-regulation and treatment retention in cocaine dependent individuals: a longitudinal study. Drug Alcohol Depend. 122 (1-2), 142–148. doi:http://dx.doi.org/10.1016/j.drugalcdep.2011.09.025.

Volkow, N.D., Chang, L., Wang, G.J., Fowler, J.S., Franceschi, D., Sedler, M., Logan, J., 2001. Loss of dopamine transporters in methamphetamine abusers recovers with protracted abstinence. J. Neurosci. 21 (23), 9414–9418.

Wetherill, R.R., Squeglia, L.M., Yang, T.T., Tapert, S.F., 2013. A longitudinal examination of adolescent response inhibition: neural differences before and after the initiation of heavy drinking. Psychopharmacology (Berl) 230 (4), 663–671. doi:http://dx.doi.org/10.1007/s00213-013-3198-2.

Whelan, R., Watts, R., Orr, C.A., Althoff, R.R., Artiges, E., Banaschewski, T., et al., 2014. Neuropsychosocial profiles of current and future adolescent alcohol misusers. Nature 512 (7513), 185–189. doi:http://dx.doi.org/10.1038/nature13402.

Woicik, P.A., Moeller, S.J., Alia-Klein, N., Maloney, T., Lukasik, T.M., Yeliosof, O., et al., 2009. The neuropsychology of cocaine addiction: recent cocaine use masks impairment. Neuropsychopharmacology 34 (5), 1112–1122.

Wong, M.M., Brower, K.J., Nigg, J.T., Zucker, R.A., 2010. Childhood sleep problems, response inhibition, and alcohol and drug outcomes in adolescence and young adulthood. Alcohol. Clin. Exp. Res. 34 (6), 1033–1044. doi:http://dx.doi.org/10.1111/j.1530-0277.2010.01178.x.

Worhunsky, P.D., Stevens, M.C., Carroll, K.M., Rounsaville, B.J., Calhoun, V.D., Pearlson, G.D., Potenza, M.N., 2013. Functional brain networks associated with cognitive control, cocaine dependence, and treatment outcome. Psychol. Addict. Behav. 27 (2), 477–488. doi:http://dx.doi.org/10.1037/a0029092.

PART

2

Drugs and Behaviours for Addiction Medicine

CHAPTER

Neuroscience of nicotine for addiction medicine: novel targets for smoking cessation medications

10

Manoranjan S. D'Souza[1]

Department of Biomedical and Pharmaceutical Sciences, The Raabe College of Pharmacy, Ohio Northern University, Ada, OH, USA

[1]Corresponding author: Tel.: +1-419-772-3950; Fax: +1-419-772-1917, e-mail address: m-dsouza@onu.edu

Abstract

Morbidity and mortality associated with tobacco smoking constitutes a significant burden on healthcare budgets all over the world. Therefore, promoting smoking cessation is an important goal of health professionals and policy makers throughout the world. Nicotine is a major psychoactive component in tobacco that is largely responsible for the widespread addiction to tobacco. A majority of the currently available FDA-approved smoking cessation medications act via neuronal nicotinic receptors. These medications are effective in approximately half of all the smokers, who want to quit and relapse among abstinent smokers continues to be high. In addition to relapse among abstinent smokers, unpleasant effects associated with nicotine withdrawal are a major motivational factor in continued tobacco smoking. Over the last two decades, animal studies have helped in identifying several neural substrates that are involved in nicotine-dependent behaviors including those associated with nicotine withdrawal and relapse to tobacco smoking. In this review, first the role of specific brain regions/circuits that are involved in nicotine dependence will be discussed. Next, the review will describe the role of specific nicotinic receptor subunits in nicotine dependence. Finally, the review will discuss the role of classical neurotransmitters (dopamine, serotonin, noradrenaline, glutamate, and γ-aminobutyric acid) as well as endogenous opioid and endocannabinoid signaling in nicotine dependence. The nicotinic and nonnicotinic neural substrates involved in nicotine-dependent behaviors can serve as possible targets for future smoking cessation medications.

Keywords

Glutamate, GABA, Nicotine withdrawal, Reinstatement, Opioids, Endocannabinoids, Serotonin, Habenula, Nucleus accumbens, Interpeduncular nucleus

Abbreviations

AMPA	amino-3-hydroxy-5-methyl-4-isoxazolepropionate/kainate
CRF	corticotropin-releasing factor
FAAH	fatty acid amide hydrolase
GABA	γ-aminobutyric acid
ICSS	intracranial self-stimulation
mGlu	metabotropic glutamate
NAcc	nucleus accumbens
nAChRs	nicotinic acetylcholine receptors
NMDA	*N*-methyl-D-aspartate
nor-BNI	norbinaltorphimine
PAMs	positive allosteric modulators
VTA	ventral tegmental area

1 INTRODUCTION

Currently, there are approximately 1.3 billion adult smoker's worldwide, making tobacco addiction one of the most prevalent addictions all over the world. Importantly, mortality attributed to tobacco smoking is estimated to rise to approximately 8 million by 2030 (WHO, 2011). Cessation of tobacco smoking can reverse some of the adverse health outcomes associated with tobacco smoking (Fagerstrom, 2002). Although a significant number of smokers are willing to quit, few succeed without professional help. In fact, most smokers, who attempt to quit on their own without professional help, will relapse within the first 6 months (Hughes et al., 2004).

Nicotine is the major psychoactive component of tobacco smoke (Stolerman and Jarvis, 1995). The effects of nicotine are mediated by neuronal nicotinic acetylcholine receptors (nAChRs) and current first-line Food and Drug Administration (FDA)-approved smoking cessation medications such as varenicline and nicotine replacement therapies target these nAChRs (Nides, 2008; Rennard and Daughton, 2014; see Table 1). However, these FDA-approved medications are not effective in all smokers who express a desire to quit. Thus, to improve the overall smoking cessation rates, it is imperative to identify neurobiological substrates that play a role in continued nicotine seeking. In this review, the different phases of nicotine dependence will be discussed. In addition, the review will describe specific neurobiological substrates that play a role in continued nicotine seeking and may serve as targets for new smoking cessation medications and improve quit rates among smokers.

2 PHASES OF NICOTINE DEPENDENCE

The reinforcing effects of nicotine play an important role in the initiation and maintenance of tobacco smoking in humans and nicotine seeking in animals. In humans, nicotine intake through tobacco smoking produces a pleasurable rush, mild euphoria,

Table 1 FDA-Approved Smoking Cessation Medications and Their Molecular Mechanism of Action

Place in Therapy	Target Receptor	Drug	Molecular Mechanism of Action
First-line	Nicotinic	Nicotine replacement products (gum, lozenge, patch, nasal spray, inhaler)	Stimulation of nicotinic acetylcholine receptors (nAChRs)
		Varenicline	Partial agonist of $\alpha4\beta2$ nAChRs
	Nicotinic and nonnicotinic	Bupropion	Weak inhibitor of dopamine/noradrenaline uptake transporters, nAChR antagonist?
Second-line	Nonnicotinic	Clonidine	$\alpha2$-adrenergic agonist
		Nortriptyline	Inhibitor of noradrenaline uptake transporters

increased arousal, decreased fatigue, and relaxation (Henningfield et al., 1985). More importantly, nicotine-dependent smokers, who quit smoking, experience unpleasant affective effects such as depressed mood, anhedonia, dysphoria, anxiety, irritability, difficulty in concentrating, and craving (Shiffman and Jarvik, 1976). In addition, some smokers may develop mild "physical" or somatic withdrawal manifestations such as bradycardia, insomnia, gastrointestinal discomfort, and weight gain (Hughes et al., 1991). These unpleasant effects are mediated by nicotine-induced neuroadaptations at distinct sites/nuclei in the brain (see below). Like in humans, withdrawal of nicotine in nicotine-dependent animals produces somatic and affective disturbances (see Table 2). Abstinent smokers often succumb to intense cravings and reinitiate smoking, which is termed as relapse. Relapse often occurs in response to stress, nicotine/tobacco smoke, and/or environmental stimuli associated with nicotine (Carmody, 1992). In animals too, stress, nicotine and nicotine-associated cues, and contexts result in reinstatement of nicotine seeking after a period of nicotine withdrawal and extinction training (Shaham et al., 2003; Stoker and Markou, 2015; see Table 2). Overall, the unpleasant effects associated with nicotine withdrawal and relapse play an important role in continued nicotine seeking. A description of animal models used to study the neurobiology of nicotine withdrawal and relapse to nicotine seeking in humans is described in Table 2. The next sections will describe the advances in our understanding of the neural mechanisms underlying the unpleasant effects of nicotine withdrawal and relapse to nicotine seeking using animal models.

3 NEUROCIRCUITRY UNDERLYING THE DEVELOPMENT OF NICOTINE DEPENDENCE

The reinforcing effects of nicotine are mediated by the mesocorticolimbic dopaminergic neurons, which originate in the ventral tegmental area (VTA) and project to several cortical and subcortical sites including the prefrontal cortex (PFC),

Table 2 Animal Models Commonly Used to Assess Different Nicotine-Dependent Behaviors

Phase of Nicotine Dependence	Nicotine-Dependent Behavior	Model	Description	Data Interpretation	References (Example)
Nicotine withdrawal	Somatic withdrawal effects	Somatic signs of nicotine withdrawal	Animals are first made dependent on nicotine using subcutaneously implanted osmotic minipumps over a period of time (7 days–4 weeks) and then subsequently nicotine withdrawal is induced by removal of the osmotic pumps (spontaneous withdrawal) or administration of a nicotinic receptor antagonist (precipitated withdrawal). Animals are observed for signs of nicotine withdrawal. Withdrawal effects are compared between nicotine- and saline-treated (control) animals. The somatic signs occurring as a result of nicotine withdrawal are scored by an independent observer	During withdrawal from nicotine, nicotine-dependent animals show somatic signs, which include rearing, jumping, wet-dog shakes, scratching, front paw tremors, ptosis, abdominal constrictions, chewing, teeth chattering, and piloerection. Nicotine-dependent animals show more pronounced and significantly greater number of nicotine withdrawal-associated somatic signs compared to saline-treated animals	Jackson et al. (2013)
	Affective (anhedonia-like state)	Nicotine withdrawal-induced anhedonia-like state using	Animals are surgically implanted with electrodes directed at the medial forebrain bundle/lateral hypothalamus. Animals are subsequently	Elevation of reward thresholds compared to baseline in nicotine-dependent animals during nicotine withdrawal is indicative of development of an	Epping-Jordan et al. (1998)

		intracranial self-stimulation (ICSS)	trained to self-administer an electric current to the brain over a wide range of intensities. Once the animals are trained, a reward threshold is determined for each animal. Reward threshold is defined as the minimum current that is perceived as rewarding by the animal. Animals are then made dependent on nicotine using osmotic minipumps as described above. Reward thresholds during nicotine withdrawal are compared to baseline reward thresholds of the animals in a nicotine-dependent state. Reward thresholds are also compared between nicotine- and saline-treated animals	aversive anhedonia-like state in animals	
	Affective (anxiety-like state)	Nicotine withdrawal-induced anxiety-like state using the elevated plus maze	Animals are made dependent on nicotine using osmotic minipumps as described above. Nicotine withdrawal is induced either by administering a nicotinic receptor antagonist or by removing the nicotine pumps. During withdrawal from nicotine, animals are placed in an elevated plus maze	During nicotine withdrawal, nicotine-dependent animals compared to saline-treated animals spend significantly less time in the open arm and demonstrate fewer open arm entries. Together, these behaviors indicate the development of an anxiety-like state	Jackson et al. (2009)

Continued

Table 2 Animal Models Commonly Used to Assess Different Nicotine-Dependent Behaviors—cont'd

Phase of Nicotine Dependence	Nicotine-Dependent Behavior	Model	Description	Data Interpretation	References (Example)
			consisting of two open arms and two closed arms. Time spent by the animals in the open arm and number of entries into the open from the closed arm are compared between animals withdrawing from nicotine versus saline-treated animals (control)		
	Affective (generalized aversion)	Nicotine withdrawal-induced aversive state using conditioned place aversion (CPA)	Animals are made dependent on nicotine using subcutaneously implanted osmotic minipumps. Several days after pump implantation (at least 7 days; sometimes up to 14 days), animals are conditioned using two distinct environments/compartments. During the conditioning procedure, animals are pretreated with the nicotinic receptor antagonist and placed in one of the compartments. During the next conditioning session, animals are pretreated with saline (control) and placed in a distinct compartment. The rationale behind this procedure is that animals begin to	Nicotine-dependent animals compared to saline-treated animals spend significantly less time in the compartment associated with the nicotinic receptor antagonist indicating the development of CPA	Jackson et al. (2009) and Shram et al. (2008)

			associate the aversive effects of nicotine withdrawal induced via administration of the nicotinic receptor antagonist with the specific compartment they are confined to. The number of conditioning sessions can vary depending on the study. On the test day, animals are given access to both compartments, and time spent by the animals in the compartment associated with the nicotinic receptor antagonist is compared to the time spent by the animals in the saline-associated compartment		
Relapse	Reinstatement of nicotine seeking	Cue-induced reinstatement of nicotine seeking	In this model, animals are surgically prepared with intravenous catheters and are trained using operant boxes to self-administer nicotine by pressing a lever (acquisition of self-administration). Once the animals have demonstrated stable nicotine self-administration behavior over 3–4 weeks, animals undergo extinction training. During extinction training, animals are placed in the same operant boxes with access to the lever,	Lever responding during the cue-induced reinstatement session is significantly greater than lever responding during the last day of extinction training, indicating cue-induced reinstatement of nicotine seeking	D'Souza and Markou (2014)

Continued

Table 2 Animal Models Commonly Used to Assess Different Nicotine-Dependent Behaviors—cont'd

Phase of Nicotine Dependence	Nicotine-Dependent Behavior	Model	Description	Data Interpretation	References (Example)
			but lever pressing does not result in delivery of nicotine or cues associated with nicotine. Animals show decreased lever responding during extinction training. Finally, during the reinstatement session, animals are presented with cues associated with nicotine and the lever, which results in increased lever responding which is termed as reinstatement of nicotine seeking. Lever presses during the reinstatement session results in presentation of cues associated with nicotine but no nicotine delivery		
		Stress-induced reinstatement of nicotine seeking	In this procedure, the acquisition of nicotine self-administration and extinction training are similar to that used in the cue-induced reinstatement procedure. However, prior to the reinstatement session, animals are first subjected to stress	Lever responding during the stress-induced reinstatement session is significantly greater than lever responding during the last day of extinction training, indicating stress-induced reinstatement of nicotine seeking	Grella et al. (2014) and Qi et al. (2015)

			using footshocks or yohimbine (an $\alpha 2$-adrenergic antagonist) and then placed in the operant chamber with access to the lever. Lever presses during the reinstatement session results in presentation of cues associated with nicotine but no nicotine delivery		
		Nicotine-induced reinstatement of nicotine seeking	In this procedure, the acquisition of nicotine self-administration and extinction training are similar to that used in the cue-induced reinstatement procedure. However, either prior or during the reinstatement session, animals are administered a single injection of nicotine and then given access to the lever. Lever presses during the reinstatement session results in presentation of cues associated with nicotine but no nicotine delivery	Lever responding during the nicotine-induced reinstatement session is significantly greater than lever responding during the last day of extinction training, indicating nicotine-induced reinstatement of nicotine seeking	Forget et al. (2009)

hippocampus, habenula, nucleus accumbens (NAcc), and amygdala (De Biasi and Dani, 2011; Koob and Volkow, 2010; Picciotto and Mineur, 2014). The activity of these mesocorticolimbic dopaminergic neurons is regulated by reciprocal glutamatergic projections from the PFC, amygdala, lateral hypothalamus, and habenula (Grace et al., 2007; Watabe-Uchida et al., 2012). In addition, the mesocorticolimbic dopaminergic neurons receive inhibitory γ-aminobutyric acid (GABA) projections from the NAcc, rostromedial tegmental nucleus, and local circuit GABA neurons (Geisler and Zahm, 2005). The mesocorticolimbic dopamine neurons also receive cholinergic afferents from the pedunculopontine tegmentum and lateral dorsal tegmentum (Jhou et al., 2009).

The withdrawal effects of nicotine are mediated by changes in dopamine levels in mesocorticolimbic brain sites such as the NAcc. Withdrawal of nicotine decreased dopamine levels in the NAcc compared to baseline in nicotine-dependent rats, possibly due to increased expression of the dopamine uptake transporter (Hadjiconstantinou et al., 2011; Hildebrand et al., 1998). Interestingly, the decrease in NAcc dopamine levels after administration of the nAChR antagonist mecamylamine to nicotine-dependent rats was more pronounced in adolescent rats compared to adult rats, suggesting that the mesocorticolimbic dopamine system in adolescent rats may be more susceptible to neuroadaptive changes compared to adult rats (Natividad et al., 2009). Nicotine withdrawal also resulted in increased expression of corticotrophin-releasing hormone mRNA in the mesocorticolimbic dopaminergic neurons located in the posterior VTA (Grieder et al., 2014). Furthermore, the same study showed that blockade of corticotropin-releasing factor (CRF) 1 receptors in the posterior VTA in mice attenuated the unpleasant affective effects associated with nicotine withdrawal. In addition to the mesocorticolimbic dopaminergic circuitry, a significant body of evidence has described the role of the habenulo-interpeduncular circuit in mediating the withdrawal effects of nicotine (Velasquez et al., 2014). In fact, during nicotine withdrawal, GABAergic neurons in the interpeduncular nucleus are activated by glutamatergic afferents from the medial habenula (Zhao-Shea et al., 2013). In summary, above-described studies suggest that the mesocorticolimbic dopaminergic neurons and the habenulo-interpeduncular circuit play an important role in mediating the unpleasant affective symptoms associated with nicotine withdrawal.

Exposure to environmental cues and contexts among abstinent smokers can result in intense cravings and reactivation of nicotine-associated memories leading to relapse. Thus, associative learning between nicotine and environmental cues plays an important role in the process of relapse (Stoker and Markou, 2015). As described above, the mesocorticolimbic dopaminergic neurons from the VTA project to a number of cortical and limbic nuclei, such as the PFC, amygdala, hippocampus, and NAcc, that are involved in the process of learning between environmental cues and nicotine, as well as emotional responses to nicotine-associated stimuli. Empirical evidence supports a role for several cortical structures including the insular cortex, medial PFC, hippocampus in the process of drug craving, and relapse (Koob and Volkow, 2010). In addition, subcortical structures such as the NAcc core and VTA, which receive excitatory inputs from the cortical structures described above, are implicated in the reinstatement of nicotine seeking (Gipson et al., 2013;

Grieder et al., 2014). In summary, even though several distinct brain nuclei/regions are involved in the different nicotine-dependent behaviors, it must be emphasized that these brain regions closely regulate each other and often form parts of overlapping circuitries. We will now describe the role of specific neurotransmitters and receptors in the above-described brain regions in nicotine withdrawal and reinstatement of nicotine seeking.

4 NEURAL SUBSTRATES UNDERLYING NICOTINE DEPENDENCE

4.1 CHOLINERGIC NEUROTRANSMISSION AND NICOTINE DEPENDENCE

The action of endogenous acetylcholine in the brain is mediated by neuronal nAChRs, which are widely distributed in the brain. Neuronal nAChRs are excitatory ligand-gated ion channels composed of five subunits, which come together to form a functional receptor (Dani and Bertrand, 2007). Most of the neuronal nAChRs are heteromeric and composed of different isoforms of alpha (α2–α9) and beta (β1–β4) subunits. Heteromeric nAChRs have a lot of diversity and can vary in their pharmacological responses based on the particular combination of alpha and beta receptor subunits and differences in stoichiometry of nAChRs (Changeux, 2010; Zoli et al., 2015). For example, the pharmacological response of a neuronal nAChR having two α4 subunits and three β2 subunits $[(\alpha4)_2(\beta2)_3]$ will be different from a nAChR having three α4 subunits and two β2 subunits $[(\alpha4)_3(\beta2)_2]$. In contrast, some neuronal nAChRs are homomeric and consist of five α7 subunits. Importantly, chronic exposure of nicotine induces neuroadaptations in expression of nAChRs resulting in either upregulation or downregulation of the nAChRs in different parts of the brain (Colombo et al., 2013). Finally, the neuronal nAChRs exist as heteroreceptors on presynaptic terminals of a number of neurotransmitters including dopamine, serotonin, glutamate, and GABA and can alter the release of these neurotransmitters when activated by nicotine/acetylcholine (Balfour, 2009). NAChR subunits also play an important role in the reinforcing effects of nicotine. However, the discussion of nAChR subunits mediating the reinforcing effects of nicotine is beyond the scope of this review, and the reader is referred to other scholarly work for a discussion on this topic (D'Souza and Markou, 2011; Li et al., 2014; Picciotto et al., 1998). In this review, we will mainly discuss the role of the nAChR subunits in the withdrawal effects of nicotine and reinstatement of nicotine seeking.

Several specific nAChR subunits mediate the unpleasant somatic and affective effects associated with nicotine withdrawal in nicotine-dependent animals. For example, knockout of either the β4, α2, α5, or α7 subunits attenuated the somatic signs of nicotine withdrawal in nicotine-dependent mice compared to their respective wild-type counterparts (Jackson et al., 2008; Salas et al., 2004, 2009). The nAChRs also mediate the unpleasant affective symptoms associated with nicotine withdrawal.

Both systemic and intra-VTA administration of dihydrobetaerythroidine (DhβE), a nAChR antagonist selective for the α4β2-containing nAChRs, elevated intracranial self-stimulation (ICSS) thresholds in nicotine-dependent rats (Bruijnzeel and Markou, 2004; Epping-Jordan et al., 1998; see Table 2). This DhβE-induced elevation of ICSS thresholds in nicotine-dependent rats suggests the development of nicotine withdrawal-induced anhedonia-like state in animals. Interestingly, mice lacking either the β4 or α7 nAChR subunits showed delayed onset of nicotine withdrawal-induced anhedonia-like state (Stoker et al., 2012a). In addition, pharmacological blockade of the α6 subunit-containing nAChRs attenuated the nicotine withdrawal-induced anxiety and conditioned place aversion in mice, suggesting a role for α6 subunit-containing nAChRs in the unpleasant affective effects of nicotine withdrawal (Jackson et al., 2009). Overall, the data suggest that the unpleasant physical signs associated with nicotine withdrawal are mediated by the nAChRs that contain either the β4, α2, α5, or α7 subunits. Furthermore, the data suggest that the unpleasant affective signs associated with nicotine withdrawal are mediated by the nAChRs that contain the β2, β4, α4, α6, or α7 subunits.

The nAChRs are also involved in the reinstatement of nicotine seeking in animals, a putative model for relapse in humans. Administration of methyllycaconitine, a selective antagonist of the α7-containing nAChRs, attenuated cue-induced reinstatement of nicotine seeking in rats (Liu, 2014). NAChRs containing the α7 subunits are often found as heteroreceptors on presynaptic glutamate terminals, and activation of these α7-containing nAChRs increases glutamate release. A recent study has shown that blockade of glutamatergic transmission via *N*-methyl-D-aspartate (NMDA) receptors attenuated cue-induced reinstatement of nicotine seeking (Gipson et al., 2013). Therefore, the decrease in cue-induced reinstatement of nicotine seeking due to blockade of the α7-containing nAChRs could be due to blockade of release of other neurotransmitters such as glutamate. Varenicline, a partial agonist at the α4β2-containing nAChRs, attenuated nicotine-induced, but not cue-induced, reinstatement of nicotine seeking in rats (O'Connor et al., 2010). Taken together, data from the above-described studies suggest that nAChRs with different subunit compositions may have a differential role in reinstatement of nicotine seeking. For example, α7-containing nAChRs possibly mediate cue-induced reinstatement of nicotine seeking, while the α4β2-containing nAChRs possibly mediate the nicotine-induced reinstatement of nicotine seeking. Future studies using optogenetics and molecular biology tools are needed to identify the role of the nAChRs in specific neural circuits that are involved in the different types of reinstatement models described above (cue- vs. nicotine-induced reinstatement of nicotine seeking). In another study, galantamine, an acetylcholinesterase inhibitor and a nonselective positive modulator of nAChRs, attenuated cue-induced reinstatement of nicotine seeking (Hopkins et al., 2012). In summary, targeting specific nAChR subunits may help alleviate unpleasant effects of nicotine withdrawal and prevent relapse among abstinent smokers. Thus, medications targeting specific nAChR subunits may generate smoking cessation medications with better efficacy compared to currently approved FDA medications.

4.2 GLUTAMATE NEUROTRANSMISSION AND NICOTINE DEPENDENCE

Glutamate is the major excitatory neurotransmitter in the mammalian brain, and its action is mediated by fast-acting ionotropic and slow-acting G protein-coupled metabotropic receptors (Conn and Pin, 1997; Hollmann and Heinemann, 1994). The ionotropic receptors include NMDA, amino-3-hydroxy-5-methyl-4-isoxazolepropionate (AMPA), and kainate receptors. The metabotropic glutamate (mGlu) receptors are classified into Group I (mGlul and mGlu5), Group II (mGlu2 and mGlu3), and Group III (mGlu4, mGlu6, mGlu7, and mGlu8) receptors depending on their signal transduction pathways, sequence homology, and pharmacological selectivity (Pin and Duvoisin, 1995). Nicotine increases glutamate release by binding to excitatory α7-containing nAChRs located on presynaptic glutamatergic terminals (Mansvelder and McGehee, 2002).

It is hypothesized that chronic nicotine exposure leads to the development of a hypoglutamatergic state, which is responsible for the unpleasant affective state associated with nicotine withdrawal (Markou, 2008). Consistent with this hypothesis, blockade of glutamatergic transmission using either the AMPA receptor antagonist (NBQX) or the mGlu5 receptor antagonist (MPEP) resulted in the elevation of ICSS thresholds in nicotine-dependent rats, suggesting development of an anhedonia-like aversive state (Kenny et al., 2003; Liechti and Markou, 2007). Interestingly, nicotine-dependent mice lacking the mGlu5 receptor showed attenuated anhedonia-like aversive state during nicotine withdrawal compared to their wild-type counterparts (Stoker et al., 2012b). The difference in findings with respect to the mGlu5 receptors between the pharmacological and genetic studies described above could be attributed to neuroadaptations that are often seen in congenital knockout mice. Overall, the data suggest that blockade of glutamatergic transmission worsens the unpleasant affective effects associated with nicotine withdrawal and thus reversal of nicotine-induced hypoglutamatergic state may help alleviate the unpleasant effects of nicotine withdrawal.

Interestingly, increased glutamate levels have been reported following exposure to nicotine-associated cues in the core subdivision of the NAcc (Gipson et al., 2013). Blockade of the NMDA-mediated glutamatergic transmission after systemic administration of ifenprodil, a NMDA receptor antagonist, attenuated cue-induced reinstatement of nicotine seeking (Gipson et al., 2013). The same study also showed that injections of a NMDA receptor antagonist selective for the NR2A subunit (TCN-201) in the NAcc core attenuated cue-induced reinstatement of nicotine seeking. In contrast to these findings, blockade of the NMDA receptors in the NAcc core by the competitive NMDA receptor antagonist LY235959 increased cue-induced reinstatement of nicotine seeking compared to saline control (D'Souza and Markou, 2014). The differential effects observed between the above-described studies could be either due to selectivity of the different compounds (LY235959 vs. TCN-201) for the different NMDA receptor subunits and/or the differential localization (synaptic vs. extrasynaptic) of the NMDA receptors containing these different subunits. More work will be required to determine the precise role of the NMDA receptors in the

NAcc core in cue-induced reinstatement of nicotine seeking. Furthermore, cue-induced reinstatement of nicotine seeking was attenuated following blockade of glutamate transmission via manipulation of the mGlu5 and mGlu2/3 receptors (Bespalov et al., 2005; Liechti et al., 2007). In addition to glutamate receptors, the levels of synaptic glutamate are regulated by uptake transporter proteins such as the cystine–glutamate antiporter (Nicholls and Attwell, 1990). *N*-acetylcysteine, which binds to the cystine–glutamate antiporter and increases levels of extrasynaptic glutamate levels, attenuated the cue-induced reinstatement of nicotine seeking (Ramirez-Nino et al., 2013). These data support a role for the cystine–glutamate exchanger in the cue-induced reinstatement of nicotine seeking. It is hypothesized that the increase in extrasynaptic glutamate transmission may help reverse the hypoglutamatergic state induced by chronic nicotine exposure. In summary, several glutamatergic receptors described above can serve as targets for smoking cessation medications. The mGlu receptors may be better targets for smoking cessation medications compared to iGlu receptors, because the mGlu receptors subtly modulate glutamate transmission, and it is hypothesized that manipulation of the mGlu receptors will produce fewer side effects compared to the iGlu receptors (D'Souza and Markou, 2012).

4.3 GABA NEUROTRANSMISSION AND NICOTINE DEPENDENCE

GABA is the major inhibitory neurotransmitter in the brain. The actions of endogenously released GABA are mediated by ionotropic ($GABA_A$ and $GABA_C$) and metabotropic ($GABA_B$) receptors. Nicotine increases GABA neurotransmission by binding to excitatory nAChRs located on presynaptic GABA neurons. Importantly, the nAChRs located on GABAergic neurons undergo rapid desensitization, and chronic nicotine exposure is associated with decreased GABA transmission (Mansvelder et al., 2002). Consistent with this hypothesis, systemic administration of the $GABA_B$ agonist baclofen attenuated the somatic signs of nicotine withdrawal in nicotine-dependent mice (Varani et al., 2014). In contrast, both the $GABA_B$ receptor agonist (CGP44532) and the $GABA_B$ receptor positive allosteric modulator (PAM) (BHF177) elevated ICSS thresholds during nicotine withdrawal in animals, indicating an exacerbation of the anhedonia-like aversive effects associated with nicotine withdrawal (Vlachou et al., 2011b). Together, these data suggest that the $GABA_B$ receptors play a differential role in somatic versus affective manifestations of nicotine withdrawal. The differences between the studies could be attributed to methodological differences (measurement of affective vs. somatic signs) and differences in species (rats vs. mice) used in the above studies. Interestingly, the $GABA_B$ receptor antagonist (CGP56433) also exacerbated the anhedonia-like effects of nicotine withdrawal (Vlachou et al., 2011b). The similar effects of the $GABA_B$ receptor agonist/PAM and $GABA_B$ receptor antagonist on the unpleasant affective effects associated with nicotine withdrawal could be due to action of these compounds at the $GABA_B$ receptors located at differential sites such as the presynaptic versus postsynaptic neurons. In addition, these effects could also be explained by the differential action of these $GABA_B$ compounds at different $GABA_B$ subunits such as the $GABA_{B1}$ versus $GABA_{B2}$. Further work will be required to fully understand the role of the $GABA_B$ receptors in the affective and somatic effects associated with nicotine withdrawal. Finally, both the $GABA_B$ receptor agonists (CGP44532) and the $GABA_B$ receptor PAM attenuated cue-induced reinstatement of nicotine seeking (Paterson et al., 2005; Vlachou et al., 2011a). Overall, the data suggest that the $GABA_B$ receptors can serve as useful targets for smoking cessation medications.

4.4 ENDOGENOUS OPIOID SIGNALING AND NICOTINE DEPENDENCE

Endogenous opioids such as β-endorphin, met- and leu-enkephalin, and dynorphins are also involved in the development of nicotine dependence (Berrendero et al., 2010). The effects of β-endorphin, enkephalin, and dynorphin are mediated by three different subtypes of opioid receptors, namely, the mu, delta, and kappa-opioid receptors, respectively (Lutz and Pfister, 1992). Nicotine increases endogenous opioid neurotransmission by binding to nAChRs located on presynaptic terminals of neurons containing opioid peptides (Hadjiconstantinou and Neff, 2011). Administration of the nonselective opioid receptor antagonist naloxone, to nicotine-dependent rats and mice, precipitated the somatic signs associated with nicotine withdrawal (Biala et al., 2005; Malin et al., 1993). Naloxone also precipitated nicotine withdrawal in a small group of nicotine-dependent smokers (Krishnan-Sarin et al., 1999). It is hypothesized that this naloxone-precipitated nicotine withdrawal syndrome could be mediated by binding of naloxone to nAChRs. In contrast, nicotine-dependent mice lacking the mu-opioid receptors or preproenkephalin gene showed attenuated somatic signs of nicotine withdrawal as compared to their respective wild-type controls (Berrendero et al., 2002, 2005). The discrepancy between these genetic and pharmacological studies described above could be due to compensatory changes in mice lacking a specific protein during development. In contrast to the mu and delta opioid receptors, blockade of the kappa opioid receptors using norbinaltorphimine (nor-BNI) attenuated the unpleasant somatic and affective nicotine withdrawal signs in nicotine-dependent animals (Jackson et al., 2010; Tejeda et al., 2012; but also see Ise et al., 2002). Blockade of the kappa opioid receptors using nor-BNI attenuated stress-induced, but not cue-induced, reinstatement of nicotine seeking (Grella et al., 2014). In addition, naltrexone attenuated cue-induced reinstatement of nicotine seeking in rats (Liu et al., 2009). Overall, opioid receptors may be useful targets for alleviation of nicotine withdrawal symptoms and prevention of relapse among abstinent smokers.

4.5 ENDOCANNABINOID SIGNALING AND NICOTINE DEPENDENCE

The effects of endogenous cannabinoids anandamide and 2 arachidonylglycerol are mediated by the CB_1 and CB_2 cannabinoid receptors. Furthermore, the levels of endogenous cannabinoids can be increased by inhibiting their degradative enzymes such as fatty acid amide hydrolase (FAAH) and monoacylglycerol lipase. Administration of delta-9-tetrahydrocannabinol (THC), the primary active constituent of marijuana, attenuated the somatic and affective signs associated with nicotine withdrawal in rats (Balerio et al., 2004). Consistent with these studies, the FAAH inhibitor (URB597), which increases endogenous cannabinoid signaling, reduced nicotine withdrawal-induced anxiety in rats (Cippitelli et al., 2011). However, knockout of either the CB_1 or CB_2 receptors did not worsen expression of somatic nicotine withdrawal signs as compared to their respective wild-type counterparts (Castane et al., 2002; Ignatowska-Jankowska et al., 2013).

Blockade of the CB_1 receptors attenuated both cue- and nicotine-induced reinstatement of nicotine seeking (Diergaarde et al., 2008; Forget et al., 2009). In contrast, reinstatement of nicotine seeking was also attenuated by increasing endocannabinoid signaling via blockade of the anandamide transporter with AM404 (Gamaleddin et al., 2013). It is not clear why the blockade of the CB_1 receptors and increase in anandamide levels had similar effects on reinstatement of nicotine seeking. It is hypothesized that CB_1 receptor antagonists may act in neurocircuits that express endocannabinoid ligands other than anandamide. In summary, the above data support a role for endocannabinoid signaling in nicotine dependence, and the CB_1 receptors and anandamide transporter can serve as possible targets for future smoking cessation medications.

4.6 ROLE OF OTHER NEUROBIOLOGICAL SUBSTRATES IN NICOTINE DEPENDENCE

The role of monoamines such as dopamine, noradrenaline, and serotonin in nicotine dependence is discussed in Table 3. Several studies have reported a role for other peptides such as melanocortin, orexin/hypocretin, galanin, CRF, neurokinins, brain-derived neurotrophic factor, and alpha-peroxisome proliferator receptors in nicotine withdrawal and nicotine seeking (Dao et al., 2014; Jackson et al., 2011; Kivinummi et al., 2011; Plaza-Zabala et al., 2012, 2013). In addition, several signaling molecules such as calcium/calmodulin-dependent protein kinases II and IV and cyclic-AMP response element-binding protein are involved in nicotine-dependent behaviors (Jackson and Damaj, 2009; Kivinummi et al., 2011). The discussion of these neuromediators/neuropeptides and signaling molecules is beyond the scope of this review, and the reader is referred to other scholarly reviews for a discussion on the above-described molecules (Barik and Wonnacott, 2009; Brunzell and Picciotto, 2009; Jackson et al., 2014; Kenny, 2011; Picciotto and Mineur, 2014).

Table 3 Role of Monoamines (Dopamine, Noradrenaline, and Serotonin) in Nicotine-Dependent Behaviors

Neurotransmitters	Nicotine-Dependent Behavior	Treatment	Effects on Behavior	References
Dopamine	Nicotine withdrawal	Weak inhibitor of dopamine transporter (Bupropion)	Decreased nicotine withdrawal-induced somatic signs	Paterson et al. (2007)
		Weak inhibitor of dopamine transporter (Bupropion)	Reversed nicotine withdrawal-induced elevation in ICSS thresholds	Paterson et al. (2007)
	Reinstatement of nicotine seeking	D_1 antagonist (SCH23390)	Decreased cue-induced reinstatement of nicotine seeking	Liu et al. (2010)
		D_2 antagonist (Eticlopride)	Decreased cue-induced reinstatement of nicotine seeking	Liu et al. (2010)
		D_3 antagonist (SB277011-A)	Decreased cue- and nicotine-induced reinstatement of nicotine seeking	Khaled et al. (2010)
		D_4 antagonist (L-745870)	Decreased cue- and nicotine-induced reinstatement of nicotine seeking	Yan et al. (2012)
Serotonin	Nicotine withdrawal	5-Hydroxytryptophan (serotonin precursor)	Decreased nicotine withdrawal-induced somatic signs	Ohmura et al. (2011)
		Selective serotonin reuptake inhibitor (SSRI) + $5HT_{1A}$ antagonist	Reversed nicotine withdrawal-induced elevation in ICSS thresholds	Harrison et al. (2001)
		$5HT_3$ antagonist (Ondansetron)	Attenuated nicotine withdrawal-induced CPA	Suzuki et al. (1997)
	Reinstatement of nicotine seeking	$5HT_{2A}$ receptor antagonist (M100907)	Decreased cue- and nicotine-induced reinstatement of nicotine seeking	Fletcher et al. (2012)
		$5HT_{2c}$ receptor agonist (Ro60-0175)	Decreased cue- and nicotine-induced reinstatement of nicotine seeking	Fletcher et al. (2012)
Noradrenaline	Reinstatement of nicotine seeking	α_1 Receptor antagonist (Prazosin)	Decreased cue- and nicotine-induced reinstatement of nicotine seeking	Forget et al. (2010)
		β Blocker (Propranolol)	Decreased cue-induced reinstatement of nicotine seeking	Chiamulera et al. (2010)

ICSS, intracranial self-stimulation; CPA, conditioned place aversion.

5 CONCLUSION AND FUTURE DIRECTIONS

Nicotine is a highly addictive substance, and tobacco smoking is a major burden on healthcare systems all over the world. In this review, several brain regions that play a role in the development of nicotine dependence have been described. Furthermore, this review has described both nicotinic and nonnicotinic neural substrates that play a role in nicotine withdrawal and nicotine seeking. Currently, available FDA-approved smoking cessation medications target neuronal nAChRs, which will continue to remain an important target for future smoking cessation medications. This review has described the role of specific subunits of nAChRs that play an important role in both nicotine withdrawal and seeking. Targeting these specific nAChR subunits may result in more efficacious smoking cessation medications compared to those currently available. However, the wide distribution of nAChRs, existence of nAChRs as heteroreceptors and autoreceptors, neuroadaptations in the functioning of nAChRs (upregulation vs. downregulation) upon chronic nicotine exposure, and the difference in pharmacological response of nAChRs to both nicotine and acetylcholine due to nAChR stoichiometry, makes targeting nAChRs for the treatment of nicotine dependence extremely challenging. Thus, in addition to nAChRs, it is imperative to look at nonnicotinic neural substrates that play a role in the development of nicotine dependence. Studies using animal models suggest an important role for other neurotransmitters like dopamine, glutamate, GABA, serotonin, noradrenaline, and signaling peptides such as opioids and endocannabinoids in the development of nicotine dependence. It remains to be seen whether the promise of several neural targets obtained from animal studies will result in the availability of clinically efficacious smoking cessation medications. In summary, the challenge of reducing the burden of tobacco smoking in society will require the discovery of new smoking cessation medications and will involve targeting both nicotinic and nonnicotinic neural substrates.

ACKNOWLEDGMENTS

This work was supported by Bower, Bennet, and Bennet Endowed Chair Research Award awarded to Dr. Manoranjan S. D'Souza by The Raabe College of Pharmacy, Ohio Northern University (ONU), Ada, Ohio. The author would also like to thank Mr. Haval Norman (Pharmacy candidate 2018, The Raabe College of Pharmacy, Ohio Northern University, ONU) for editorial comments on the manuscript.

Disclosures: Manoranjan D'Souza reports no financial conflicts of interests.

REFERENCES

Balerio, G.N., et al., 2004. Delta9-tetrahydrocannabinol decreases somatic and motivational manifestations of nicotine withdrawal in mice. Eur. J. Neurosci. 20, 2737–2748.

Balfour, D.J., 2009. The neuronal pathways mediating the behavioral and addictive properties of nicotine. Handb. Exp. Pharmacol. 192, 209–233.

Barik, J., Wonnacott, S., 2009. Molecular and cellular mechanisms of action of nicotine in the CNS. Handb. Exp. Pharmacol. 192, 173–207.

Berrendero, F., Kieffer, B.L., Maldonado, R., 2002. Attenuation of nicotine-induced antinociception, rewarding effects, and dependence in mu-opioid receptor knock-out mice. J. Neurosci. 22, 10935–10940.

Berrendero, F., et al., 2005. Nicotine-induced antinociception, rewarding effects, and physical dependence are decreased in mice lacking the preproenkephalin gene. J. Neurosci. 25, 1103–1112.

Berrendero, F., et al., 2010. Neurobiological mechanisms involved in nicotine dependence and reward: participation of the endogenous opioid system. Neurosci. Biobehav. Rev. 35, 220–231.

Bespalov, A.Y., et al., 2005. Metabotropic glutamate receptor (mGluR5) antagonist MPEP attenuated cue- and schedule-induced reinstatement of nicotine self-administration behavior in rats. Neuropharmacology 49 (Suppl. 1), 167–178.

Biala, G., Budzynska, B., Kruk, M., 2005. Naloxone precipitates nicotine abstinence syndrome and attenuates nicotine-induced antinociception in mice. Pharmacol. Rep. 57, 755–760.

Bruijnzeel, A.W., Markou, A., 2004. Adaptations in cholinergic transmission in the ventral tegmental area associated with the affective signs of nicotine withdrawal in rats. Neuropharmacology 47, 572–579.

Brunzell, D.H., Picciotto, M.R., 2009. Molecular mechanisms underlying the motivational effects of nicotine. Nebr. Symp. Motiv. 55, 17–30.

Carmody, T.P., 1992. Preventing relapse in the treatment of nicotine addiction: current issues and future directions. J. Psychoactive Drugs 24, 131–158.

Castane, A., et al., 2002. Lack of CB1 cannabinoid receptors modifies nicotine behavioural responses, but not nicotine abstinence. Neuropharmacology 43, 857–867.

Changeux, J.P., 2010. Allosteric receptors: from electric organ to cognition. Annu. Rev. Pharmacol. Toxicol. 50, 1–38.

Chiamulera, C., et al., 2010. Propranolol transiently inhibits reinstatement of nicotine-seeking behaviour in rats. J. Psychopharmacol. 24, 389–395.

Cippitelli, A., et al., 2011. Endocannabinoid regulation of acute and protracted nicotine withdrawal: effect of FAAH inhibition. PLoS One 6, e28142.

Colombo, S.F., et al., 2013. Biogenesis, trafficking and up-regulation of nicotinic ACh receptors. Biochem. Pharmacol. 86, 1063–1073.

Conn, P.J., Pin, J.P., 1997. Pharmacology and functions of metabotropic glutamate receptors. Annu. Rev. Pharmacol. Toxicol. 37, 205–237.

Dani, J.A., Bertrand, D., 2007. Nicotinic acetylcholine receptors and nicotinic cholinergic mechanisms of the central nervous system. Annu. Rev. Pharmacol. Toxicol. 47, 699–729.

Dao, D.Q., et al., 2014. Nicotine enhances excitability of medial habenular neurons via facilitation of neurokinin signaling. J. Neurosci. 34, 4273–4284.

De Biasi, M., Dani, J.A., 2011. Reward, addiction, withdrawal to nicotine. Annu. Rev. Neurosci. 34, 105–130.

Diergaarde, L., et al., 2008. Contextual renewal of nicotine seeking in rats and its suppression by the cannabinoid-1 receptor antagonist Rimonabant (SR141716A). Neuropharmacology 55, 712–716.

D'Souza, M.S., Markou, A., 2011. Neuronal mechanisms underlying development of nicotine dependence: implications for novel smoking-cessation treatments. Addict. Sci. Clin. Pract. 6, 4–16.

D'Souza, M.S., Markou, A., 2012. The "Stop" & "Go" of nicotine dependence: role of GABA and glutamate. In: Pierce, R.C., Kenny, P.J., (Eds.), Addiction: A Neurobiological Perspective. Cold Spring Harbor Press, New York, pp. 251–268.

D'Souza, M.S., Markou, A., 2014. Differential role of N-methyl-D-aspartate receptor-mediated glutamate transmission in the nucleus accumbens shell and core in nicotine seeking in rats. Eur. J. Neurosci. 39, 1314–1322.

Epping-Jordan, M.P., et al., 1998. Dramatic decreases in brain reward function during nicotine withdrawal. Nature 393, 76–79.

Fagerstrom, K., 2002. The epidemiology of smoking: health consequences and benefits of cessation. Drugs 62 (Suppl. 2), 1–9.

Fletcher, P.J., Rizos, Z., Noble, K., Soko, A.D., Silenieks, L.B., Le, A.D., Higgins, G.A., 2012. Effects of the 5-HT2C receptor agonist Ro60-0175 and the 5-HT2A receptor antagonist M100907 on nicotine self-administration and reinstatement. Neuropharmacology 62, 2288–2298.

Forget, B., Coen, K.M., Le Foll, B., 2009. Inhibition of fatty acid amide hydrolase reduces reinstatement of nicotine seeking but not break point for nicotine self-administration—comparison with CB(1) receptor blockade. Psychopharmacology (Berl.) 205, 613–624.

Forget, B., et al., 2010. Noradrenergic alpha1 receptors as a novel target for the treatment of nicotine addiction. Neuropsychopharmacology 35, 1751–1760.

Gamaleddin, I., et al., 2013. AM404 attenuates reinstatement of nicotine seeking induced by nicotine-associated cues and nicotine priming but does not affect nicotine- and food-taking. J. Psychopharmacol. 27, 564–571.

Geisler, S., Zahm, D.S., 2005. Afferents of the ventral tegmental area in the rat-anatomical substratum for integrative functions. J. Comp. Neurol. 490, 270–294.

Gipson, C.D., et al., 2013. Reinstatement of nicotine seeking is mediated by glutamatergic plasticity. Proc. Natl. Acad. Sci. U.S.A. 110, 9124–9129.

Grace, A.A., et al., 2007. Regulation of firing of dopaminergic neurons and control of goal-directed behaviors. Trends Neurosci. 30, 220–227.

Grella, S.L., et al., 2014. Role of the kappa-opioid receptor system in stress-induced reinstatement of nicotine seeking in rats. Behav. Brain Res. 265, 188–197.

Grieder, T.E., et al., 2014. VTA CRF neurons mediate the aversive effects of nicotine withdrawal and promote intake escalation. Nat. Neurosci. 17, 1751–1758.

Hadjiconstantinou, M., Neff, N.H., 2011. Nicotine and endogenous opioids: neurochemical and pharmacological evidence. Neuropharmacology 60, 1209–1220.

Hadjiconstantinou, M., et al., 2011. Enhanced dopamine transporter function in striatum during nicotine withdrawal. Synapse 65, 91–98.

Harrison, A.A., Liem, Y.T., Markou, A., 2001. Fluoxetine combined with a serotonin-1A receptor antagonist reversed reward deficits observed during nicotine and amphetamine withdrawal in rats. Neuropsychopharmacology 25, 55–71.

Henningfield, J.E., Miyasato, K., Jasinski, D.R., 1985. Abuse liability and pharmacodynamic characteristics of intravenous and inhaled nicotine. J. Pharmacol. Exp. Ther. 234, 1–12.

Hildebrand, B.E., et al., 1998. Reduced dopamine output in the nucleus accumbens but not in the medial prefrontal cortex in rats displaying a mecamylamine-precipitated nicotine withdrawal syndrome. Brain Res. 779, 214–225.

Hollmann, M., Heinemann, S., 1994. Cloned glutamate receptors. Annu. Rev. Neurosci. 17, 31–108.

Hopkins, T.J., et al., 2012. Galantamine, an acetylcholinesterase inhibitor and positive allosteric modulator of nicotinic acetylcholine receptors, attenuates nicotine taking and seeking in rats. Neuropsychopharmacology 37, 2310–2321.

Hughes, J.R., et al., 1991. Symptoms of tobacco withdrawal. A replication and extension. Arch. Gen. Psychiatry 48, 52–59.

Hughes, J.R., Keely, J., Naud, S., 2004. Shape of the relapse curve and long-term abstinence among untreated smokers. Addiction 99, 29–38.

Ignatowska-Jankowska, B.M., et al., 2013. The cannabinoid CB2 receptor is necessary for nicotine-conditioned place preference, but not other behavioral effects of nicotine in mice. Psychopharmacology (Berl.) 229, 591–601.

Ise, Y., et al., 2002. Modulation of kappa-opioidergic systems on mecamylamine-precipitated nicotine-withdrawal aversion in rats. Neurosci. Lett. 323, 164–166.

Jackson, K.J., Damaj, M.I., 2009. L-type calcium channels and calcium/calmodulin-dependent kinase II differentially mediate behaviors associated with nicotine withdrawal in mice. J. Pharmacol. Exp. Ther. 330, 152–161.

Jackson, K.J., et al., 2008. Differential role of nicotinic acetylcholine receptor subunits in physical and affective nicotine withdrawal signs. J. Pharmacol. Exp. Ther. 325, 302–312.

Jackson, K.J., et al., 2009. The role of alpha6-containing nicotinic acetylcholine receptors in nicotine reward and withdrawal. J. Pharmacol. Exp. Ther. 331, 547–554.

Jackson, K.J., et al., 2010. Effect of the selective kappa-opioid receptor antagonist JDTic on nicotine antinociception, reward, and withdrawal in the mouse. Psychopharmacology (Berl.) 210, 285–294.

Jackson, K.J., et al., 2011. The neuropeptide galanin and variants in the GalR1 gene are associated with nicotine dependence. Neuropsychopharmacology 36, 2339–2348.

Jackson, K.J., et al., 2013. The $\alpha3\beta4^*$ nicotinic acetylcholine receptor subtype mediates nicotine reward and physical nicotine withdrawal signs independently of the $\alpha5$ subunit in the mouse. Neuropharmacology 70, 228–235.

Jackson, K.J., et al., 2014. New mechanisms and perspectives in nicotine withdrawal. Neuropharmacology 96, 223–234.

Jhou, T.C., et al., 2009. The mesopontine rostromedial tegmental nucleus: a structure targeted by the lateral habenula that projects to the ventral tegmental area of Tsai and substantia nigra compacta. J. Comp. Neurol. 513, 566–596.

Kenny, P.J., 2011. Tobacco dependence, the insular cortex and the hypocretin connection. Pharmacol. Biochem. Behav. 97, 700–707.

Kenny, P.J., Gasparini, F., Markou, A., 2003. Group II metabotropic and alpha-amino-3-hydroxy-5-methyl-4-isoxazole propionate (AMPA)/kainate glutamate receptors regulate the deficit in brain reward function associated with nicotine withdrawal in rats. J. Pharmacol. Exp. Ther. 306, 1068–1076.

Khaled, M.A., et al., 2010. The selective dopamine D3 receptor antagonist SB 277011-A, but not the partial agonist BP 897, blocks cue-induced reinstatement of nicotine-seeking. Int. J. Neuropsychopharmacol. 13, 181–190.

Kivinummi, T., et al., 2011. Alterations in BDNF and phospho-CREB levels following chronic oral nicotine treatment and its withdrawal in dopaminergic brain areas of mice. Neurosci. Lett. 491, 108–112.

Koob, G.F., Volkow, N.D., 2010. Neurocircuitry of addiction. Neuropsychopharmacology 35, 217–238.

Krishnan-Sarin, S., Rosen, M.I., O'Malley, S.S., 1999. Naloxone challenge in smokers. Preliminary evidence of an opioid component in nicotine dependence. Arch. Gen. Psychiatry 56, 663–668.

Li, X., et al., 2014. Involvement of glutamatergic and GABAergic systems in nicotine dependence: implications for novel pharmacotherapies for smoking cessation. Neuropharmacology 76 (Pt. B), 554–565.

Liechti, M.E., Markou, A., 2007. Interactive effects of the mGlu5 receptor antagonist MPEP and the mGlu2/3 receptor antagonist LY341495 on nicotine self-administration and reward deficits associated with nicotine withdrawal in rats. Eur. J. Pharmacol. 554, 164–174.

Liechti, M.E., et al., 2007. Metabotropic glutamate 2/3 receptors in the ventral tegmental area and the nucleus accumbens shell are involved in behaviors relating to nicotine dependence. J. Neurosci. 27, 9077–9085.

Liu, X., 2014. Effects of blockade of alpha4beta2 and alpha7 nicotinic acetylcholine receptors on cue-induced reinstatement of nicotine-seeking behaviour in rats. Int. J. Neuropsychopharmacol. 17, 105–116.

Liu, X., et al., 2009. Naltrexone attenuation of conditioned but not primary reinforcement of nicotine in rats. Psychopharmacology (Berl.) 202, 589–598.

Liu, X., et al., 2010. Effects of dopamine antagonists on drug cue-induced reinstatement of nicotine-seeking behavior in rats. Behav. Pharmacol. 21, 153–160.

Lutz, R.A., Pfister, H.P., 1992. Opioid receptors and their pharmacological profiles. J. Recept. Res. 12, 267–286.

Malin, D.H., et al., 1993. Naloxone precipitates nicotine abstinence syndrome in the rat. Psychopharmacology (Berl.) 112, 339–342.

Mansvelder, H.D., McGehee, D.S., 2002. Cellular and synaptic mechanisms of nicotine addiction. J. Neurobiol. 53, 606–617.

Mansvelder, H.D., Keath, J.R., McGehee, D.S., 2002. Synaptic mechanisms underlie nicotine-induced excitability of brain reward areas. Neuron 33, 905–919.

Markou, A., 2008. Review. Neurobiology of nicotine dependence. Philos. Trans. R. Soc. Lond. B Biol. Sci. 363, 3159–3168.

Natividad, L.A., et al., 2009. Nicotine withdrawal produces a decrease in extracellular levels of dopamine in the nucleus accumbens that is lower in adolescent versus adult male rats. Synapse 64, 136–145.

Nicholls, D., Attwell, D., 1990. The release and uptake of excitatory amino acids. Trends Pharmacol. Sci. 11, 462–468.

Nides, M., 2008. Update on pharmacologic options for smoking cessation treatment. Am. J. Med. 121, S20–S31.

O'Connor, E.C., et al., 2010. The alpha4beta2 nicotinic acetylcholine-receptor partial agonist varenicline inhibits both nicotine self-administration following repeated dosing and reinstatement of nicotine seeking in rats. Psychopharmacology (Berl.) 208, 365–376.

Ohmura, Y., et al., 2011. 5-Hydroxytryptophan attenuates somatic signs of nicotine withdrawal. J. Pharmacol. Sci. 117, 121–124.

Paterson, N.E., Balfour, D.J., Markou, A., 2007. Chronic bupropion attenuated the anhedonic component of nicotine withdrawal in rats via inhibition of dopamine reuptake in the nucleus accumbens shell. Eur J Neurosci. 25, 3099–3108.

Paterson, N.E., Froestl, W., Markou, A., 2005. Repeated administration of the GABAB receptor agonist CGP44532 decreased nicotine self-administration, and acute administration decreased cue-induced reinstatement of nicotine-seeking in rats. Neuropsychopharmacology 30, 119–128.

Picciotto, M.R., Mineur, Y.S., 2014. Molecules and circuits involved in nicotine addiction: the many faces of smoking. Neuropharmacology 76 (Pt. B), 545–553.

Picciotto, M.R., et al., 1998. Acetylcholine receptors containing the beta2 subunit are involved in the reinforcing properties of nicotine. Nature 391, 173–177.

Pin, J.P., Duvoisin, R., 1995. The metabotropic glutamate receptors: structure and functions. Neuropharmacology 34, 1–26.

Plaza-Zabala, A., et al., 2012. Hypocretin/orexin signaling in the hypothalamic paraventricular nucleus is essential for the expression of nicotine withdrawal. Biol. Psychiatry 71, 214–223.

Plaza-Zabala, A., et al., 2013. A role for hypocretin/orexin receptor-1 in cue-induced reinstatement of nicotine-seeking behavior. Neuropsychopharmacology 38, 1724–1736.

Qi, X., et al., 2015. A critical role for the melanocortin 4 receptor in stress-induced relapse to nicotine seeking in rats. Addict. Biol. 20, 324–335.

Ramirez-Nino, A.M., D'Souza, M.S., Markou, A., 2013. N-acetylcysteine decreased nicotine self-administration and cue-induced reinstatement of nicotine seeking in rats: comparison with the effects of N-acetylcysteine on food responding and food seeking. Psychopharmacology (Berl.) 225, 473–482.

Rennard, S.I., Daughton, D.M., 2014. Smoking cessation. Clin. Chest Med. 35, 165–176.

Salas, R., Pieri, F., De Biasi, M., 2004. Decreased signs of nicotine withdrawal in mice null for the beta4 nicotinic acetylcholine receptor subunit. J. Neurosci. 24, 10035–10039.

Salas, R., et al., 2009. Nicotinic receptors in the habenulo-interpeduncular system are necessary for nicotine withdrawal in mice. J. Neurosci. 29, 3014–3018.

Shaham, Y., et al., 2003. The reinstatement model of drug relapse: history, methodology and major findings. Psychopharmacology (Berl.) 168, 3–20.

Shiffman, S.M., Jarvik, M.E., 1976. Smoking withdrawal symptoms in two weeks of abstinence. Psychopharmacology (Berl.) 50, 35–39.

Shram, M.J., et al., 2008. Interactions between age and the aversive effects of nicotine withdrawal under mecamylamine-precipitated and spontaneous conditions in male Wistar rats. Psychopharmacology (Berl.) 198, 181–190.

Stoker, A.K., Markou, A., 2015. Neurobiological bases of cue and nicotine-induced reinstatement of nicotine seeking: implications for the development of smoking cessation medications. Curr. Top. Behav. Neurosci. 24, 125–154.

Stoker, A.K., Olivier, B., Markou, A., 2012a. Role of alpha7- and beta4-containing nicotinic acetylcholine receptors in the affective and somatic aspects of nicotine withdrawal: studies in knockout mice. Behav. Genet. 42, 423–436.

Stoker, A.K., Olivier, B., Markou, A., 2012b. Involvement of metabotropic glutamate receptor 5 in brain reward deficits associated with cocaine and nicotine withdrawal and somatic signs of nicotine withdrawal. Psychopharmacology (Berl.) 221, 317–327.

Stolerman, I.P., Jarvis, M.J., 1995. The scientific case that nicotine is addictive. Psychopharmacology (Berl.) 117, 2–10. discussion 14–20.

Suzuki, T., et al., 1997. Attenuation of mecamylamine-precipitated nicotine-withdrawal aversion by the 5-HT3 receptor antagonist ondansetron. Life Sci. 61, PL249–254.

Tejeda, H.A., et al., 2012. Dysregulation of kappa-opioid receptor systems by chronic nicotine modulate the nicotine withdrawal syndrome in an age-dependent manner. Psychopharmacology (Berl.) 224, 289–301.

Varani, A.P., et al., 2014. Attenuation by baclofen of nicotine rewarding properties and nicotine withdrawal manifestations. Psychopharmacology (Berl.) 231, 3031–3040.

Velasquez, K.M., Molfese, D.L., Salas, R., 2014. The role of the habenula in drug addiction. Front. Hum. Neurosci. 8, 174.

Vlachou, S., et al., 2011a. Repeated administration of the GABAB receptor positive modulator BHF177 decreased nicotine self-administration, and acute administration decreased cue-induced reinstatement of nicotine seeking in rats. Psychopharmacology (Berl.) 215, 117–128.

Vlachou, S., et al., 2011b. Both GABA(B) receptor activation and blockade exacerbated anhedonic aspects of nicotine withdrawal in rats. Eur. J. Pharmacol. 655, 52–58.

Watabe-Uchida, M., et al., 2012. Whole-brain mapping of direct inputs to midbrain dopamine neurons. Neuron 74, 858–873.

WHO, 2011. WHO Report on the Global Tobacco Epidemic, 2011: Warning About the Dangers of Tobacco. WHO website [online], http://www.who.int/tobacco/global_report/2011/en (2011).

Yan, Y., et al., 2012. Blockade of dopamine d4 receptors attenuates reinstatement of extinguished nicotine-seeking behavior in rats. Neuropsychopharmacology 37, 685–696.

Zhao-Shea, R., et al., 2013. Activation of GABAergic neurons in the interpeduncular nucleus triggers physical nicotine withdrawal symptoms. Curr. Biol. 23, 2327–2335.

Zoli, M., Pistillo, F., Gotti, C., 2015. Diversity of native nicotinic receptor subtypes in mammalian brain. Neuropharmacology 96 (Pt. B), 302–311.

CHAPTER

Neuroscience of alcohol for addiction medicine: Neurobiological targets for prevention and intervention in adolescents

11

Anita Cservenka*, Bonnie J. Nagel*,†,1

**Departments of Psychiatry, Oregon Health & Science University, Portland, OR, USA*

†Behavioral Neuroscience, Oregon Health & Science University, Portland, OR, USA

[1]Corresponding author: Tel.: +1-503-4944612; Fax: +1-503-4185774, e-mail address: nagelb@ohsu.edu

Abstract

Structural and functional neuroimaging studies indicate that heavy alcohol use during adolescence may be neurotoxic to the brain. This chapter reviews the neuroimaging findings in cross-sectional and longitudinal studies of adolescent heavy alcohol users. These youth exhibit reductions in prefrontal, hippocampal, and cerebellar brain volume, decreased fronto-parietal, and increased frontolimbic white matter integrity, as well as alterations in blood oxygen level-dependent response during working memory, inhibitory control, verbal encoding, decision making, and reward processing—some of which appear to differ between males and females. Although some exist, additional longitudinal studies will significantly advance addiction medicine by aiding prevention scientists and treatment providers to develop neurobiologically informed ways of strengthening neural networks prior to and after the onset of heavy alcohol use, thereby promoting healthy cognitive functioning across the adolescent period.

Keywords

Alcohol, Adolescence, Brain volume, White matter, fMRI

1 INTRODUCTION

Human neuroscience research examining the impact of alcohol use on the brain has made tremendous progress over the last 20 years. Advances in neuroimaging technology have allowed researchers to more carefully examine the effects of alcohol use

Progress in Brain Research, Volume 223, ISSN 0079-6123, http://dx.doi.org/10.1016/bs.pbr.2015.07.027

on brain structure and brain functioning. This work has provided a better understanding of the brain systems most susceptible to the neurotoxic effects of alcohol and could ultimately be used to neurobiologically inform treatment and rehabilitation efforts. Prevention scientists may also benefit from this research, as neuroimaging studies may foster translational efforts focused on behavioral or cognitive interventions that strengthen brain networks shown to be vulnerable to the effects of alcohol.

An important population of study in the field of neuroimaging and alcohol use is adolescents. The significance and advantages of studying youth to inform addiction medicine are multifold. First, adolescence is a time of protracted brain maturation (Giedd et al., 1999; Gogtay et al., 2004), during which affective and reward-related neurocircuitry, including the mesolimbic system, is believed to mature prior to top-down regulatory brain regions, such as the prefrontal cortex (PFC; Mills et al., 2014; Van Leijenhorst et al., 2010). The ongoing maturation of the adolescent brain renders it particularly vulnerable to neurotoxic substances, such as alcohol. Second, adolescence is a time of alcohol use initiation, with alcohol being the most widely used intoxicating substance among youth. According to the 2014 Monitoring the Future Survey (Johnston et al., 2015), while rates of adolescent alcohol are at their lowest since initiation of the survey, alcohol use still increases substantially between the 8th and 12th grades, such that over a quarter of youth have consumed alcohol by 8th grade, and two-thirds of youth have consumed alcohol by the end of high school, with 50% reporting being drunk by this time. Even more concerning, 20% of 12th graders reported binge drinking (consuming five or more drinks in one setting) in the 2-week period prior to the survey. Many risk factors for increased alcohol use have been reported during adolescence, including age at first drink (Dawson et al., 2008), elevations in sensation seeking (Macpherson et al., 2010), cumulative life stress (Casement et al., 2015), family history of alcoholism (FHP; Lieb et al., 2002), and externalizing and internalizing symptoms (King et al., 2004). Thus, ongoing brain maturation, incidence of binge drinking, and risk factors for alcohol abuse during adolescence are all important motives to study the consequences of alcohol's effects on brain structure and functioning during this developmental period. In this chapter, we outline the major findings that have been reported through neuroimaging studies of alcohol use during adolescence and summarize areas of research that can inform addiction medicine. We also emphasize the need for additional efforts for the study of sex differences in the consequences and risk factors for alcohol abuse during adolescence, as well as longitudinal studies of within-subject change in brain structure, functioning, and neurocognition in alcohol-using youth.

2 BRAIN VOLUME

2.1 HIPPOCAMPUS

The hippocampus is a brain region critical for encoding new memories and has been shown to be vulnerable to the effects of alcohol in animal models (Broadwater et al., 2014) and adults with alcohol use disorders (AUDs; Beresford et al., 2006), which makes it an important brain area of inquiry for research on alcohol's neurotoxic effects during adolescence. Three studies of adolescents with AUDs have examined hippocampal morphology (De Bellis et al., 2000; Medina et al., 2007; Nagel et al., 2005). The first of these studies found that adolescents with AUDs had reduced bilateral hippocampal volume compared with healthy controls, with longer duration of alcohol use and earlier age of AUD onset related to smaller hippocampal volumes (De Bellis et al., 2000). However, these findings were in a sample of AUD youth who had comorbid psychiatric diagnoses, which could have confounded results, given demonstrated hippocampal abnormalities in adolescents with psychopathology alone, such as depression (Hulvershorn et al., 2011). A study thereafter in AUD youth without comorbid psychiatric diagnoses also found reductions in hippocampal volume, limited to the left hemisphere, and relative to controls (Nagel et al., 2005). Interestingly, no "dose–response" relationship was found in this study, suggesting only an association between the diagnosis of AUD and the hippocampal volume. It is also possible that premorbid differences in hippocampal volume may have been present between groups, prior to heavy alcohol consumption, suggesting reduced hippocampal volume as a risk factor for AUDs in these adolescents. Finally, a subsequent study replicated the findings of reduced left hippocampal volume in adolescent alcohol users relative to controls or youth who used alcohol and marijuana (Medina et al., 2007). This study also reported that increased right to left asymmetry of the hippocampus in adolescent alcohol users was associated with the severity of their alcohol abuse or dependence. Interestingly, while left to right asymmetry of the hippocampus was related to verbal learning performance in controls, this relationship was absent in alcohol users, suggesting a direct neurocognitive correlate of hippocampal morphometry that is aberrant in alcohol-using youth.

Given the overlap of findings among these three studies, more work is needed to confirm these results, with additional studies that focus on neurocognitive, alcohol dose, and diagnostic severity relationships with hippocampal volume in alcohol-using adolescents, as well as identification of premorbid risk-related abnormalities. This appears to be a promising avenue of research for treatment studies that could focus on improving memory encoding in adolescent alcohol users, or strengthening memory functioning in youth at risk for future alcohol abuse.

2.2 FRONTAL LOBE

The PFC is one of the latest regions to undergo synaptic pruning and is responsible for many higher-order executive functions that are developing during adolescence, such as decision making, inhibitory control, working memory, selective attention, and goal flexibility (for review, see Casey et al., 2000). Smaller overall PFC and PFC white matter volumes have been demonstrated in youth with AUDs relative to controls (De Bellis et al., 2005). Additionally, number of drinks in the largest drinking episode was negatively correlated with PFC gray matter volume in these youth. While gray matter volume reduction is characteristic of adolescent brain maturation and may reflect normative synaptic pruning, excessive reduction may represent an atypical trajectory of brain development. It will be important for future studies to find neurocognitive correlates of this reduction in overall PFC volume in youth with AUDs and to determine whether these results translate to studies of heavy alcohol-using teens who do not meet criteria for an AUD diagnosis.

Another study that examined PFC volume in adolescents with AUDs divided the PFC into posterior, anterior dorsal, and anterior ventral regions to understand specific PFC subregions that may be more affected in adolescents with AUDs (Medina et al., 2008). The authors found an interesting sex-by-group interaction, such that adolescent females with AUDs showed reductions in PFC gray and white matter volume relative to female controls, driven by reduced anterior dorsal and anterior ventral PFC volumes, while males with AUDs showed the opposite effect, such that they had larger PFC gray and white matter volumes relative to male controls, in the same subregions. This striking interaction draws attention to an important area of inquiry in the study of alcohol use's effects on the adolescent brain—that is, are there sex differences in the effects that alcohol has on brain structure, or different neural risk features for adolescent alcohol use between the sexes?

Another sex-by-group interaction, reporting contrary patterns to the study above, was shown in a study examining PFC cortical thickness, one contributor to gray matter volume, in adolescent binge drinkers (Squeglia et al., 2012b). In this study, male binge drinkers had reduced PFC cortical thickness relative to nondrinking male controls, but female binge drinkers had thicker prefrontal cortices compared with nondrinking female controls. Increased prefrontal cortical thickness in females was significantly related to poorer performance on a variety of neuropsychological tests, including tests of attention, visuospatial processing, and inhibitory control. Future studies with larger sample sizes will be needed to carefully study these interactions, which could have major implications for addiction medicine, as health care providers may need to design sex-specific treatment programs for adolescent alcohol abusers.

The orbitofrontal cortex (OFC) is another frontal lobe region implicated in addiction, as it has structural and functional connectivity with reward-related brain regions, such as the nucleus accumbens (NAcc). A volumetric analysis of this region in a large sample of juvenile justice-involved adolescents found that recent drinking frequency was negatively related to OFC volume (Thayer et al., 2012). It is uncertain whether these would be long term, persisting effects, or if abstinence from alcohol use may diminish some of these effects, suggesting potential recovery of brain regions.

One longitudinal investigation of alcohol-using adolescents with neuroimaging data at both baseline and after alcohol use initiation found that cortical thickness in the right middle frontal gyrus showed significantly larger decreases in youth who had started using alcohol versus those who had not begun using (Luciana et al., 2013). No premorbid differences in frontal lobe cortical thickness were present between alcohol users and nonusers, further substantiating the possibility of an alcohol-related effect. It will be important for future longitudinal studies to continue to dissociate premorbid differences from those related to alcohol-induced changes in brain development, particularly in groups matched over time for other potentially confounding influences.

Another longitudinal study of current and future adolescent binge drinkers demonstrated that ventromedial PFC and inferior frontal gyrus gray matter volume were both smaller in current binge drinkers compared with controls, while smaller bilateral superior frontal gyri volume was predictive of future binge drinking (Whelan et al., 2014). These findings suggest that there may be dissociable regions of the PFC reflecting vulnerability toward the consequences of alcohol use versus ability to predict future onset of heavy use.

2.3 CEREBELLUM

A frequently reported site of alcohol-related brain damage in adults with AUDs has been the cerebellum, and associated cerebellar motor deficits have been seen in this population. In addition to its role in motor skills and balance, the cerebellum is also integrated with frontal lobe circuitry, and frontocerebellar pathways are thought to be critical to executive functioning (Schmahmann and Pandya, 1997). In a study by De Bellis et al. (2005), where youth with AUDs had comorbid psychiatric disorders, cerebellar volume was reduced in males with AUDs compared with controls—a finding that was not present in females with AUDs. This then begs the question, is reduced cerebellar volume in males with AUDs related to motor or cognitive deficits in those youth? Also, how much does the presence of other psychiatric disorders account for these findings?

A study of cerebellar volume in binge-drinking youth without AUD diagnoses found reduced cerebellar volume in binge drinkers relative to controls across both male and female youth and reported a striking negative relationship between peak number of drinks consumed in the past 3 months and cerebellar volume (Lisdahl et al., 2013). This suggests a strong possibility, that in high doses, alcohol may have a neurotoxic effect on the cellular architecture of the brain and that the cerebellum should be a continued region of interest for future neuroimaging studies.

2.4 SUBCORTICAL STRUCTURES: BASAL GANGLIA, THALAMUS

Fewer studies in adolescent alcohol users have reported on brain morphometric alterations of other brain regions, such as the basal ganglia and the thalamus. While De Bellis et al. (2005) found no significant volumetric differences in the thalamus between AUD youth and controls, a more recent study found that male youth with AUDs and no comorbid psychiatric disorders had smaller thalamic volumes compared with male controls, while female youth with AUDs had larger thalamic volumes compared with female controls (Fein et al., 2013). This study was conducted with voxel-based morphometry to measure gray matter density, while the study by De Bellis et al. (2005) performed manual tracing of the thalamus. These differences in methodology and the presence or absence of comorbidities could have accounted for these discrepant findings. Further, the putamen is a motor structure of the basal ganglia implicated in habit formation, and this area showed the same sex-by-group interaction as reported for the thalamus, suggesting that basal ganglia structure may also be affected by heavy alcohol use during adolescence (Fein et al., 2013). A recent longitudinal study reported basal ganglia volume change in binge drinkers in the caudate, such that greater number of days of alcohol consumption was correlated with larger reductions of caudate volume (Squeglia et al., 2014). Other regions, such as the left ventral diencephalon, brain stem, and left inferior and middle temporal gyrus, also showed brain volume reductions in youth who initiated heavy alcohol use, but correlations with alcohol use were not present for these regions.

Finally, a study of the NAcc, the major dopamine output region of the mesolimbic pathway, showed that past 3-month drinking frequency was *positively* associated with the volume of this region (Thayer et al., 2012). This finding stands in contrast to previously observed volumetric reductions in executive functioning brain circuitry, such as the PFC and cerebellum, in drinking adolescents. This could indicate that heavy alcohol use leads to increases in volume of reward-related areas, but reductions in volume of cognitive control regions, thereby altering neurocircuitry in both bottom-up and top-down brain systems.

2.5 INSULA

The insula is a structure that has also been the focus of addiction research, as it is implicated in internal emotional and bodily states that could be relevant for processing feelings associated with pleasure derived from or craving associated with substance use (Chung and Clark, 2014). This structure was the focus of a longitudinal investigation in adolescents, a majority of whom had used alcohol and marijuana and were in a treatment program for substance use (Chung and Clark, 2014). A mediation model found a significant indirect effect between insular white matter volume and binge-drinking frequency at baseline and 1 year into the study that was mediated through enhancement motives for alcohol use. This suggests that the insula may be a particularly good target for future studies aimed at understanding the neural correlates of motivations for alcohol use that are linked with internal states.

3 WHITE MATTER MICROSTRUCTURE

3.1 ALCOHOL

Diffusion tensor imaging is able to indirectly infer information about water diffusion along axons, with more restricted diffusion thought to reflect greater anisotropic movement of water molecules along white matter fiber tracts. Higher fractional anisotropy (FA) and lower mean diffusivity (MD) are usually indicators of greater axonal integrity and myelination, with higher FA and lower MD values seen across development as white matter matures. A number of studies have specifically focused on the effects of alcohol on white matter integrity during adolescence, while others have had more mixed samples of adolescent substance users, including multisubstance or substance use disorder (SUD) histories, such as regular comorbid marijuana and tobacco use.

In a study of binge-drinking adolescents, McQueeny et al. (2009) reported widespread reductions in FA in many white matter tracts, including the superior longitudinal fasciculus (SLF), corpus callosum, corona radiata, internal/external capsules, commissural fibers, and the corticospinal tract. These results indicate that alcohol may have global effects on white matter integrity that are not limited to specific tracts. Many of the reductions in FA were related to greater prevalence of withdrawal/hangover symptoms in these youth. However, two other studies of youth with AUDs found higher FA in limbic tracts (Cardenas et al., 2013) and the corpus callosum (De Bellis et al., 2008). The former study suggested that higher FA in white matter tracts connecting reward and affect-related regions may be a risk marker for drinking or perhaps greater integrity of these tracts reinforces alcohol use. In contrast to those two studies, an investigation of juvenile justice-involved adolescents found that youth with high scores on the Alcohol Use Disorders Identification Test had reduced FA in the SLF, which was related to higher impulsivity (Thayer et al., 2013). Thus, it is plausible that lower FA may be present in white matter tracts involved with regulatory functions, such as impulse control, while higher FA might be seen in youth with AUDs in affect and reward-related connections, which could be driving their propensity to consume alcohol.

One longitudinal investigation found that FA increased in the inferior fronto-occipital fasciculus and dorsal caudate from baseline to follow-up in adolescents who abstained from alcohol use for approximately 2 years, and this was driven by decreases in radial diffusivity (diffusion of water perpendicular to the primary direction of diffusion along the white matter tracts; Luciana et al., 2013). These developmental changes were not observed in the alcohol-using group, suggesting that white matter maturation may have been blunted in these youth. It will be important to determine the neurocognitive correlates of altered patterns of white matter maturation during adolescence, especially with long-term continued alcohol use.

3.2 ALCOHOL AND MARIJUANA

Since a large proportion of adolescents use alcohol and marijuana, it becomes difficult to dissociate the effects of these substances on white matter integrity. In a study of heavy alcohol and marijuana users compared with controls, widespread reductions in FA were present across 10 different fiber tracts, including frontoparietal and temporal lobe circuitry, in adolescent substance users compared with controls (Bava et al., 2009). A few white matter tracts showed increased FA in users relative to controls, and it is uncertain if these might be compensatory neuroadaptations due to widespread reductions in white matter integrity among other regions. A follow-up study in the same group of combined substance users and controls found that users had poorer performance on processing speed, attention, and working memory that were associated with lower FA in regions previously identified as having reduced white matter integrity, such as the right inferior longitudinal fasciculus and superior temporal lobe pathways (Bava et al., 2010). On the other hand, potential compensatory changes might be present in occipital tracts where users had higher FA, which was related to better visuomotor and verbal learning performance.

A study by Jacobus et al. (2009) used a three-group design with binge drinkers, combined binge drinkers and marijuana users, and controls, to better dissociate the effects of different substances on white matter integrity. While binge drinkers had reductions in FA relative to controls in many long-range association tracts, such as the superior and inferior longitudinal fasciculi, these same reductions in combined alcohol and marijuana users were not present when compared with controls. These surprising findings suggest that while binge drinking may be damaging to white matter integrity, the combined effects of using multiple substances are not always additive and may have different consequences for the maturation of white matter during adolescence.

In some contrast to the findings above, the same research group conducted a longitudinal study with 16 adolescents and found that those youth who became heavy alcohol and marijuana users by the time of their second study visit, 3 years later, had lower FA compared with their baseline line scan, a decrease not present in those who only initiated heavy alcohol use (Jacobus et al., 2013b). These results should be interpreted with caution, as the sample size for this study was fairly small; however, the within individual change may be more informative than the above-referenced cross-sectional design.

A longitudinal study of a larger sample of adolescent alcohol and marijuana users found that escalation of substance use over a 1.5-year period was predictive of MD in white matter tracts at follow-up, such as the SLF (Bava et al., 2013). Another longitudinal study that followed a group of controls, binge drinkers, and combined binge and marijuana users over 3 years found reduced FA at the third neuroimaging visit in users compared with controls (Jacobus et al., 2013a). Both substance-using groups showed reductions in FA over the 3-year period, so in this case, binge drinking alone as well as combined alcohol and marijuana use appeared to compromise FA. It is important to note that these youth were in emerging adulthood (ages 19–22) at the time of the third scan, so discrepancies between this finding and other studies (Jacobus et al., 2009) could have been due to the age of the participants at the time of the scan, as the effects of binge drinking and marijuana use on white matter microstructure may depend on brain maturity.

Sex differences in alcohol's effects on white matter microstructure also merit attention, but few studies have specifically examined these differences. One study, which replicated former results suggesting reduced integrity of SLF white matter in adolescent alcohol users, also found a significant sex-by-group interaction, such that females with SUDs had lower FA in this tract than males with SUDs, while female controls had higher FA in this tract than male controls (Thatcher et al., 2010). These results should be interpreted with caution due to relatively small sample sizes, and specificity to alcohol is uncertain given overlapping abuse and/or dependence of multiple substances, as well as diagnoses of externalizing and internalizing disorders in the SUD youth. It will be important for future studies to examine sex differences in binge drinkers or youth with AUDs with no other significant substance use or psychopathology.

4 BRAIN FUNCTION

4.1 WORKING MEMORY

The most frequently studied neurocognitive function in alcohol-using adolescents using functional magnetic resonance imaging (fMRI) has been working memory, an executive function that improves across development and engages frontoparietal circuitry, which matures during this time period (Crone et al., 2006). Working memory-related brain response in adolescents who met criteria for AUDs was first examined by Tapert et al. (2004b) using a visual working memory task. No differences in behavioral performance were present, but increased parietal and decreased cerebellar, occipital, and temporal activity were seen in youth with AUDs. Additionally, many brain regions showed both positive and negative relationships between alcohol consumption variables, such as number of drinks consumed and withdrawal symptoms. However, a study that examined sex-by-group interactions, albeit using a different task, found very different patterns of brain response between female and male youth with AUDs, which could have been masked in the former findings. For example, in bilateral superior frontal gyri, females with AUDs showed decreased spatial working memory (SWM)-related brain response, while males showed positive brain activity in these regions (Caldwell et al., 2005). Yet, in other areas, such as the superior temporal lobes, females with AUDs showed greater response relative to males with AUDs or the female controls. These complex interactions suggest that heavy alcohol use may exert its influence on brain response very differently between the sexes and among brain regions. Another study of SWM-related brain response in adolescent binge drinkers showed that female binge drinkers had reduced SWM-related brain activity in frontal lobe areas, such as the superior frontal gyrus, compared with female controls, but male binge drinkers only showed elevated activity relative to male controls in half of the regions in which interactions were found (Squeglia et al., 2011). These findings could indicate a female sensitivity to alcohol's neurotoxic effects when consumed in high quantities, especially since poorer attention and working memory scores were related to reduced response in frontal and cerebellar regions.

Another study by Tapert et al. (2004a) included youth with varying alcohol experience and administered a questionnaire to examine self-reported effects of alcohol during initial use of the substance. Higher number of drinks needed to experience effects was both positively and negatively related to working memory-related brain response in superior frontal and superior parietal regions, respectively, which were among many regions correlated with this measure. These results suggest that different brain regions may be engaged by youth with varying levels of response to alcohol's effects and could help establish neural markers of risk in adolescents who report low levels of response to alcohol or neural markers that are protective in youth who report less drinks needed to experience subjective effects.

The first longitudinal study of brain response in heavy alcohol users during a working memory task examined group differences in heavy alcohol users versus controls and also looked at baseline differences between these youth who later transitioned into heavy alcohol use versus those who continued to remain abstinent across the 3-year period of the study. Greater frontoparietal activity was present in heavy alcohol-using adolescents compared with controls, but at baseline, opposite patterns were seen, such that in multiple regions, including the inferior parietal lobe, brain response was lower in heavy drinkers compared with controls (Squeglia et al., 2012a). It is possible that while normal neurodevelopmental trajectories are taking place in healthy adolescents, altered neuromaturation in heavy alcohol users is leading to different patterns of brain response and that baseline neural phenotypes that predict who goes on to use alcohol heavily may be very informative to preventing heavy alcohol use in at-risk youth.

4.2 INHIBITORY CONTROL

Poor impulse control has been associated with alcohol abuse, and the neural correlates of inhibitory control have been examined in adolescents who go on to use alcohol heavily. Using a longitudinal study design, Wetherill et al. (2013) examined nogo versus go brain response in two groups of youth at baseline and 3 years later. The controls remained low users, while the other group emerged into heavy drinking by the time of the second study visit. Group-by-time interactions showed that at baseline youth who later emerged into heavy alcohol use had lower middle frontal, inferior parietal, and cerebellar tonsil brain activity, but higher brain response in these regions at follow-up compared with controls. This complements other findings that showed similar group-by-time interactions during a working memory task (Squeglia et al., 2012a), which suggests overlapping patterns of brain activity during executive functioning tasks at baseline and follow-up in youth who emerge into heavy alcohol use. These results are also supported by another study of inhibitory control where only baseline data were collected, but similar findings were seen showing less inhibitory control brain activity in PFC, parietal, and temporal regions in youth who later emerged into heavy alcohol use compared with their peers (Norman et al., 2011).

4.3 VERBAL ENCODING

An fMRI study of binge-drinking youth showed increased frontoparietal activity compared with controls during verbal encoding of word pairs, and no significant left hippocampal activity while encoding words (Schweinsburg et al., 2010), a region where volumetric reductions had been found previously in adolescents with AUD (Nagel et al., 2005). The elevated frontoparietal response could indicate compensatory changes in neural energetics or network efficiency in youth who use alcohol heavily. Superior frontal and inferior parietal brain response was also elevated in binge drinkers in a four-group design, with controls, users of alcohol or marijuana, and adolescents who used both substances (Schweinsburg et al., 2011). Polysubstance users' brain activity resembled that of controls, more so than use of either substance alone, suggesting that use patterns and types of substances used are important considerations for future studies even when groups are matched on task performance.

4.4 DECISION MAKING AND REWARD PROCESSING

Adolescent heavy alcohol use may be associated with immature decision-making abilities that lead to maladaptive risk taking. The Iowa gambling task is a decision-making task that has been used in an fMRI study of binge drinkers and showed that heavy alcohol users make more disadvantageous decisions on the task than nonusers (Xiao et al., 2013). Significantly, elevated insular and amygdalar response was present in binge drinkers during the task compared to controls, and insular response was positively related to drinking problems and urgency scores, a measure of impulsive behavior driven by negative emotions. These findings, similar to altered insular white matter volume in heavy alcohol users, may indicate that bottom-up affect-driven decision making may be related to risky behaviors, such as binge drinking.

A comparison of current binge drinkers and controls that identified brain areas that best differentiated the two groups via a machine learning analysis found that ventromedial PFC and inferior frontal gyrus activity were reduced during reward anticipation and consumption in binge drinkers compared with controls (Whelan et al., 2014). While bottom-up decision making-related brain activity may be enhanced, these findings suggest that reward evaluation may be compromised in heavy alcohol users.

5 CONCLUSION AND FUTURE DIRECTIONS

A wealth of findings from structural (Tables 1 and 2) and functional (Table 3) neuroimaging studies of adolescent alcohol users indicates that a diagnosis of an AUD or heavy alcohol use, in the absence of an AUD diagnosis, impacts brain volume, cortical thickness, white matter microstructure, and brain activity. Reduced hippocampal, PFC, and cerebellar volumes have all been reported in adolescent alcohol users, while reward-related ventral striatal volumes have been shown to be increased. Similarly, white matter integrity appears reduced in long-range association tracts connecting frontal and parietal regions, but increased in frontolimbic tracts. Working memory-related and inhibitory control brain activity appears to be reduced prior to heavy alcohol use in those who go on to binge drink, while being greater than or at the level of controls in binge drinkers after heavy alcohol use. Binge drinkers also show compensatory activity in frontoparietal regions during verbal encoding, in the absence of hippocampal activity. Bottom-up emotion processing brain areas may be hyperactive during affective decision making, while ventromedial PFC and inferior frontal gyrus activity may be compromised during reward evaluation.

For many of these structural and functional brain imaging studies, sex-by-group interactions are present, indicating a need for future research to examine these effects

Table 1 Brain Volume and Cortical Thickness Findings in Adolescent Alcohol Users

Structure	Study	Age Range (Years)	Population	Main Findings
Hippocampus	De Bellis et al. (2000)	13.5–21.0	$N=12$ AUD; $N=24$ C	↓L and R hippocampal volume in A
	Nagel et al. (2005)	15–17	$N=14$ AUD; $N=17$ C	↓L hippocampal volume in A
	Medina et al. (2007)	15–18	$N=16$ A; $N=26$ A+M; $N=21$ C	↓L hippocampal volume and R>L asymmetry in A
Frontal lobe	De Bellis et al. (2005)	13–21	$N=14$ AUD; $N=28$ C	↓PFC and PFC white matter volume in A, correlated with alcohol use
	Medina et al. (2008)	15–17	$N=14$ AUD, $N=17$ C	A females ↓avPFC volume, A males ↑avPFC volume
	Squeglia et al. (2012b)	16–19	$N=29$ binge; $N=30$ C	A females thicker L PFC cortices, A males thinner L PFC cortices
	Thayer et al. (2012)	14–18	$N=167$ JJ adolescents	OFC volume negatively associated with drinking frequency
	Luciana et al. (2013)	16–22 (at follow-up)	$N=30$ A; $N=25$ C	A ↑R MFG cortical thickness loss at follow-up
	Whelan et al. (2014)	14 year olds	$N=115$ binge; $N=150$ C	A ↓bilateral vmPFC, R inferior, and L MFG volume
Cerebellum	De Bellis et al. (2005)	13–21	$N=14$ AUD; $N=28$ C	A males ↓volume
	Lisdahl et al. (2013)	16–19	$N=46$ Binge; $N=50$ C	Peak binge drinks in past 3 months negatively related to cerebellar gray matter
Thalamus	De Bellis et al. (2005)	13–21	$N=14$ AUD; $N=28$ C	↔Between A and C
	Fein et al. (2013)	12–16	$N=64$ AUD; $N=64$ C	A males ↓thalamus volume; A females ↑thalamus volume

Table 1 Brain Volume and Cortical Thickness Findings in Adolescent Alcohol Users—cont'd

Structure	Study	Age Range (Years)	Population	Main Findings
Basal ganglia	Thayer et al. (2012)	14–18	N=167 JJ adolescents	NAcc volume positively associated with past 3-month frequency of drinking
	Fein et al. (2013)	12–16	N=64 AUD; N=64 C	A males ↓putamen volume; A females ↑putamen volume
	Squeglia et al. (2014)	15–21 (at follow-up)	N=20 heavy drinkers; N=20 C	More lifetime alcohol use negatively associated with left caudate volume reduction in A
Insula	Chung and Clark (2014)	14–18	N=30 drinkers (14 with AUD)	Enhancement motives for alcohol use mediated the relationship between insula white matter volume and frequency of past year binge drinking

A, alcohol users; A+M, alcohol and marijuana users; AUD, alcohol use disorder; avPFC, anterior ventral prefrontal cortex; C, controls; JJ, juvenile justice-involved adolescents; L, left; MFG, middle frontal gyrus; NAcc, nucleus accumbens; OFC, orbitofrontal cortex; PFC, prefrontal cortex; R, right; vmPFC, ventromedial prefrontal cortex; ↓, smaller or decreased; ↑, larger or increased; ↔, no difference.

more closely to develop sex-specific prevention and treatment strategies for adolescent alcohol abusers. Furthermore, while comorbid alcohol and marijuana use may limit our understanding of alcohol's specific actions on brain structure and function, multigroup study designs can aid with interpretation.

Importantly, since a majority of the previous research has been cross-sectional, it is possible that many differences observed between heavy alcohol users and controls may be premorbid and present prior to the onset of heavy alcohol use. This suggestion is already supported by longitudinal studies indicating baseline differences in brain response between adolescents who do and do not go on to use alcohol heavily (Norman et al., 2011; Wetherill et al., 2013). Furthermore, neuroimaging studies of at-risk youth with a FHP have shown overlapping findings to those described in this chapter. For example, volumetric studies have found smaller parahippocampal gray matter density in FHP nonalcohol-abusing adults (Sjoerds et al., 2013), while smaller hippocampal volume has been reported in multiple studies of alcohol-abusing adolescents (De Bellis et al., 2000; Medina et al., 2007; Nagel et al., 2005). This begs the question of whether premorbid risk-related difference in hippocampal volume may

Table 2 White Matter Microstructure Findings in Adolescent Alcohol Users

White Matter Microstructure	Study	Age Range (Years)	Population	Main Findings
Alcohol	De Bellis et al. (2008)	13.3–19.3	*N*=32 AUD; *N*=28 C	A ↓MD, ↑FA in isthmus region of corpus callosum; ↑FA in rostral body of corpus callosum in A
	McQueeny et al. (2009)	16–19	*N*=14 binge; *N*=14 C	A ↓FA across 18 white matter tracts (i.e., association, commissural, projection fibers); some associated with hangover symptoms and BAC
	Thayer et al. (2013)	14–18	*N*=125 JJ adolescents	High AUDIT group have ↓FA in PCR, R SLF; ↑FA in R ACR
	Luciana et al. (2013)	16–22 (at follow-up)	*N*=30 A; *N*=25 C	A ↓FA increase in caudate and inferior fronto-occipital fasciculus at follow-up
	Cardenas et al. (2013)	Not reported	*N*=50 AUD; *N*=50 C	↑FA in limbic white matter tracts
Alcohol + marijuana	Bava et al. (2009)	16–19	*N*=36 A+M; *N*=36 C	↓FA in A+M in 10 regions (many frontoparietal); ↑FA in occipital, SLF, internal capsule
	Jacobus et al. (2009)	16–19	*N*=14 binge; *N*=14 B+M; *N*=14 C	↓FA in A across eight white matter tracts (i.e., L SCR); in 4 of these polysubstance users ↑FA than A
	Bava et al. (2010)	16–19	*N*=36 A+M; *N*=36 C	↓FA in temporal regions associated with poorer working memory, processing speed, attention; ↑FA related to better verbal learning and visuomotor skills

Table 2 White Matter Microstructure Findings in Adolescent Alcohol Users—cont'd

White Matter Microstructure	Study	Age Range (Years)	Population	Main Findings
	Thatcher et al. (2010)	14–18	*N*=24 adolescents entering treatment; *N*=12 C	↓FA in treatment group in SLF; females with SUD ↓FA than males with SUD; C females ↑FA than C males
	Bava et al. (2013)	17–22 (at follow-up)	*N*=41 A+M; *N*=51 C	↓FA in A in R corpus callosum, prefrontal thalamic fibers, PCR; ↑MD and RD in many tracts, including SLF, PCR, prefrontal thalamic fibers
	Jacobus et al. (2013a)	19–22 (at follow-up)	*N*=17 binge; *N*=21 B+M; *N*=16 C	↓FA in A and A+M across 14 white matter tracts (i.e., corpus callosum, ACR, SLF, UF, SCR, internal capsule)
	Jacobus et al. (2013b)	19–22 (at follow-up)	*N*=8 heavy alcohol; *N*=8 heavy alcohol +M	↓FA from baseline to follow-up in A +M in 12 tracts (association, projection, interhemispheric tracts); six tracts ↑FA for A and ↓FA for A+M

A, alcohol users; A+M, alcohol and marijuana users; ACR, anterior corona radiata; AUD, alcohol use disorder; AUDIT, alcohol use disorders identification test; BAC, blood alcohol content; C, controls; FA, fractional anisotropy; JJ, juvenile justice-involved adolescents; L, left; M, marijuana users; MD, mean diffusivity; PCR, posterior corona radiata; PFC, prefrontal cortex; R, right; RD, radial diffusivity; SLF, superior longitudinal fasciculus; SUD, substance use disorder; UF, uncinate fasciculus; ↓, smaller or decreased; ↑, larger or increased.

Table 3 Brain Activity Findings in Adolescent Alcohol Users

fMRI	Study	Age Range (Years)	Population	Main Findings
Working memory	Tapert et al. (2004a)	15–17	$N=35$ alcohol users	Self-reported response to alcohol positively related to superior and middle frontal gyrus activity
	Tapert et al. (2004b)	15–17	$N=15$ AUD; $N=19$ C	A ↑parietal activity, ↓precentral and cerebellar response
	Caldwell et al. (2005)	14–17	$N=18$ AUD; $N=21$ C	A x sex interaction in SFG; A females ↓than A males or C females
	Squeglia et al. (2011)	16–19	$N=40$ binge; $N=55$ C	A females ↓, A males ↑activity across frontal, cingulate, temporal, cerebellar cortices
	Squeglia et al. (2012a)	12–16 (Study 2)	$N=20$ heavy drinkers; $N=20$ C	↑Frontal and parietal activity from baseline to follow-up in A, ↓in C
Inhibitory control	Norman et al. (2011)	12–14 (baseline)	$N=21$ transitioned to heavy use; $N=17$ C	↓Activation at baseline in transitioners in frontoparietal and frontostriatal areas during nogo
	Wetherill et al. (2013)	12–17 (baseline) 15–21 (follow-up)	$N=20$ binge; $N=20$ C	↓Activation at baseline in transitioners in frontoparietal, subcortical, cerebellar regions during nogo; ↑activation in frontoparietal, cerebellar regions at follow-up
Verbal encoding	Schweinsburg et al. (2010)	16–18	$N=12$ binge; $N=12$ C	↑Activation in R SFG, bilateral posterior parietal cortex, ↓activation in occipital cortex during novel encoding in A
	Schweinsburg et al. (2011)	16–18	$N=16$ binge; $N=28$ B+M; $N=8$ M; $N=22$ C	↓Inferior frontal; ↑dorsal frontal and parietal in A; polysubstance users resembled controls

Table 3 Brain Activity Findings in Adolescent Alcohol Users—cont'd

fMRI	Study	Age Range (Years)	Population	Main Findings
Decision making/ reward	Xiao et al. (2013)	16–18	*N*=14 binge; *N*=14 C	↑L amygdala and bilateral insula activity in A during affective decision making
	Whelan et al. (2014)	14 year olds	*N*=115 binge; *N*=150 C	↓Activation in A in vmPFC, L IFG, putamen, and hippocampus during reward anticipation; ↓activation in vmPFC, L IFG, R hippocampus during reward receipt

A, alcohol users; AUD, alcohol use disorder; B+M, binge drinkers and marijuana users; C, controls; IFG, inferior frontal gyrus; L, left; M, marijuana users; R, right; SFG, superior frontal gyrus; vmPFC, ventromedial prefrontal cortex; ↓, smaller or decreased; ↑, larger or increased.

have contributed to previous hippocampal findings in alcohol-using youth. Additionally, numerous studies have reported reduced FA across white matter tracts critical for subserving executive functioning, such as the SLF, in both heavy alcohol users (Jacobus et al., 2013a; McQueeny et al., 2009; Thatcher et al., 2010) and FHP youth compared with controls (Acheson et al., 2014; Herting et al., 2010), suggesting premorbid white matter microstructure deficits may heighten vulnerability toward alcohol abuse. On the other hand, it is possible that some differences between groups represent developmental delays in brain maturation in adolescent alcohol users or that some of these changes are reversible over time if alcohol use ceases. Ongoing longitudinal research studies are critical to answer these questions and inform the addiction medicine prevention and treatment community about neural risk factors for alcohol use as well as the consequences of heavy alcohol use on neural structure and functioning. Additionally, studies that can provide direct neurocognitive correlates of these structural or functional phenotypes can be helpful in identifying strategies where interventions are necessary to aid healthy cognitive functioning and brain maturation during adolescence.

REFERENCES

Acheson, A., Wijtenburg, S.A., Rowland, L.M., Winkler, A.M., Gaston, F., Mathias, C.W., Fox, P.T., Lovallo, W.R., Wright, S.N., Hong, L.E., Dougherty, D.M., Kochunov, P., 2014. Assessment of whole brain white matter integrity in youths and young adults with a family history of substance-use disorders. Hum. Brain Mapp. 35, 5401–5413.

Bava, S., Frank, L.R., McQueeny, T., Schweinsburg, B.C., Schweinsburg, A.D., Tapert, S.F., 2009. Altered white matter microstructure in adolescent substance users. Psychiatry Res. 173, 228–237.

Bava, S., Jacobus, J., Mahmood, O., Yang, T.T., Tapert, S.F., 2010. Neurocognitive correlates of white matter quality in adolescent substance users. Brain Cogn. 72, 347–354.

Bava, S., Jacobus, J., Thayer, R.E., Tapert, S.F., 2013. Longitudinal changes in white matter integrity among adolescent substance users. Alcohol. Clin. Exp. Res. 37 (Suppl. 1), E181–E189.

Beresford, T.P., Arciniegas, D.B., Alfers, J., Clapp, L., Martin, B., Du, Y., Liu, D., Shen, D., Davatzikos, C., 2006. Hippocampus volume loss due to chronic heavy drinking. Alcohol. Clin. Exp. Res. 30, 1866–1870.

Broadwater, M.A., Liu, W., Crews, F.T., Spear, L.P., 2014. Persistent loss of hippocampal neurogenesis and increased cell death following adolescent, but not adult, chronic ethanol exposure. Dev. Neurosci. 36, 297–305.

Caldwell, L.C., Schweinsburg, A.D., Nagel, B.J., Barlett, V.C., Brown, S.A., Tapert, S.F., 2005. Gender and adolescent alcohol use disorders on BOLD (blood oxygen level dependent) response to spatial working memory. Alcohol Alcohol. 40, 194–200.

Cardenas, V.A., Greenstein, D., Fouche, J.P., Ferrett, H., Cuzen, N., Stein, D.J., Fein, G., 2013. Not lesser but Greater fractional anisotropy in adolescents with alcohol use disorders. Neuroimage Clin. 2, 804–809.

Casement, M.D., Shaw, D.S., Sitnick, S.L., Musselman, S.C., Forbes, E.E., 2015. Life stress in adolescence predicts early adult reward-related brain function and alcohol dependence. Soc. Cogn. Affect. Neurosci. 10, 416–423.

Casey, B.J., Giedd, J.N., Thomas, K.M., 2000. Structural and functional brain development and its relation to cognitive development. Biol. Psychol. 54, 241–257.

Chung, T., Clark, D.B., 2014. Insula white matter volume linked to binge drinking frequency through enhancement motives in treated adolescents. Alcohol. Clin. Exp. Res. 38, 1932–1940.

Crone, E.A., Wendelken, C., Donohue, S., Van Leijenhorst, L., Bunge, S.A., 2006. Neurocognitive development of the ability to manipulate information in working memory. Proc. Natl. Acad. Sci. U.S.A. 103, 9315–9320.

Dawson, D.A., Goldstein, R.B., Chou, S.P., Ruan, W.J., Grant, B.F., 2008. Age at first drink and the first incidence of adult-onset DSM-IV alcohol use disorders. Alcohol. Clin. Exp. Res. 32, 2149–2160.

De Bellis, M.D., Clark, D.B., Beers, S.R., Soloff, P.H., Boring, A.M., Hall, J., Kersh, A., Keshavan, M.S., 2000. Hippocampal volume in adolescent-onset alcohol use disorders. Am. J. Psychiatry 157, 737–744.

De Bellis, M.D., Narasimhan, A., Thatcher, D.L., Keshavan, M.S., Soloff, P., Clark, D.B., 2005. Prefrontal cortex, thalamus, and cerebellar volumes in adolescents and young adults with adolescent-onset alcohol use disorders and comorbid mental disorders. Alcohol. Clin. Exp. Res. 29, 1590–1600.

De Bellis, M.D., Van Voorhees, E., Hooper, S.R., Gibler, N., Nelson, L., Hege, S.G., Payne, M.E., MacFall, J., 2008. Diffusion tensor measures of the corpus callosum in adolescents with adolescent onset alcohol use disorders. Alcohol. Clin. Exp. Res. 32, 395–404.

Fein, G., Greenstein, D., Cardenas, V.A., Cuzen, N.L., Fouche, J.P., Ferrett, H., Thomas, K., Stein, D.J., 2013. Cortical and subcortical volumes in adolescents with alcohol dependence but without substance or psychiatric comorbidities. Psychiatry Res. Neuroimaging 214, 1–8.

Giedd, J.N., Blumenthal, J., Jeffries, N.O., Castellanos, F.X., Liu, H., Zijdenbos, A., Paus, T., Evans, A.C., Rapoport, J.L., 1999. Brain development during childhood and adolescence: a longitudinal MRI study. Nat. Neurosci. 2, 861–863.

Gogtay, N., Giedd, J.N., Lusk, L., Hayashi, K.M., Greenstein, D., Vaituzis, A.C., Nugent 3rd, T.F., Herman, D.H., Clasen, L.S., Toga, A.W., Rapoport, J.L., Thompson, P.M., 2004. Dynamic mapping of human cortical development during childhood through early adulthood. Proc. Natl. Acad. Sci. U.S.A. 101, 8174–8179.

Herting, M.M., Schwartz, D., Mitchell, S.H., Nagel, B.J., 2010. Delay discounting behavior and white matter microstructure abnormalities in youth with a family history of alcoholism. Alcohol. Clin. Exp. Res. 34, 1590–1602.

Hulvershorn, L.A., Cullen, K., Anand, A., 2011. Toward dysfunctional connectivity: a review of neuroimaging findings in pediatric major depressive disorder. Brain Imaging Behav. 5, 307–328.

Jacobus, J., McQueeny, T., Bava, S., Schweinsburg, B.C., Frank, L.R., Yang, T.T., Tapert, S.F., 2009. White matter integrity in adolescents with histories of marijuana use and binge drinking. Neurotoxicol. Teratol. 31, 349–355.

Jacobus, J., Squeglia, L.M., Bava, S., Tapert, S.F., 2013a. White matter characterization of adolescent binge drinking with and without co-occurring marijuana use: a 3-year investigation. Psychiatry Res. 214, 374–381.

Jacobus, J., Squeglia, L.M., Infante, M.A., Bava, S., Tapert, S.F., 2013b. White matter integrity pre- and post marijuana and alcohol initiation in adolescence. Brain Sci. 3, 396–414.

Johnston, L.D., O'Malley, P.M., Meich, R.A., Bachman, J.G., Schulenberg, J.E., 2015. Monitoring the Future National Survey Results on Drug Use: 2014 Overview, Key Findings on Adolescent Drug Use. Institute for Social Research, The University of Michigan, Ann Arbor, MI.

King, S.M., Iacono, W.G., McGue, M., 2004. Childhood externalizing and internalizing psychopathology in the prediction of early substance use. Addiction 99, 1548–1559.

Lieb, R., Merikangas, K.R., Hofler, M., Pfister, H., Isensee, B., Wittchen, H.U., 2002. Parental alcohol use disorders and alcohol use and disorders in offspring: a community study. Psychol. Med. 32, 63–78.

Lisdahl, K.M., Thayer, R., Squeglia, L.M., McQueeny, T.M., Tapert, S.F., 2013. Recent binge drinking predicts smaller cerebellar volumes in adolescents. Psychiatry Res. 211, 17–23.

Luciana, M., Collins, P.F., Muetzel, R.L., Lim, K.O., 2013. Effects of alcohol use initiation on brain structure in typically developing adolescents. Am. J. Drug Alcohol Abuse 39, 345–355.

Macpherson, L., Magidson, J.F., Reynolds, E.K., Kahler, C.W., Lejuez, C.W., 2010. Changes in sensation seeking and risk-taking propensity predict increases in alcohol use among early adolescents. Alcohol. Clin. Exp. Res. 34, 1400–1408.

McQueeny, T., Schweinsburg, B.C., Schweinsburg, A.D., Jacobus, J., Bava, S., Frank, L.R., Tapert, S.F., 2009. Altered white matter integrity in adolescent binge drinkers. Alcohol. Clin. Exp. Res. 33, 1278–1285.

Medina, K.L., Schweinsburg, A.D., Cohen-Zion, M., Nagel, B.J., Tapert, S.F., 2007. Effects of alcohol and combined marijuana and alcohol use during adolescence on hippocampal volume and asymmetry. Neurotoxicol. Teratol. 29, 141–152.

Medina, K.L., McQueeny, T., Nagel, B.J., Hanson, K.L., Schweinsburg, A.D., Tapert, S.F., 2008. Prefrontal cortex volumes in adolescents with alcohol use disorders: unique gender effects. Alcohol. Clin. Exp. Res. 32, 386–394.

Mills, K.L., Goddings, A.L., Clasen, L.S., Giedd, J.N., Blakemore, S.J., 2014. The developmental mismatch in structural brain maturation during adolescence. Dev. Neurosci. 36, 147–160.

Nagel, B.J., Schweinsburg, A.D., Phan, V., Tapert, S.F., 2005. Reduced hippocampal volume among adolescents with alcohol use disorders without psychiatric comorbidity. Psychiatry Res. Neuroimaging 139, 181–190.

Norman, A.L., Pulido, C., Squeglia, L.M., Spadoni, A.D., Paulus, M.P., Tapert, S.F., 2011. Neural activation during inhibition predicts initiation of substance use in adolescence. Drug Alcohol Depend. 119, 216–223.

Schmahmann, J.D., Pandya, D.N., 1997. Anatomic organization of the basilar pontine projections from prefrontal cortices in rhesus monkey. J. Neurosci. 17, 438–458.

Schweinsburg, A.D., McQueeny, T., Nagel, B.J., Eyler, L.T., Tapert, S.F., 2010. A preliminary study of functional magnetic resonance imaging response during verbal encoding among adolescent binge drinkers. Alcohol 44, 111–117.

Schweinsburg, A.D., Schweinsburg, B.C., Nagel, B.J., Eyler, L.T., Tapert, S.F., 2011. Neural correlates of verbal learning in adolescent alcohol and marijuana users. Addiction 106, 564–573.

Sjoerds, Z., Van Tol, M.J., Van den Brink, W., Van der Wee, N.J., Van Buchem, M.A., Aleman, A., Penninx, B.W., Veltman, D.J., 2013. Family history of alcohol dependence and gray matter abnormalities in non-alcoholic adults. World J. Biol. Psychiatry 14, 565–573.

Squeglia, L.M., Schweinsburg, A.D., Pulido, C., Tapert, S.F., 2011. Adolescent binge drinking linked to abnormal spatial working memory brain activation: differential gender effects. Alcohol. Clin. Exp. Res. 35, 1831–1841.

Squeglia, L.M., Pulido, C., Wetherill, R.R., Jacobus, J., Brown, G.G., Tapert, S.F., 2012a. Brain response to working memory over three years of adolescence: influence of initiating heavy drinking. J. Stud. Alcohol Drugs 73, 749–760.

Squeglia, L.M., Sorg, S.F., Schweinsburg, A.D., Wetherill, R.R., Pulido, C., Tapert, S.F., 2012b. Binge drinking differentially affects adolescent male and female brain morphometry. Psychopharmacology (Berl.) 220, 529–539.

Squeglia, L.M., Rinker, D.A., Bartsch, H., Castro, N., Chung, Y., Dale, A.M., Jernigan, T.L., Tapert, S.F., 2014. Brain volume reductions in adolescent heavy drinkers. Dev. Cogn. Neurosci. 9, 117–125.

Tapert, S.F., Pulido, C., Paulus, M.P., Schuckit, M.A., Burke, C., 2004a. Level of response to alcohol and brain response during visual working memory. J. Stud. Alcohol 65, 692–700.

Tapert, S.F., Schweinsburg, A.D., Barlett, V.C., Brown, S.A., Frank, L.R., Brown, G.G., Meloy, M.J., 2004b. Blood oxygen level dependent response and spatial working memory in adolescents with alcohol use disorders. Alcohol. Clin. Exp. Res. 28, 1577–1586.

Thatcher, D.L., Pajtek, S., Chung, T., Terwilliger, R.A., Clark, D.B., 2010. Gender differences in the relationship between white matter organization and adolescent substance use disorders. Drug Alcohol Depend. 110, 55–61.

Thayer, R.E., Crotwell, S.M., Callahan, T.J., Hutchison, K.E., Bryan, A.D., 2012. Nucleus accumbens volume is associated with frequency of alcohol use among juvenile justice-involved adolescents. Brain Sci. 2, 605–618.

Thayer, R.E., Callahan, T.J., Weiland, B.J., Hutchison, K.E., Bryan, A.D., 2013. Associations between fractional anisotropy and problematic alcohol use in juvenile justice-involved adolescents. Am. J. Drug Alcohol Abuse 39, 365–371.

Van Leijenhorst, L., Moor, B.G., Op de Macks, Z.A., Rombouts, S.A., Westenberg, P.M., Crone, E.A., 2010. Adolescent risky decision-making: neurocognitive development of reward and control regions. Neuroimage 51, 345–355.

Wetherill, R.R., Squeglia, L.M., Yang, T.T., Tapert, S.F., 2013. A longitudinal examination of adolescent response inhibition: neural differences before and after the initiation of heavy drinking. Psychopharmacology (Berl.) 230, 663–671.

Whelan, R., Watts, R., Orr, C.A., Althoff, R.R., Artiges, E., Banaschewski, T., Barker, G.J., Bokde, A.L., Buchel, C., Carvalho, F.M., Conrod, P.J., Flor, H., Fauth-Buhler, M., Frouin, V., Gallinat, J., Gan, G., Gowland, P., Heinz, A., Ittermann, B., Lawrence, C., Mann, K., Martinot, J.L., Nees, F., Ortiz, N., Paillere-Martinot, M.L., Paus, T., Pausova, Z., Rietschel, M., Robbins, T.W., Smolka, M.N., Strohle, A., Schumann, G., Garavan, H., 2014. Neuropsychosocial profiles of current and future adolescent alcohol misusers. Nature 512, 185–189.

Xiao, L., Bechara, A., Gong, Q., Huang, X., Li, X., Xue, G., Wong, S., Lu, Z.L., Palmer, P., Wei, Y., Jia, Y., Johnson, C.A., 2013. Abnormal affective decision making revealed in adolescent binge drinkers using a functional magnetic resonance imaging study. Psychol. Addict. Behav. 27, 443–454.

CHAPTER

Neuroscience of opiates for addiction medicine: From stress-responsive systems to behavior

12

Yan Zhou[*,1], **Francesco Leri**[†]

The Laboratory of the Biology of Addictive Diseases, The Rockefeller University, New York, NY, USA

[†]*Department of Psychology, University of Guelph, Guelph, ON, Canada*

[1]*Corresponding author: Tel.: +1-212-3278248; Fax: +1-212-3278574, e-mail address: yan.zhou@rockefeller.edu*

Abstract

Opiate addiction, similarly to addiction to other psychoactive drugs, is chronic relapsing brain disease caused by drug-induced short-term and long-term neuroadaptations at the molecular, cellular, and behavioral levels. Preclinical research in laboratory animals has found important interactions between opiate exposure and stress-responsive systems. In this review, we will discuss the dysregulation of several stress-responsive systems in opiate addiction: vasopressin and its receptor system, endogenous opioid systems (including proopiomelanocortin/mu opioid receptor and dynorphin/kappa opioid receptor), orexin and its receptor system, and the hypothalamic–pituitary–adrenal axis. A more complete understanding of how opiates alter these stress systems, through further laboratory-based studies, is required to identify novel and effective pharmacological targets for the long-term treatment of heroin addiction.

Keywords

Stress, HPA axis, Vasopressin, V1b receptor, Dynorphin, POMC, Opioid receptor, Orexin, Heroin addiction

Abbreviation

AVP arginine vasopressin
CPP conditioned place preference
CRF corticotropin-releasing factor
CRF-R1 CRF type I receptor

Progress in Brain Research, Volume 223, ISSN 0079-6123, http://dx.doi.org/10.1016/bs.pbr.2015.09.001

eGFP	enhanced green fluorescent protein
HPA	hypothalamic–pituitary–adrenal
KOP-r	kappa opioid receptor
MOP-r	mu opioid receptor
NAc	nucleus accumbens
nor-BNI	nor-binaltorphimine
POMC	proopiomelanocortin
pPVN	parvocellular division of PVN
PVN	paraventricular nucleus
V1b	AVP type 1b receptor
VTA	ventral tegmental area

1 INTRODUCTION

Opiate addiction is a major global public health problem, and there is still a need for alternative medications that could enhance the effectiveness of existing treatments (e.g., methadone maintenance treatment) in reducing opiate abuse, manage withdrawal, and prevent relapse. Addiction to heroin, a potent agonist at opioid receptors, is often characterized by a drug user's initial use resulting in tolerance and eventually severe withdrawal symptoms during periods of abstinence. The reward pathways and withdrawal symptoms, and the desire to avoid them often result in relapse and reescalation of drug use. The length of time a person spends in each of the different developing stages of opiate addiction varies by individual; stress plays a major role in initiation, maintenance, and withdrawal, and elevates drug craving (Koob and Kreek, 2007).

Recent studies in laboratory animals have implicated the dysregulation of several brain stress-responsive systems and hypothalamic–pituitary–adrenal (HPA) axis in the acquisition of opiate self-administration behaviors and progression toward opiate dependence. For example, environmental stressors modulate the effects of drugs on the acquisition of drug self-administration behavior, locomotor activity, and reinstatement of self-administration after extinction (e.g., Breese et al., 2011; Heilig et al., 2010; Koob and Kreek, 2007; Schank et al., 2012; Shalev et al., 2010; Sinha et al., 2011; Spanagel et al., 2014). Atypical stress responsivity is one of the critical factors influencing individual vulnerability to drug relapse. This review will discuss the roles of four important systems to current addiction research.

The arginine vasopressin (AVP) system has been studied in neuroendocrinology and drug addiction. The HPA axis in rodents is directly influenced by the AVP system. This neuropeptide system is profoundly altered by opiates in rodent models, which is discussed in detail in this review. Of interest is the impact of drugs of abuse, such as heroin, alcohol, and cocaine, on AVP and V1b (AVP type 1b receptor) system and their potential roles in taking and seeking behaviors. Specific brain areas such as the medial amygdala, paraventricular nucleus (PVN) of the hypothalamus, and anterior pituitary will be discussed as well as the functions of AVP and its V1b receptors which have been identified and elucidated.

The endogenous opioid systems include the proopiomelanocortin (POMC)/mu opioid receptor (MOP-r) and dynorphin/kappa opioid receptor (KOP-r) systems. The new data on potential roles of POMC and dynorphin in opiate-related seeking behaviors and the control of the HPA axis will be discussed herein as the second topic.

The third is the stress-responsive orexin (or hypocretin) system. Most of the lateral hypothalamic orexin neurons coexpress dynorphin. Orexin has actions in reward-related areas of the brain, such as the nucleus accumbens (NAc) and ventral tegmental area (VTA) with implications in rewarding and addictive-like behaviors.

Finally, there is substantial evidence demonstrating that opiates alter the HPA axis, and in turn the abnormal HPA activity may contribute to the development of opiate addiction and relapse to drug use. In our discussion of the four systems, we will provide an overview of recent research on opiate addiction, with specific emphasis on preclinical laboratory-based research to elucidate the neurobiology of opiate addiction.

2 AVP AND V1B SYSTEMS

The two central G protein-coupled AVP receptor subtypes V1a and V1b are highly expressed in the rat extended amygdala. The V1b receptors are expressed prominently in the amygdala, hypothalamus, hippocampus, and anterior pituitary. Recent studies suggest that increased AVP neuronal activity in the amygdala represents a step in the neurobiology of stress-related behaviors in rodent models: (a) acute stress increases extracellular AVP levels in the rat amygdala (Wigger et al., 2004) and (b) activation of AVP V1b receptors underlies anxiety-like and depression-like behaviors (Griebel et al., 2002; Roper et al., 2011; Salome et al., 2006; Serradeil-Le Gal et al., 2002).

In order to examine the role of AVP in heroin addiction, the expression of the AVP gene in the rat amygdala and hypothalamus was studied after chronic intermittent escalating-dose heroin or during early and late spontaneous withdrawal. AVP mRNA levels expressed in the medial amygdala were increased during early heroin withdrawal (Zhou et al., 2008a). To further study the AVP system in response to stress, we designed experiments using unpredictable foot shock, a widely used stimulus to elicit stress responses in rodents. Foot-shock stress was found to increase AVP mRNA levels in the medial amygdala in rats trained to self-administer heroin (in comparison with the nonshock rats self-administering heroin and the rats with saline self-administration), suggesting that AVP sensitivity to stress is increased after voluntary heroin exposure. It was also investigated whether blockade of central AVP receptors (V1a or V1b receptor) would attenuate the reinstatement of heroin-seeking behavior by foot-shock stress, and HPA hormonal responses to foot shock. The selective V1b receptor antagonist SSR149415 (but not the relatively selective V1a antagonist) dose-dependently attenuated foot shock-induced reinstatement of heroin seeking and lowered foot shock-induced HPA activation (Zhou et al., 2008a).

Additionally, increased AVP mRNA levels in the medial amygdala were also found after acute withdrawal from chronic cocaine exposure (Zhou et al., 2005). Together, these data suggest that stress-responsive AVP/V1b receptor systems in the medial amygdala may be critical components of the neural circuitry underlying the effect of aversive emotional states, such as drug withdrawal, on drug-seeking behavior (Brown and Lawrence, 2009; D'Souza and Markou, 2010; Zhou et al., 2008a).

Additionally, to investigate the involvement of AVP and V1b in alcohol drinking behavior, we used genetically selected Sardinian alcohol-preferring (sP) rats (Colombo et al., 2006), in collaboration with the Colombo laboratory at C.N.R. Institute of Neuroscience at the University of Cagliari in Italy, and found that pharmacological blockade of V1b receptor reduced alcohol consumption in sP rats (Zhou et al., 2011a). This is consistent with the observation that the selective V1b receptor antagonist SSR149415 dose-dependently reduced excessive levels of alcohol self-administration in alcohol-dependent Wistar rats (induced by chronic intermittent alcohol vapor exposure) without affecting the limited levels of alcohol drinking in nondependent rats (Edwards et al., 2012).

Interestingly, higher basal levels of AVP mRNA are found in the PVN of Indiana alcohol-preferring and high-alcohol drinking rats, when compared to alcohol-nonpreferring and low-alcohol drinking rats. Our recent studies on selectively bred Sardinian alcohol-preferring versus Sardinian alcohol-nonpreferring rats showed similar results: Sardinian alcohol-preferring rats display higher basal AVP mRNA levels in the PVN and medial amygdala than Sardinian alcohol-nonpreferring rats, and voluntary alcohol consumption decreased the AVP mRNA levels in both the PVN and medial amygdala of Sardinian alcohol-preferring rats (Zhou et al., 2011a).

Earlier studies have shown that AVP in the parvocellular division of PVN (pPVN) does not contribute to the acute stimulatory effects of cocaine on HPA activity. However, we recently found persistent elevations of both peripheral plasma ACTH levels and AVP mRNA levels in the pPVN of the rats after 14 days of protracted cocaine withdrawal, and V1b antagonists attenuated cocaine withdrawal-induced HPA activation (Zhou et al., 2011b). Interestingly, in AVP-enhanced green fluorescent protein (eGFP) transgenic mice, cocaine withdrawal increased the number of pPVN AVP neurons expressing GFP, further confirming that enhanced pPVN AVP gene expression is associated with persistent elevations of basal HPA activity. These results indicate that AVP and its receptor system are involved in chronic stress and may be an attractive therapeutic target for treating anxiety and depressive symptoms associated with withdrawal from drugs of abuse. Hence, it may be important to explore the value of both the V1b and V1a receptor antagonists in the management of opiate, cocaine, or alcohol abuse and relapse (Bisagno and Cadet, 2014; Koob, 2008; Koshimizu et al., 2012; See and Waters, 2010). Also, though AVP is a potent modulator of stress, further studies should clarify the specific involvement of AVP and V1b receptor systems in relapse-like behaviors after chronic withdrawal from drugs.

3 ENDOGENOUS OPIOID SYSTEMS

3.1 POMC SYSTEMS

The POMC gene encodes a prohormone expressed at significant levels in the pituitary or specific brain regions. Cell-specific posttranslational processing of the POMC prohormone generates a variety of biologically active peptides. In the pituitary, anterior lobe corticotrophs release ACTH, whereas intermediate lobe melanotrophs further process POMC to produce N-acetylated forms of α-melanocyte-stimulating hormone and β-endorphin. In the brain, a specialized population of neurons in the hypothalamic arcuate nucleus expresses POMC, giving rise to α-, β-, and γ-melanocortins and the potent opioid peptide beta-endorphin (e.g., Cowley et al., 2001; Rubinstein et al., 1996). Beta-endorphin, primarily acting on MOP-r, is mainly expressed in the hypothalamic POMC neurons, the principal cluster of POMC/beta-endorphin neurons in the brain. It is known that activation of MOP-r by beta-endorphin results in positive affective reactions and is rewarding in rats, possibly by modulation of DA cells in the VTA and associated release of DA in the NAc (e.g., Spanagel et al., 1991). Hence, it is hardly surprising that beta-endorphin is involved in the motivational and affective actions of most drugs of abuse. Dopamine D_2-like receptors are involved in regulation of POMC gene expression in the hypothalamus (Zhou et al., 2004). The selective D_2 antagonist sulpiride increased POMC mRNA levels in the hypothalamus, indicating that D_2-like receptors exert tonic inhibitory effect on hypothalamic POMC gene expression. Blockade by a selective D_1-like receptor antagonist SCH23390 had no effect on hypothalamic POMC mRNA levels, suggesting a specific role for the D_2 receptors in hypothalamic POMC expression.

Heroin, morphine, and other short-acting opiates regulate the activity of endogenous opioid systems. In rats, chronic intermittent heroin administration by experimenters (Zhou et al., 2013c) or chronic heroin self-administration (Zhou et al., 2015) resulted in decreased POMC mRNA levels in the hypothalamus during acute (1 day) withdrawal or after 9 days of withdrawal, respectively, suggesting a hypothesis that long-term exposure to opiates leads to relative deficiency in beta-endorphin system (Negus and Rice, 2009). In parallel, there is an upregulation of MOP-r mRNA expression levels after acute opiate withdrawal in NAc, caudate-putamen, and lateral hypothalamus, indicating a compensatory increase in MOP-r biosynthesis due to the less beta-endorphin (Le Merrer et al., 2009; Zhou et al., 2006). The presumed relative deficiency in endogenous beta-endorphin (as reflected by decreased POMC mRNA levels in the hypothalamus) could lead to a hyperactive HPA axis and corresponding increases in hormone secretion during spontaneous opiate withdrawal. Indeed, increases in HPA hormonal levels have been observed in rats during acute spontaneous opiate withdrawal (Ignar and Kuhn, 1990; Martinez et al., 1990; Zhou et al., 2006).

POMC-derived peptides, especially beta-endorphin, are also distributed in the dopaminergic mesocorticolimbic regions, including the NAc, VTA, and frontal cortex. In addition to the arcuate nucleus, POMC mRNA has also been detected in the

NAc and dorsal striatum at relatively low levels. Therefore, the demonstration of POMC neuron distribution in the NAc region is an essential issue concerning the neural networks containing POMC mRNA and derived peptides in the NAc. Using POMC-eGFP transgenic mice in which POMC-expressing neurons were labeled with eGFP and enhanced by immunohistochemistry procedures, we found that POMC-eGFP-expressing neurons are present in modest amounts in the NAc core, shell, and dorsal striatum of POMC-eGFP mice (Zhou et al., 2013a).

We also measured POMC mRNA levels in these two subdivisions of NAc and the dorsal striatum of Sardinian alcohol-preferring (sP) rats exposed to 17-day alcohol drinking and found that voluntary consumption of high amounts of alcohol by sP rats was associated with increases in POMC mRNA levels in the NAc shell, but not NAc core. This result suggests that voluntary alcohol drinking modulates POMC mRNA expression in the POMC neuron populations in the NAc shell. The POMC neurons in the shell are a region long considered to mediate processes of reward and reinforcement (e.g., Di Chiara, 2002), and our results support their role in alcohol intake in this model (Zhou et al., 2013a). Several studies have also demonstrated that alcohol, cocaine, or cannabinoids are self-administered directly into the NAc shell, but not the core (e.g., Rodd-Henricks et al., 2002). Therefore, the shell (not the core) may be the region in which alcohol and other drugs of abuse contribute to reinforcing effects mediated through the POMC neuronal activation. The potential involvement of NAc POMC in heroin- or cocaine-related behaviors needs further investigation.

3.2 DYNORPHIN SYSTEMS

Activation of the dynorphin/KOP-r systems has been implicated in the negative reinforcement aspects of opiate, alcohol, and psychostimulant addictions (Bodnar, 2014; Koob and Kreek, 2007). As there is more recent progress on the dynorphin/KOP-r systems in alcohol research, we highlight some relevant findings below. In rats, acute administration of KOP-r agonists attenuates alcohol self-administration and decreases alcohol-induced conditioned place preference (CPP), while the selective KOP-r antagonist nor-binaltorphimine (nor-BNI) increased alcohol drinking in rats with high basal levels of alcohol consumption. KOP-r agonists inhibit GABAergic synaptic responses and alcohol effects in the central nucleus of amygdala and bed nucleus of the stria terminalis (Li et al., 2012), which may be the mechanism by which these agonists can modulate responses to alcohol and possibly other addictive drugs. Recently, KOP-r antagonists have also been reported to attenuate alcohol-seeking behavior induced by stress in mice (Sperling et al., 2010) and to reduce alcohol consumption in alcohol-dependent rats (Walker and Koob, 2008).

Dynorphin and KOP-r systems in the NAc shell have been implicated in the modulation of drug escalation and seeking behaviors (Nealey et al., 2011; Schlosburg et al., 2013; Zhou et al., 2013b). In fact, KOP-r antagonists administered in the shell reduce alcohol consumption in alcohol-dependent rats (Nealey et al., 2011) as well as escalation of heroin intake in dependent rats (Schlosburg et al., 2013). Interestingly, heroin seeking induced by stress is associated with increased dynorphin mRNA levels in this region (Zhou et al., 2013b).

Laboratory studies in humans have found that yohimbine increased subjective anxiety in normal subjects and drug craving in abstinent opiate addicts (Stine et al., 2002). Yohimbine enhances central noradrenergic activity by acting as an antagonist at alpha-2 adrenergic autoreceptors and reinstates heroin, methamphetamine, cocaine, alcohol, and food seeking. Of interest, the pretreatment with KOP-r antagonist nor-BNI blocked the heroin-seeking behavior induced by food deprivation stress (Sedki et al., 2015), or yohimbine stress (Zhou et al., 2013b), and attenuated escalation of heroin taking (Schlosburg et al., 2013). In line with these findings, blockade of KOP-r also attenuated naltrexone-precipitated withdrawal and conditioned place aversion in morphine-dependent rats, suggesting that KOP-r activation is involved in opiate withdrawal, its aversive consequences, and probably consequent relapse (Kelsey et al., 2015).

In tissue plasminogen activator (tPA, endogenous microglia activator) knockout mice, basal levels of dynorphin (but not KOP-r or enkephalin) mRNA are increased in the NAc (Maiya et al., 2009). It has been demonstrated that NAc dialysate dopamine levels are decreased after stimulation of KOP-r ((Spanagel et al., 1990). On the basis of this evidence, it has been hypothesized that increased basal levels of dynorphin in the NAc may lead to altered dopaminergic signaling after drug exposure. Indeed, cocaine-induced CPP and locomotor sensitization, as well as cocaine withdrawal-induced stress responses (CRF mRNA increases in the amygdala), are attenuated in tPA knockout mice (Maiya et al., 2009; Zhou et al., 2010). Our work also indicates a potential involvement of microglial systems in regulation of NAc dynorphin/KOP-r responsivity to drug reward, as tPA deletion resulted in a deficiency of endogenous microglia activation (Maiya et al., 2009).

As mentioned above, we have learned a great deal from selective animal models regarding POMC/MOP-r and dynorphin/KOP-r systems, focusing on the neurobiology of heroin addiction. The MOP-r and KOP-r are attractive therapeutic targets for treating opiate addiction.

4 OREXIN AND ITS RECEPTORS

The orexins are expressed in the lateral hypothalamus, perifornical area, and dorsomedial hypothalamus, with extensive projections in the brain (de Lecea et al., 1998). Orexin A acts at orexin type 1 and 2 receptors (OX1R and OX2R), and orexin B acts on OX2R exclusively. Hypothalamic orexins are involved in the regulation of sleep–wakefulness, arousal, feeding, and stress. Orexin receptor blockade in the VTA after acute morphine administration, for example, attenuated an increase in extracellular dopamine levels in the NAc (Narita et al., 2006), and orexins and orexin receptor interactions have been found to trigger morphine-motivated behaviors (Harris et al., 2005; Sharf et al., 2010; Smith and Aston-Jones, 2012), suggesting that orexins have a role in the modulation of drug reward and drug-seeking behaviors

(Baimel et al., 2015; Boutrel et al., 2010; Calipari and España, 2012; Yeoh et al., 2014), and OX1R antagonists could have medication potential in opioids of abuse (Harris et al., 2005).

During the aversive state of acute opiate withdrawal, orexin neuronal activity and mRNA levels are increased in the lateral hypothalamus, indicating that the increased orexin neuronal activity could contribute to negative affective states in opiate withdrawal (Georgescu et al., 2003). Because most of the lateral hypothalamic orexin neurons coexpress the dynorphin gene (Chou et al., 2001), we examined levels of both the orexin and dynorphin mRNAs in the lateral hypothalamus, and found that orexin mRNA levels were increased in rat lateral hypothalamus, but the dynorphin mRNA levels remain unaltered in the lateral hypothalamus in acute withdrawal from chronic escalating-dose morphine (Zhou et al., 2006). To investigate whether this observation held true with other drugs of abuse, we extended our research to cocaine. We were primarily interested in investigating whether orexin or dynorphin mRNA levels in rat lateral hypothalamus or medial hypothalamus (perifornical and dorsomedial areas) are altered by acute cocaine withdrawal (Zhou et al., 2008b). Similar to morphine withdrawal, acute withdrawal from chronic escalating-dose cocaine administration also resulted in increased orexin mRNA levels in the lateral hypothalamus, but not the medial hypothalamus. In contrast to the opiate withdrawal, acute cocaine withdrawal also increased the dynorphin mRNA levels in the lateral hypothalamus (Zhou et al., 2008b). These results suggest that the orexin expression alteration in response to drug withdrawal seems lateral hypothalamus-specific. Orexin and dynorphin in the lateral hypothalamus may differentially contribute to the enhanced negative affective states in opiate and cocaine withdrawal.

Modafinil is a stimulant used to treat narcolepsy and other disorders involving excessive sleepiness. Because of its long-acting stimulatory action on orexin, central monoamine, and glutamate systems (Minzenberg and Carter, 2008), modafinil was investigated as a potential "substitution" treatment for cocaine dependence, withdrawal, and relapse. We employed Sprague–Dawley rats trained to self-administer cocaine to explore whether chronic administration of modafinil (0, 25, or 50 mg/kg, IP, twice daily) could alter spontaneous cocaine seeking observed during extinction, and reinstatement of cocaine seeking precipitated by foot-shock stress (Leri et al., 2009). In parallel, two additional experiments were performed to assess the effect of cocaine sensitization (20 mg/kg × 5 days) on the stimulatory and rewarding (CPP) effects of modafinil (50 and 100 mg/kg). In the reinstatement study, it was found that modafinil did not alter responding during extinction or during the test of reinstatement induced by exposure to foot-shock stress. In the locomotion cross-sensitization experiment, it was found that the stimulatory properties of modafinil were significantly increased by chronic cocaine exposure, but cocaine sensitization did not enhance the rewarding properties of modafinil. Finally, in all experiments, modafinil enhanced the effects of cocaine exposure on elevations of orexin mRNA expression in the lateral hypothalamus. Interestingly, it has been reported that cue-induced relapse could be blocked by OX1 receptor antagonist SB334867 (Kallupi et al., 2010). Therefore, these results in laboratory rats do not support the effectiveness of modafinil in preventing relapse to cocaine seeking caused by stress, and further suggest that modafinil may have enhanced stimulatory properties in cocaine addicts. Moreover, the blockade of orexin system rather than its activation could be an efficacious strategy for the treatment of cocaine abuse and, because modafinil tends to enhance mRNA production of orexin, it is possible that this drug may be a safe pharmacological approach only the alleviation of cocaine withdrawal (Leri et al., 2009).

5 HPA AXIS

It has been reported that stress-induced elevation of HPA activity predicts relapse to drug use and amounts of subsequent use (e.g., Sinha et al., 2006). Vulnerability to drug abuse is enhanced by stress, and the HPA response to stress seems one of the critical factors influencing individual vulnerability to drug abuse. The HPA axis is a well-studied stress-responsive system in animals and humans. Stress, triggered by internal or external stimuli, increases corticotropin-releasing factor (CRF) and AVP, causing their release into the pituitary portal circulation from terminals of hypothalamic PVN. Activation of CRF type I receptor (CRF-R1) and V1b stimulates the release of the POMC peptides (Vale et al., 1981). Two major POMC gene products are ACTH and beta-endorphin (39 and 31 amino acids, respectively). Acting primarily on the adrenal cortex, ACTH releases corticosterone in rodents. Glucocorticoids, which are essential for life in mammals, act in a negative feedback mood by decreasing biosynthesis, release, and function of CRF and AVP in the hypothalamus and of CRF-R1 and V1b receptors in the anterior pituitary. Glucocorticoids also act to reduce the processing and release of POMC peptides in the anterior pituitary.

In early studies, an acute opiate challenge had a stimulatory effect on plasma ACTH and corticosterone release in opiate-naïve mice and rats. After chronic opiate administration from 1 to 2 weeks, however, the HPA axis was not activated by heroin or morphine but rather it was suppressed (Zhou et al., 2008a, 2013c). Chronic exposure to short-acting opiates acts as chronic stressor by virtue of the withdrawal that occurs between every exposure, and it may alter the responsivity of the HPA axis, as do many other stressors (e.g., Houshyar et al., 2003). Indeed, a low dose of heroin challenge decreased ACTH levels in rats during chronic withdrawal after HPA hormonal levels returned to baseline, which is in contrast to the stimulatory effect of acute heroin in opiate-naïve animals (Zhou et al., 2013c). The effects of opiates on HPA activity also depend on the presence or absence of external stressors, although the mechanisms responsible for this interaction are not well understood. In support of this concept, it was found that while either acute morphine or stress alone increased ACTH levels as an independent stimulus, morphine decreases plasma ACTH levels elevated by the stress (Zhou et al., 2013c). Together, animal studies demonstrate that morphine, heroin, or other short-acting opiates reduce HPA activity caused by stress, indicating that opioids act in a counter-regulatory role in modulating HPA stress responsivity under stress conditions.

Recent findings demonstrated that in comparison with saline or water controls, an abnormal HPA activity (including an enhanced dexamethasone–CRF response and increased basal hormonal levels) was found in mice and rats during chronic alcohol or cocaine withdrawal with an increased AVP mRNA level or a decreased CRF mRNA level (Pang et al., 2013; Zhou et al., 2011b). However, few studies have specifically addressed the involvement of AVP and V1b receptor systems in the HPA modulation and relapse-like behavior after chronic withdrawal from long-term opiate exposure, though AVP is a potent modulator of brain stress-responsive systems and HPA axis.

Studies in animals and humans have also demonstrated that beta-endorphin and dynorphin exert tonic inhibition and stimulation of HPA activity by acting on the MOP-r and the KOP-r, respectively. For instance, beta-endorphin acting on the MOP-r exerts tonic inhibition of CRF and then of the HPA axis in rodents. KOP-r agonists stimulate plasma corticosterone levels in rats. The stimulatory effects of the KOP-r agonists on the HPA axis were blocked by selective KOP-r antagonist nor-BNI. Consistent with the evidence that dynorphin/KOP-r modulates stress via the HPA axis, recent work further found that the yohimbine-induced or food restriction stress-induced HPA activation was blunted by selective KOP-r antagonist nor-BNI, providing evidence that there is an involvement of endogenous dynorphin/KOP-r system in modulation of HPA activity (Allen et al., 2013; Zhou et al., 2013b).

6 CONCLUSION AND FUTURE DIRECTIONS

As shown in this review, there has been substantial progress in understanding how exposure to drug of abuse interacts with stress-responsive brain systems in order to regulate addictive behaviors. The endogenous opioid systems (including POMC/MOP-r and dynorphin/KOP-r systems) clearly play a major role in heroin addiction, and specific gene alterations may contribute to stress responsivity and may affect vulnerability to develop heroin addiction and to relapse. Other stress-responsive systems mentioned above (including the vasopressin with its V1b receptors, orexin with its receptors, and the HPA axis) are also potentially involved in opiate addiction. By targeting multiple neurotransmitter pathways, combination medications are likely to have enhanced efficacy over the traditional single-medication approach. Indeed, the above-mentioned neurobiological studies have found supporting observations, given that several stress-responsive systems are profoundly altered by chronic heroin exposure. For example, though naltrexone is an available therapy for alcohol dependence, this and other single-pathway targeted pharmacotherapies (e.g., acamprosate on NMDA receptors) have only shown modest therapeutic value over placebo, indicating a need for new compounds with greater efficacy (Müller et al., 2014). Perhaps, given the complexity of the interactions, a potential useful strategy may be to combine compounds that target different neuropharmacological mechanisms (i.e., naltrexone + acamprosate, or methadone + modafinil) (e.g., Heyser et al., 2003).

REFERENCES

Allen, C.P., Zhou, Y., Leri, F., 2013. Effect of food restriction on cocaine locomotor sensitization in Sprague–Dawley rats: role of kappa opioid receptors. Psychopharmacology (Berl) 226, 571–578.

Baimel, C., Bartlett, S.E., Chiou, L.C., Lawrence, A.J., Muschamp, J.W., Patkar, O., Tung, L.W., Borgland, S.L., 2015. Orexin/hypocretin role in reward: implications for opioid and other addictions. Br. J. Pharmacol. 172, 334–348.

Bisagno, V., Cadet, J.L., 2014. Stress, sex, and addiction: potential roles of corticotropin-releasing factor, oxytocin, and arginine-vasopressin. Behav. Pharmacol. 25, 445–457.

Bodnar, R.J., 2014. Endogenous opiates and behavior: 2013. Peptides 62, 67–136.

Boutrel, B., Cannella, N., de Lecea, L., 2010. The role of hypocretin in driving arousal and goal-oriented behaviors. Brain Res. 1314, 103–111.

Breese, G.R., Sinha, R., Heilig, M., 2011. Chronic alcohol neuroadaptation and stress contribute to susceptibility for alcohol craving and relapse. Pharmacol. Ther. 129, 149–171.

Brown, R.M., Lawrence, A.J., 2009. Neurochemistry underlying relapse to opiate seeking behavior. Neurochem. Res. 34, 1876–1887.

Calipari, E.S., España, R.A., 2012. Hypocretin/orexin regulation of dopamine signaling: implications for reward and reinforcement mechanisms. Front. Behav. Neurosci. 6, 54.

Chou, T.C., Lee, C.E., Lu, J., Elmquist, J.K., Hara, J., Willie, J.T., 2001. Orexin (hypocretin) neurons contain dynorphin. J. Neurosci. 21, RC168.

Colombo, G., Lobina, C., Carai, M.A.M., Gessa, G.L., 2006. Phenotypic characterization of genetically selected Sardinian alcohol-preferring (sP) and -non preferring (sNP) rats. Addict. Biol. 11, 324–338.

Cowley, M.A., Smart, J.L., Rubinstein, M., Cerdan, M.G., Diano, S., Horvath, T.L., Cone, R.D., Low, M.J., 2001. Leptin activates anorexigenic POMC neurons through a neural network in the arcuate nucleus. Nature 411, 480–484.

de Lecea, L., Kilduff, T.S., Peyron, C., Gao, X., Foye, P.E., Danielson, P.E., 1998. The hypocretins: hypothalamus-specific peptides with neuroexcitatory activity. Proc. Natl. Acad. Sci. U.S.A. 95, 322–327.

Di Chiara, G., 2002. Nucleus accumbens shell and core dopamine: differential role in behavior and addiction. Behav. Brain Res. 137, 75–114.

D'Souza, M.S., Markou, A., 2010. Neural substrates of psychostimulant withdrawal-induced anhedonia. Curr. Top. Behav. Neurosci. 3, 119–178.

Edwards, S., Guerrero, M., Ghoneim, O.M., Roberts, E., Koob, G.F., 2012. Evidence that vasopressin V1b receptors mediate the transition to excessive drinking in ethanol-dependent rats. Addict. Biol. 17, 76–85.

Georgescu, D., Zachariou, V., Barrot, M., Mieda, M., Willie, J.T., Eisch, A.J., Yanagisawa, M., Nestler, E.J., Dileone, R.J., 2003. Involvement of the lateral hypothalamic peptide orexin in morphine dependence and withdrawal. J. Neurosci. 23, 3106–3111.

Griebel, G., Simiand, J., Serradeil-Le Gal, C., Wagnon, J., Pascal, M., Scatton, B., 2002. Anxiolytic- and antidepressant-like effects of the non-peptide vasopressin V1b receptor antagonist, SSR149415, suggest an innovative approach for the treatment of stress-related disorders. Proc. Natl. Acad. Sci. U.S.A. 99, 6370–6375.

Harris, G.C., Wimmer, M., Aston-Jones, G., 2005. A role for lateral hypothalamic orexin neurons in reward seeking. Nature 437, 556–559.

Heilig, M., Egli, M., Crabbe, J.C., Becker, H.C., 2010. Acute withdrawal, protracted abstinence and negative affect in alcoholism: are they linked? Addict. Biol. 15, 169–184.

Heyser, C.J., Moc, K., Koob, G.F., 2003. Effects of naltrexone alone and in combination with acamprosate on the alcohol deprivation effect in rats. Neuropsychopharmacology 28, 1463–1471.

Houshyar, H., Gomez, F., Manalo, S., Bhargava, A., Dallman, M., 2003. Intermittent morphine administration induces dependence and is a chronic stressor in rats. Neuropsychopharmacology 28, 1960–1971.

Ignar, D.M., Kuhn, C.M., 1990. Effects of specific mu and kappa opiate tolerance and abstinence on hypothalamo-pituitary-adrenal axis secretion in the rat. J. Pharmacol. Exp. Ther. 255, 1287–1295.

Kallupi, M., Cannella, N., Economidou, D., Ubaldi, M., Ruggeri, B., Weiss, F., Massi, M., Marugan, J., Heilig, M., Bonnavion, P., de Lecea, L., Ciccocioppo, R., 2010. Neuropeptide S facilitates cue-induced relapse to cocaine seeking through activation of the hypothalamic hypocretin system. Proc. Natl. Acad. Sci. U.S.A. 107, 19567–19572.

Kelsey, J.E., Verhaak, A.M., Schierberl, K.C., 2015. The kappa-opioid receptor antagonist, nor-binaltorphimine (nor-BNI), decreases morphine withdrawal and the consequent conditioned place aversion in rats. Behav. Brain Res. 283, 16–21.

Koob, G., Kreek, M.J., 2007. Stress, dysregulation of drug reward pathways, and the transition to drug dependence. Am. J. Psychiatry 164, 1149–1159.

Koob, G.F., 2008. A role for brain stress systems in addiction. Neuron 59, 11–34.

Koshimizu, T.A., Nakamura, K., Egashira, N., Hiroyama, M., Nonoguchi, H., Tanoue, A., 2012. Vasopressin V1a and V1b receptors: from molecules to physiological systems. Physiol. Rev. 92, 1813–1864.

Le Merrer, J., Becker, J.A., Befort, K., Kieffer, B.L., 2009. Reward processing by the opioid system in the brain. Physiol. Rev. 89, 1379–1412.

Leri, F., Zhou, Y., Kreek, M.J., Jacklin, D., 2009. Effect of modafinil on stress-induced cocaine seeking and cocaine cross-sensitization in laboratory rats (Abstract). In: CPDD 71st Annual Meeting (Reno, Sparks), Nevada.

Li, C., Pleil, K.E., Stamatakis, A.M., Busan, S., Vong, L., Lowell, B.B., Stuber, G.D., Kash, T.L., 2012. Presynaptic inhibition of gamma-aminobutyric acid release in the bed nucleus of the stria terminalis by kappa opioid receptor signaling. Biol. Psychiatry 71, 725–732.

Maiya, R., Zhou, Y., Norris, E., Kreek, M.J., Strickland, S., 2009. Tissue plasminogen activator modulates the cellular and behavioral response to cocaine. Proc. Natl. Acad. Sci. U.S.A. 106, 1983–1988.

Martinez, J.A., Vargas, M.L., Fuente, T., Garcia, J.D.R., Milanes, M.V., 1990. Plasma beta-endorphin and cortisol levels in morphine-tolerant rats and in naloxone-induced withdrawal. Eur. J. Pharmacol. 182, 117–123.

Minzenberg, M.J., Carter, C.S., 2008. Modafinil: a review of neurochemical actions and effects on cognition. Neuropsychopharmacology 33, 1477–1502.

Müller, C.A., Geisel, O., Banas, R., Heinz, A., 2014. Current pharmacological treatment approaches for alcohol dependence. Expert Opin. Pharmacother. 15, 471–481.

Narita, M., Nagumo, Y., Hashimoto, S., Narita, M., Khotib, J., Miyatake, M., 2006. Direct involvement of orexinergic systems in the activation of the mesolimbic dopamine pathway and related behaviors induced by morphine. J. Neurosci. 26, 398–405.

Nealey, K.A., Smith, A.W., Davis, S.M., Smith, D.G., Walker, B.M., 2011. κ-Opioid receptors are implicated in the increased potency of intra-accumbens nalmefene in ethanol-dependent rats. Neuropharmacology 61, 35–42.

Negus, S.S., Rice, K.C., 2009. Mechanisms of withdrawal-associated increases in heroin self-administration: pharmacologic modulation of heroin vs food choice in heroin-dependent rhesus monkeys. Neuropsychopharmacology 34, 899–911.

Pang, T.Y., Du, X., Catchlove, W.A., Renoir, T., Lawrence, A.J., Hannan, A.J., 2013. Positive environmental modification of depressive phenotype and abnormal hypothalamic-pituitary-adrenal axis activity in female C57BL/6J mice during abstinence from chronic ethanol consumption. Front. Pharmacol. 4, 1–9. Article 93.

Rodd-Henricks, Z.A., McKinzie, D.L., Li, T.K., Murphy, J.M., McBride, W.J., 2002. Cocaine is self-administered into the shell but not the core of the nucleus accumbens of Wistar rats. J. Pharmacol. Exp. Ther. 303, 1216–1226.

Roper, J.A., O'Carroll, A., Young III, W.S., Lolait, S.J., 2011. The vasopressin AVPr1b receptor: molecular and pharmacological studies. Stress 14, 98–115.

Rubinstein, M., Mogil, J.S., Japon, M., Chan, E.C., Allen, R.G., Low, M.J., 1996. Absence of opioid stress-induced analgesia in mice lacking beta-endorphin by site-directed mutagenesis. Proc. Natl. Acad. Sci. U.S.A. 93, 3995–4000.

Salome, N., Stemmelin, J., Cohen, C., Griebel, G., 2006. Differential roles of amygdaloid nuclei in the anxiolytic- and antidepressant-like effects of the V1b receptor antagonist, SSR149415, in rats. Psychopharmacology (Berl) 187, 237–244.

Schank, J.R., Ryabinin, A.E., Giardino, W.J., Ciccocioppo, R., Heilig, M., 2012. Stress-related neuropeptides and addictive behaviors: beyond the usual suspects. Neuron 76, 192–208.

Schlosburg, J.E., Whitfield Jr., T.W., Park, P.E., Crawford, E.F., George, O., Vendruscolo, L.F., Koob, G.F., 2013. Long-term antagonism of κ opioid receptors prevents escalation of and increased motivation for heroin intake. J. Neurosci. 33, 19384–19392.

Sedki, F., Eigenmann, K., Gelinas, J., Schouela, N., Courchesne, S., Shalev, U., 2015. A role for kappa-, but not mu-opioid, receptor activation in acute food deprivation-induced reinstatement of heroin seeking in rats. Addict. Biol. 20, 423–432.

See, R.E., Waters, R.P., 2010. Pharmacologically-induced stress: a cross-species probe for translational research in drug addiction and relapse. Am. J. Transl. Res. 3, 81–89.

Serradeil-Le Gal, C., Wagnon, J., Simiand, J., Griebel, G., Lacour, C., Guillon, G., Barberis, C., Brossard, G., Pascal, M., Soubrie, P., Nisato, D., Pascal, M., Pruss, R., Scatton, B., Maffrand, J.P., Le Fur, G., 2002. Characterization of (2S,4R)-1-[5-chloro-1-[(2,4-dimethoxyphenyl)sulfonyl]-3-(2-methoxy-phenyl)-2-oxo-2,3-dihydro-1H-indol-3-yl]-4-hydroxy-N,N-dimethyl-2-pyrrolidine carboxamide (SSR149415), a selective and oral active vasopressin V1b receptor antagonist. J. Pharmacol. Exp. Ther. 300, 1122–1130.

Shalev, U., Erb, S., Shaham, Y., 2010. Role of CRF and other neuropeptides in stress-induced reinstatement of drug seeking. Brain Res. 1314, 15–28.

Sharf, R., Sarhan, M., Dileone, R.J., 2010. Role of orexin/hypocretin in dependence and addiction. Brain Res. 1314, 130–138.

Sinha, R., Garcia, M., Paliwal, P., Kreek, M.J., Rounsaville, B.J., 2006. Stress-induced cocaine craving and hypothalamic-pituitary-adrenal responses are predictive of cocaine relapse outcomes. Arch. Gen. Psychiatry 63, 324–331.

Sinha, R., Shaham, Y., Heilig, M., 2011. Translational and reverse translational research on the role of stress in drug craving and relapse. Psychopharmacology (Berl) 218, 69–82.

Smith, R.J., Aston-Jones, G., 2012. Orexin/hypocretin 1 receptor antagonist reduces heroin self-administration and cue-induced heroin seeking. Eur. J. Neurosci. 35, 798–804.

Spanagel, R., Herz, A., Shippenberg, T.S., 1990. The effects of opioid peptides on dopamine release in the nucleus accumbens: an in vivo microdialysis study. J. Neurochem. 55, 1734–1740.

Spanagel, R., Herz, A., Bals-Kubik, R., Shippenberg, T.S., 1991. Beta-endorphin-induced locomotor stimulation and reinforcement are associated with an increase in dopamine release in the nucleus accumbens. Psychopharmacology (Berl) 104, 51–56.

Spanagel, R., Noori, H.R., Heilig, M., 2014. Stress and alcohol interactions: animal studies and clinical significance. Trends Neurosci. 37, 219–227.

Sperling, R.E., Gomes, S.M., Sypek, E.I., Carey, A.N., McLaughlin, J.P., 2010. Endogenous kappa-opioid mediation of stress-induced potentiation of ethanol-conditioned place preference and self-administration. Psychopharmacology (Berl) 210, 199–209.

Stine, S., Southwick, S., Petrakis, I., Kosten, T., Charney, D., Krystal, J., 2002. Yohimbine-induced withdrawal and anxiety symptoms in opioid dependent patients. Biol. Psychiatry 51, 642–651.

Vale, W., Spiess, J., Rivier, C., Rivier, J., 1981. Characterization of a 41-residue ovine hypothalamic peptide that stimulates secretion of corticotropin and beta-endorphin. Science 213, 1394–1397.

Walker, B.M., Koob, G.F., 2008. Pharmacological evidence for a motivational role of kappa-opioid systems in ethanol dependence. Neuropsychopharmacology 33, 643–652.

Wigger, A., Sánchez, M.M., Mathys, K.C., Ebner, K., Frank, E., Liu, D., Kresse, A., Neumann, I.D., Holsboer, F., Plotsky, P.M., Landgraf, R., 2004. Alterations in central neuropeptide expression, release, and receptor binding in rats bred for high anxiety: critical role of vasopressin. Neuropsychopharmacology 29, 1–14.

Yeoh, J.W., Campbell, E.J., James, M.H., Graham, B.A., Dayas, C.V., 2014. Orexin antagonists for neuropsychiatric disease: progress and potential pitfalls. Front. Neurosci. 8, 36.

Zhou, Y., Spangler, R., Yuferov, V.P., Schlussmann, S.D., Ho, A., Kreek, M.J., 2004. Effects of selective D1- or D2-like dopamine receptor antagonists with acute "binge" pattern cocaine on corticotropin-releasing hormone and proopiomelanocortin mRNA levels in the hypothalamus. Mol. Brain Res. 130, 61–67.

Zhou, Y., Bendor, J.T., Yuferov, V., Schlussman, S.D., Ho, A., Kreek, M.J., 2005. Amygdalar vasopressin mRNA increases in acute cocaine withdrawal: evidence for opioid receptor modulation. Neuroscience 134, 1391–1397.

Zhou, Y., Bendor, J., Hofmann, L., Randesi, M., Ho, A., Kreek, M.J., 2006. Mu opioid receptor and orexin/hypocretin mRNA levels in the lateral hypothalamus and striatum are enhanced by morphine withdrawal. J. Endocrinol. 191, 137–145.

Zhou, Y., Leri, F., Cummins, E., Hoeschele, M., Kreek, M.J., 2008a. Involvement of arginine vasopressin and V1b receptor in heroin withdrawal and heroin seeking precipitated by stress and by heroin. Neuropsychopharmacology 33, 226–236.

Zhou, Y., Cui, C.L., Schlussman, S.D., Choi, J.C., Ho, A., Han, J.S., Kreek, M.J., 2008b. Effects of cocaine place conditioning, chronic escalating-dose "binge" pattern cocaine administration and acute withdrawal on orexin/hypocretin and preprodynorphin gene expressions in lateral hypothalamus of Fischer and Sprague–Dawley rats. Neuroscience 153, 1225–1234.

Zhou, Y., Maiya, R., Norris, E., Kreek, M.J., Strickland, S., 2010. Involvement of tissue plasminogen activator in stress responsivity during acute cocaine withdrawal in mice. Stress 13, 481–490.

Zhou, Y., Colombo, G., Carai, M.A., Ho, A., Gessa, G.L., Kreek, M.J., 2011a. Involvement of arginine vasopressin and V1b receptor in alcohol drinking in Sardinian alcohol-preferring rats. Alcohol. Clin. Exp. Res. 35, 1876–1883.

Zhou, Y., Litvin, Y., Piras, A.P., Pfaff, D.W., Kreek, M.J., 2011b. Persistent increase in hypothalamic arginine vasopressin gene expression during protracted withdrawal from chronic escalating-dose cocaine in rodents. Neuropsychopharmacology 36, 2062–2075.

Zhou, Y., Colombo, G., Niikura, K., Carai, M.A.M., Femenía, T., García-Gutiérrez, M.S., Manzanares, J., Ho, A., Gessa, G.L., Kreek, M.J., 2013a. Voluntary alcohol drinking enhances proopiomelanocortin (POMC) gene expression in nucleus accumbens shell and hypothalamus of Sardinian alcohol-preferring rats. Alcohol. Clin. Exp. Res. 37, E131–E140.

Zhou, Y., Leri, F., Grella, S., Aldrich, J., Kreek, M.J., 2013b. Involvement of dynorphin and kappa opioid receptor in yohimbine-induced reinstatement of heroin seeking in rats. Synapse 67, 358–361.

Zhou, Y., Leri, F., Ho, A., Kreek, M.J., 2013c. Suppression of hypothalamic-pituitary-adrenal axis by acute heroin challenge in rats during acute and chronic withdrawal from chronic heroin administration. Neurochem. Res. 38, 1850–1860.

Zhou, Y., Leri, F., Cummins, E., Kreek, M.J., 2015. Individual differences in gene expression of vasopressin, D2 receptor, POMC and orexin: vulnerability to relapse to heroin seeking in rats. Physiol. Behav. 139, 127–135.

CHAPTER

13

Opioid neuroscience for addiction medicine: From animal models to FDA approval for alcohol addiction

Wade Berrettini[1]

Karl E Rickles Professor of Psychiatry, Center for Neurobiology and Behavior, Perelman School of Medicine, University of Pennsylvania, Philadelphia, PA, USA

[1]Corresponding author: Tel.: +1-215-898-0092; Fax: +1-215-573-2041, e-mail address: wadeb@mail.med.upenn.edu

Abstract

Alcohol addiction is one of the most common and devastating diseases in the world. Given the tremendous heterogeneity of alcohol-addicted individuals, it is unlikely that one medication will help nearly all patients. Thus, there is a clear need to develop predictors of response to existing medications. Naltrexone is a mu opioid receptor antagonist which has been approved in the United States for treatment of alcohol addiction since 1994. It has limited efficacy, in part due to noncompliance, but many patients do not respond despite high levels of compliance. There are reports that a mis-sense single-nucleotide polymorphism (rs179919 or A118G) in the mu opioid receptor gene predicts a favorable response to naltrexone if an individual carries a "G" allele. This chapter will review the evidence for this hypothesis. The data suggest that the "G" allele has a complex role in alcohol addiction, increasing the rewarding valence of alcohol. Whether the G allele increases risk for alcoholism and whether it predisposes to a beneficial naltrexone response among alcohol-addicted persons must await additional research with large sample sizes of multiple ethnicities in prospective clinical trials.

Keywords

Opioids, Alcohol addiction, Naltrexone, Pharmacogenetics, mu opioid receptor

1 INTRODUCTION: THE ROLE OF OPIOIDS IN ALCOHOL REWARD

Ventral tegmental neurons release dopamine at nerve terminals in ventral striatum and medial prefrontal cortex. Activation of this circuit is a common element of abused drugs, including alcohol (e.g., Di Chiara and Imperato, 1988; for review see Koob and Volkow, 2010). Thus, alcohol shares in common with nicotine, cocaine, amphetamine, morphine, etc., this property of enhancing dopaminergic transmission in ventral striatum and medial prefrontal cortex. Both animal model and human studies are in agreement on this point (Boileau et al., 2003; Gilman et al., 2008; Spanagel, 2009). This release of dopamine in the ventral striatum and medial prefrontal cortex is partially enhanced by stimulation of mu opioid receptors (for which endorphin is the primary ligand) located on inhibitory GABAergic interneurons in the ventral tegmental area. The GABAergic interneurons inhibit the dopaminergic ventral tegmental neurons, whose activation signals reward. Thus, mu opioid receptor agonists enhance the likelihood of ventral tegmental dopaminergic neuron activation (and the experience of reward) by lessening the tonic inhibition of the associated GABAergic interneurons (Johnson and North, 1992; Spanagel et al., 1992; Tanda and DiChiara, 1998).

Given this circuitry, it has been consistently shown that endogenous opioids play a role in ethanol reinforcement in various animal paradigms. Endorphin elevations after alcohol are seen in discrete reward regions of the hypothalamus (Popp and Erickson, 1998), ventral tegmentum, and ventral striatum (Rasmussen et al., 1998). It is important to note that endorphin-deficient rats continue to self-administer alcohol, indicating that endorphin is not the sole mechanism of alcohol reward (Grahame et al., 1998). The importance of mu opioid receptor activation as a mechanism for alcohol reward is underscored by the fact that alcohol consumption in alcohol-preferring rats is persistently reduced after inactivating mu opioid receptors in the ventral striatum (Myers and Robinson, 1999). Similarly, decreased alcohol self-administration is observed in primates after pretreatment with opioid antagonists (Altshuler et al., 1980). C57Bl/6J mice, an inbred strain which prefers alcohol, have increased endorphin release in the hypothalamus after alcohol administration (De Waele et al., 1992). Alcohol-preferring rats have high levels of opioid gene mRNA species in the hypothalamus, prefrontal cortex, and mediodorsal nucleus of the thalamus (Marinelli et al., 2000), as well as increased mu opioid receptor density in the ventral striatum and medial prefrontal cortex.

2 CLINICAL STUDIES OF NALTREXONE IN ALCOHOLISM

The development of a substantial body of evidence, in the 1980s, that naltrexone (an orally active mu opioid receptor antagonist) diminished alcohol self-administration in animal models (Altshuler et al., 1980; Kiianmaa et al., 1983; Myers et al., 1986;

Volpicelli et al., 1986) led to the first use of naltrexone in alcohol-addicted populations in a controlled clinical trial (Volpicelli et al., 1992), the promising outcome of which was immediately confirmed in a second controlled clinical trial (O'Malley et al., 1992). Naltrexone was found to reduce alcohol craving and relapse to heavy drinking (operationally defined as five or more drinks/day for a man, four or more for a woman), but did not reduce abstinence rates. On the basis of these two controlled trials, naltrexone was approved by the FDA, in the absence of the usual pharmaceutical industry interest.

In the intervening 20 years, there have been more than 30 clinical trials of naltrexone in alcohol addiction (for review, see Bouza et al., 2004; Pettinati et al., 2006; Srisurapanont and Jarusuraisin, 2005). While the majority of these clinical trials demonstrate efficacy of naltrexone in reducing risk for relapse to heavy drinking, the effect size is small, with many patients having no benefit. This has resulted in multiple reports in which the naltrexone arm outcomes are not significantly better than the placebo arm outcomes (e.g., Krystal et al., 2001). This is an expected outcome, given the tremendous heterogeneity of clinical alcohol addiction. It is likely that important clinical characteristics, such as compliance, severity and duration of alcohol addiction, comorbidity (both medical and psychiatric), and/or attendance at psychosocial treatment, may influence outcomes.

In this situation, multiple investigators have attempted to define clinical characteristics which might enhance the probability of naltrexone response. Some clinical measures have shown promise in characterizing a naltrexone responder: high alcohol craving (Chick et al., 2000; Monterosso et al., 2001; O'Malley et al., 2002) and strong family history of alcohol addiction (Monterosso et al., 2001), but family history of alcohol addiction did not predict response to naltrexone in the combine multicenter trial (Capone et al., 2011). Alcohol addicts who experience greater euphoria after alcohol may have a better response to naltrexone (Volpicelli et al., 1995).

3 A118G OPRM1 MIS-SENSE SINGLE-NUCLEOTIDE POLYMORPHISM: MOLECULAR AND CELLULAR EFFECTS

A common mis-sense single-nucleotide polymorphism (rs1799971) in the first exon of the mu opioid receptor gene, OPRM1, was described by Bergen et al. (1997), A118G, or N40G, reflecting the fact that the A allele encodes asparagine, while the minor G allele encodes aspartate. The A (asparagine) allele is thought to be N-glycosylated (Huang et al., 2012), whereas this is not possible for the G (aspartate) allele, as there is no free amino group. Subsequent study (e.g., Crowley et al., 2003; Gelernter et al., 1999; Szeto et al., 2001; Tan et al., 2003) revealed large ethnic differences in allele frequencies (see Table 1).

This allele has been the subject of multiple molecular investigations to determine its functional consequences, in terms of gene expression, protein translation, receptor signaling, and receptor density. Initially, Bond et al. (1998) reported that the minor "G" allele mu opioid receptor resulted in decreased affinity for binding to

Table 1 Frequency of G Allele for A118G SNP in Ethnic Groups

Ethnic Group	Frequency G (%)	Ethnic Group	Frequency G (%)
African	1	Korean	31
African-American	3	Chinese	35
Swedish	11	Malaysian	43
European-American	15	Indian	47

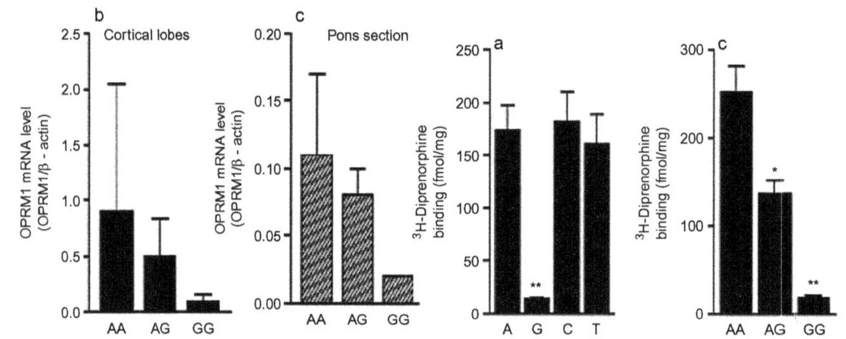

FIGURE 1

Transcriptional and translational efficiency of the 118G allele is markedly limited, compared to the A allele. **P*<0.05; ***P*<0.01.

beta-endorphin, compared to the common "A" allele receptor. There was no change in binding affinity for alkaloid ligands. This result has not been confirmed in subsequent investigations (Beyer et al., 2004; Ramchandani et al., 2011). In one such study, transfected HEK293 cells (a fibroblastoid cell type) were used (Beyer et al., 2004), but the 118G allele did not differ in binding affinity for beta-endorphin, compared to 118A. Beyer et al. (2004) also reported that the 118G allele was not different from the 118A allele in rate of desensitization, internalization, or resensitization, but 118G had decreased transcription, compared to 118A. Ramchandani et al. (2011) also did not report differences in kinetics of binding of beta-endorphin to the 118G, compared to 118A. Mahmoud et al. (2011), using a whole-cell patch clamp technique in acutely dissociated trigeminal ganglion neurons, reported that morphine was fivefold less active at the "G" allele receptor form in activating a Ca^{2+} channel. There was no such difference for fentanyl. Zhang et al. (2005) conducted allelic imbalance studies in postmortem human brain, revealing a marked decrease in 118G allele mRNA (see Fig. 1). In a second experiment, they showed *in vitro* evidence of a marked decreased translation of the 118G mRNA (see Fig. 4; Zhang et al., 2005).

4 A118G OPRM1 MIS-SENSE SINGLE-NUCLEOTIDE POLYMORPHISM: ANIMAL MODEL STUDIES

In the murine OPRM1 gene, there is no equivalent of the A118G naturally occurring variation. A homologous variation (A112G, with the A allele encoding asparagines and the G allele encoding aspartate, as in the human OPRM1 gene) was created by bacterial artificial chromosome engineering and murine transgenic techniques by Mague et al. (2009). They reported decreased transcription and translation of the G allele in transgenic C57Bl/6 mouse brain (see Fig. 2), a result congruous with the human postmortem brain *ex vivo* results of Zhang et al. (2005), as well as the *in vitro* results of Beyer et al. (2004). There was a blunted locomotor response to morphine in the 112G mice, as well as decreased morphine conditioned place preference (CPP) in 112G female mice, the latter being a sexually dimorphic response, with 112G males showing the expected CPP response to morphine.

Two other forms of transgenic mice were produced, using homologous recombination to replace the murine OPRM1 exon 1 with one of the two forms (118A and 118G) of human OPRM1 exon 1 (Ramchandani et al., 2011). These investigators conducted *in vivo* microdialysis experiments in the ventral striatum, demonstrating that the 118G mice had the expected elevations in dopamine release after alcohol, while the 118A mice had no significant increase over baseline (see Fig. 6). These data suggest that the "G" allele conveys an increased rewarding valence to alcohol, compared to the "A" allele (Fig. 3).

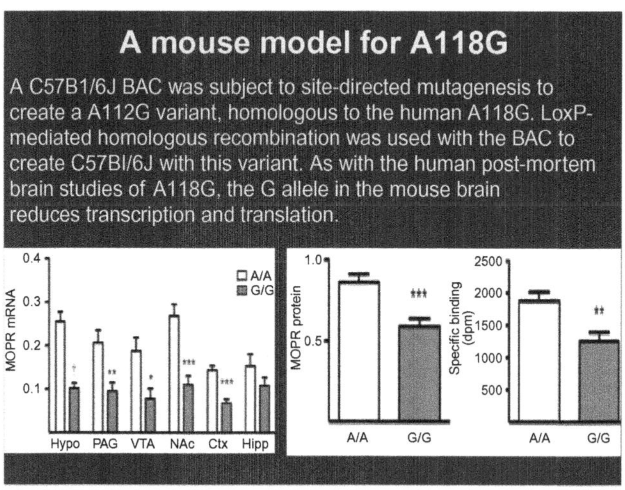

FIGURE 2

A mouse homologue of the 118G allele also shows decreased transcriptional and translational efficiency. **P*<0.05; ***P*<0.01; ****P*<0.001; †*P*<0.00001.

Mague et al. (2009).

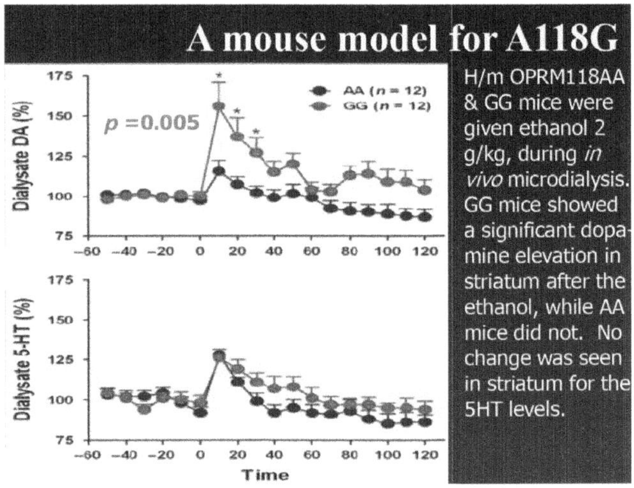

FIGURE 3

In vivo microdialysis proves that a mouse homologue of the 118G allele confers increased alcohol-induced dopamine release in the ventral striatum and presumably increased reward.

Ramchandani et al. (2011).

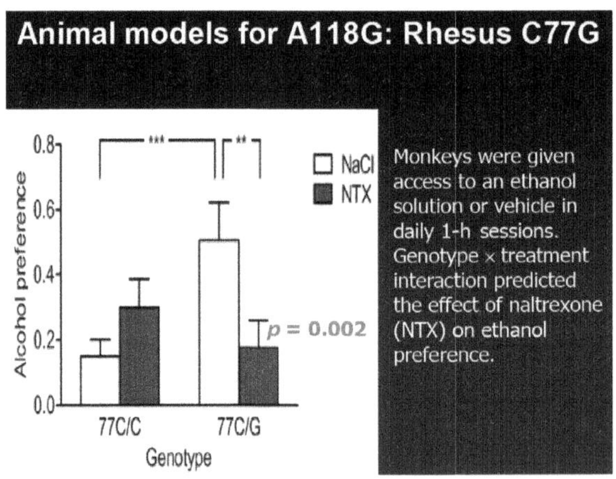

FIGURE 4

The rhesus homologue of the 118G, 77G, is associated with increased alcohol consumption, which is attenuated by naltrexone treatment.

Barr et al. (2010).

There have been several studies of a similar SNP in the rhesus monkey, the C77G, which results in a homologous amino acid change, asparagine to aspartate (Barr et al., 2007, 2010; Vallender et al., 2010). Both groups report that the G allele monkeys consume significantly more alcohol than the CC monkeys. Further, both groups note that naltrexone significantly decreases alcohol intake in the GG monkeys (Fig. 4).

However, there is scant evidence that the G allele increases alcohol consumption in the general population.

These reports, taken together, are consistent with the hypothesis that the 118G allele (or its equivalent in mouse and primate) conveys a greater rewarding effect of alcohol, a difference which is inhibited by naltrexone. These studies are remarkably consistent, given the species, paradigm, technical, and molecular engineering differences among these studies.

5 A118G OPRM1 MIS-SENSE SINGLE-NUCLEOTIDE POLYMORPHISM: HUMAN PHARMACOGENETIC STUDIES OF ALCOHOL

There have been several pharmacogenetic reports of the A118G SNP in human laboratory experiments involving alcohol (Ramchandani et al., 2011; Ray and Hutchison, 2004, 2007; Ray et al., 2010; Setiawan et al., 2011). In a laboratory investigation of the A118G pharmacogenetics of alcohol reward, Ray and Hutchison (2004, 2007) demonstrated that the G allele carriers experienced significantly greater euphoria after standard oral doses of alcohol (while controlling for breath alcohol concentration), compared to AA persons. Further, naltrexone significantly blunted the euphoria in the G allele carriers and was without effect in the AA group (see Fig. 5).

In agreement with this result, Ramchandani reported that G allele carriers had a greater striatal release of dopamine after alcohol (using a raclopride PET scan

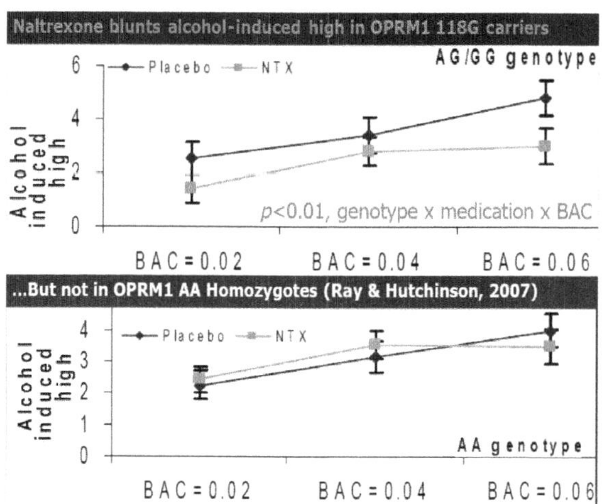

FIGURE 5

118G allele carriers have a greater euphoria response to alcohol, compared to homozygous A persons; naltrexone blunts the alcohol-induced euphoria for 118G carriers, but has no such effect in persons who are homozygous A.

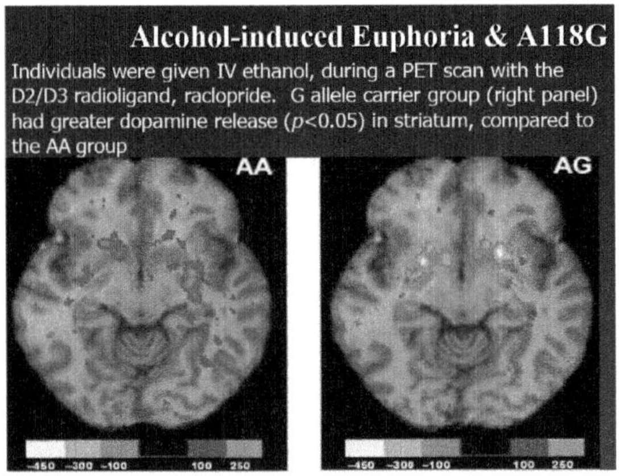

FIGURE 6

Raclopride positron emission tomograph (PET) scans reveal that 118G carriers have increase ventral striatum alcohol-induced dopamine release, compared to persons who are homozygous A.

Ramchandani et al. (2011).

technique), compared to AA participants (see Fig. 6). In a more naturalistic approach, Ray et al. (2010) studied drinking habits of social drinkers over a 5-day period, analyzing subjective responses to alcohol by A118G genotype. G allele carriers reported more significantly more "vigor" less negative mood after drinking, compared to the AA group. Similarly, Setiawan et al. (2011) studied the subjective response to alcohol in social drinkers after a dose of naltrexone. Naltrexone significantly decreased the ethanol-induced "euphoria" to a priming dose of alcohol in subjects with the G allele, compared to AA participants.

Ashenhurst et al. (2012) found evidence for a OPRD1 SNP (rs4654327) influencing the response to naltrexone in the presence of alcohol, such that carriers of the A allele at this locus reported greater naltrexone-induced blunting of alcohol stimulation and alcohol craving compared to GG homozygotes. Further, TT homozygotes reported lower naltrexone-influenced alcohol sedation as compared to carriers of the C allele at the OPRK1 SNP, rs997917. These studies indicate that multiple opioid receptors may influence the response of individuals to alcohol in the presence of naltrexone. The genetic complexity of the naltrexone–alcohol interaction is further illustrated by a human lab study of alcohol-dependent subjects, in which there was a statistically significant increased stimulation and positive mood among OPRM1 G allele carriers who were dopamine transporter (DAT) VNTR 10–10 homozygotes, compared with other genotype groups (Ray et al., 2014). Lastly, in a neuroimaging study of alcohol-dependent individuals, Schacht et al. (2013) reported a three-way interaction between medication and A118G and DAT VNTR genotypes on ventral striatum alcohol-induced activation, such that, among G allele carriers, DAT 10–10 homozygotes had less activation after naltrexone than 9-repeat-allele carriers.

Taken together, these human laboratory studies of the A118G variant on effect of alcohol are remarkably consistent, with the clear conclusion that the G allele permits people to experience alcohol in a more rewarding manner, compared to AA individuals. It is also notable that naltrexone is able to blunt this euphoria in G allele carriers, but not in AA persons. This latter observation is consistent with subjective reports of the effect of naltrexone in clinical trials for alcohol addiction, in which the medication attenuated alcohol-induced euphoria among responders (Volpicelli et al., 1995).

6 PHARMACOGENETIC STUDIES OF NALTREXONE CLINICAL TRIALS FOR ALCOHOL ADDICTION

There have been multiple pharmacogenetic studies of naltrexone clinical trials for alcohol addiction published in the last decade. The first such publication (Oslin et al., 2003) was a retrospective analysis of three naltrexone trials of similar design, two conducted at the University of Pennsylvania and one at the University of Connecticut. Compliance was monitored by riboflavin testing and by pill counts. Eighty-two patients (71 of European descent) who were randomized to naltrexone and 59 randomized to placebo (all of European descent) in one of three randomized placebo-controlled clinical trials of naltrexone were genotyped at the $A_{+118}G$ (Asn40Asp) and $C_{+17}T$ (Ala6Val) SNPs in the mu opioid gene (OPRM1). The association between genotype and drinking outcomes was measured over 12 weeks of treatment. For purposes of examining the pharmacogenetics of naltrexone response, the analysis was limited to those subjects with well-defined outcome data who had at minimum 6 weeks exposure to the medication. The primary drinking outcome considered was relapse to heavy drinking ($\geq$5 drinks in a single day for men or $\geq$4 drinks for women). This definition of heavy drinking was the primary outcome for each of the trials. The timeline follow-back method was employed (along with self-report) to measure alcohol consumption (Sobell and Sobell, 1992). There was a significantly greater proportion of naltrexone-treated subjects with the G allele variant who did not return to heavy drinking (no relapse) compared to those with those homozygous for the A allele (Wald = 4.04, 1 df, OR = 3.47 (95% CI: 1.03–11.67), and $p = 0.045$; see Table 2).

Table 2 A118G Genotype and Good Outcome in Naltrexone Studies of Pharmacotherapy for Alcohol Addiction

	Oslin et al. (2003)		Anton et al. (2008)	
Genotype at A118G	**Naltrexone (%)**	**Placebo (%)**	**Naltrexone (%)**	**Placebo (%)**
G allele carriers	85[a]	55	89[b]	54
Homozygous A	56	46	56	50

[a] *p = 0.04, odds ratio = 3.5.*
[b] *p = 0.005, genotype medication interaction; odds ratio = 5.8.*

This finding was confirmed in a larger multisite study of naltrexone, acamprosate, and placebo for alcohol addiction (Anton et al., 2008). Alcohol-addicted subjects were treated for 16 weeks with 100 mg of naltrexone. All participants received medical management alone or with combined behavioral intervention.

When considering only those patients receiving medical management alone, there was an significant effect of naltrexone on "good outcome" among the 118G carriers, while there was no such effect for the patients receiving naltrexone who were homozygous A118 (see Table 2). However, there was no such effect in the naltrexone group receiving medical management with combined behavioral intervention. The combined behavioral intervention was delivered by licensed behavioral health specialists in up to 20 flexible participant need–adjusted 50-min sessions. Combined behavioral intervention, an intensive and specific alcohol intervention, may have compensated for the placebo effect, thereby suppressing the chances of observing a main effect of naltrexone or a genetic interaction. The data presented by Anton et al. (2008) are consistent with this thinking. A gene medication interaction may be observable only in patients who can show obvious benefit from the medication over placebo.

In a small Korean study of naltrexone in alcohol addiction (Kim et al., 2009), subjects adherent to naltrexone treatment with one or two copies of the Asp40 allele took a significantly longer time than the Asn40 group to relapse to heavy drinking ($p = 0.014$). Although not significant, the Asn40 group treated with naltrexone had a 10.6 times greater relapse rate than the Asp40 variant group. There was no effect on abstinence.

In the Veterans Administration multisite study of naltrexone in alcohol addiction, Gelernter et al. (2007) reported that the 118G allele did not predict outcome among 149 participants in the naltrexone group and 64 in the placebo group. There are several possible explanations for this result. First, the efficacy of naltrexone is certainly influenced by compliance, and the compliant population was defined as those who opened the medication bottle a minimum of 50% of the time, so that medication compliance was defined liberally. Second, it is likely that high levels of comorbidity influence response to naltrexone. The study population had substantial rates of recurrent unipolar illness, antisocial personality, and anxiety disorders and had severe alcohol addiction of long duration. These factors might overwhelm any genetic predisposition to respond to naltrexone. Third, the study had limited power: for example, there were only nine 118G carriers in the placebo group.

Coller et al. (2011) recently reported the results of a naltrexone and cognitive behavioral therapy trial in 100 Australian alcohol-addicted persons. They reported an overall effect of naltrexone on relapse to heavy drinking, but no influence of the A188G variants. The absence of a control group makes this study less ideal, as does the small sample size, with 68 study completers.

Taken together, the A118G clinical trials in naltrexone treatment for alcohol addiction remain promising, but there are clear unanswered question, including the influence of counseling, compliance, and comorbidity on outcome. Available depot formulations of naltrexone may reduce noncompliance, but the influence of comorbidity and counseling may be more difficult to resolve. It will be necessary to conduct pharmacogenetic alcohol addiction naltrexone trials, for which participants are randomized by A118G genotype into the naltrexone or placebo arm to reduce possible sources of bias. These trials should be characterized by:

(1) large size (at least ~100 persons per arm, including oversampling of G allele carriers) to ensure adequate power;
(2) rigorous assessment of compliance;
(3) randomization stratified by genotype;
(4) careful assessment of comorbidity;
(5) modest psychotherapeutic intervention, so as to mirror "real world" clinical practice.

Only one such study has been published which has these characteristics (Oslin et al., 2015), and there was no influence of the G allele on outcome among those randomized to naltrexone.

7 CONCLUSION AND FUTURE DIRECTIONS

There are extensive data, across species, to suggest that the 118G form of the mu opioid receptor is characterized by decreased transcription and translation. There are convincing data, from murine, primate, and human laboratory studies, that the 118G (or its species-specific homologue) variant permits alcohol to have a greater rewarding valence, leading to increased alcohol consumption. Further, the human and rhesus data are equally convincing that naltrexone is able to blunt this greater rewarding signal. Lastly, the possibility that A118G alleles can be used clinically to identify alcohol-addicted persons with a greater probability to have a beneficial response to naltrexone is a hypothesis that deserves testing on a large scale, with the characteristics noted above.

REFERENCES

Altshuler, H.L., Phillips, P.A., Feinhandler, D.A., 1980. Alteration of ethanol self-administration by naltrexone. Life Sci. 26, 679–688.

Anton, R., Oroszi, G., O'Malley, S., Couper, D., Swift, R., Pettinati, H., Goldman, D., 2008. μ opioid receptor Asn40Asp predicts naltrexone response. Arch. Gen. Psychiatry 65 (11), 135–144.

Ashenhurst, J.R., Bujarski, S., Ray, L.A., 2012. Delta and kappa opioid receptor polymorphisms influence the effects of naltrexone on subjective responses to alcohol. Pharmacol. Biochem. Behav. 103, 253–259.

Barr, C.S., Schwandt, M., Lindell, S.G., Chen, S.A., Goldman, D., Suomi, S.J., Higley, J.D., Heilig, M., 2007. Association of a functional polymorphism in the mu-opioid receptor gene with alcohol response and consumption in male rhesus macaques. Arch. Gen. Psychiatry 64, 369–376.

Barr, C.S., Chen, S.A., Schwandt, M.L., Lindell, S.G., Sun, H., Suomi, S.J., Heilig, M., 2010. Suppression of alcohol preference by naltrexone in the rhesus macaque: a critical role of genetic variation at the mu-opioid receptor gene locus. Biol. Psychiatry 67, 78–80.

Bergen, A.W., Kokoszka, J., Peterson, R., et al., 1997. Mu opioid receptor gene variants: lack of association with alcohol dependence. Mol. Psychiatry 2 (6), 490–494.

Beyer, A., Koch, T., Schröder, H., Schulz, S., Höllt, V., 2004. Effect of the A118G polymorphism on binding affinity, potency and agonist-mediated endocytosis, desensitization, and resensitization of the human mu-opioid receptor. J. Neurochem. 89, 553–560.

Boileau, I., et al., 2003. Alcohol promotes dopamine release in the human nucleus accumbens. Synapse 49, 226–231.

Bond, C., LaForge, K.S., Tian, M., Melia, D., Zhang, S., Borg, L., Gong, J., Schluger, J., Strong, J.A., Leal, S.M., Tischfield, J.A., Kreek, M.J., Yu, L., et al., 1998. Single-nucleotide polymorphism in the human mu opioid receptor gene alters beta-endorphin binding and activity: possible implications for opiate addiction. Proc. Natl. Acad. Sci. U.S.A. 95 (16), 9608–9613.

Bouza, C., Angeles, M., Muñoz, A., Amate, J.M., 2004. Efficacy and safety of naltrexone and acamprosate in the treatment of alcohol dependence: a systematic review. Addiction 99, 811–828.

Capone, C., Kahler, C.W., Swift, R.M., O'Malley, S.S., 2011. Does family history of alcoholism moderate naltrexone's effects on alcohol use? J. Stud. Alcohol Drugs 72, 135–140.

Chick, J., Anton, R., Checinski, K., Croop, R., Drummond, D.C., Farmer, R., Labriola, D., Marshall, J., Moncrieff, J., Morgan, M.Y., Peters, T., Ritson, B., 2000. A multicenter double-blind randomized trial of naltrexone in the treatment of alcohol dependence or abuse. Alcohol Alcohol. 35, 587–593.

Coller, J.K., Cahill, S., Edmonds, C., Farquharson, A.L., Longo, M., Minniti, R., Sullivan, T., Somogyi, A.A., Whilte, J.M., 2011. OPRM1 A118G genotype fails to predict the effectiveness of naltrexone treatment for alcohol dependence. Pharmacogenet. Genomics 21, 902–905.

Crowley, J.J., Oslin, D.W., Patkar, A.A., Gottheil, E., DeMaria Jr., P.A., O'Brien, C.P., Berrettini, W.H., Grice, D.E., 2003. A genetic association study of the mu opioid receptor and severe opioid dependence. Psychiatr. Genet. 13, 169–173.

De Waele, J.P., Papachristou, D.N., Gianoulakis, C., 1992. The alcohol-preferring C57BL/6 mice present an enhanced sensitivity of the hypothalamic beta-endorphin system to ethanol than the alcohol-avoiding DBA/2 mice. J. Pharmacol. Exp. Ther. 261 (2), 788–794.

Di Chiara, G., Imperato, A., 1988. Drugs abused by humans preferentially increase synaptic dopamine concentrations in the mesolimbic system of freely moving rats. Proc. Natl. Acad. Sci. U.S.A. 85, 5274–5278.

Gelernter, J., Kranzler, H., Cubells, J., 1999. Genetics of two m opioid receptor gene (OPRM1) exon I polymorphisms: population studies, and allele frequencies in alcohol- and drug-dependent subjects. Mol. Psychiatry 4, 476–483.

Gelernter, J., Gueorguieva, R., Kranzler, H.R., et al., 2007. Opioid receptor gene (OPRM1, OPRK1, and OPRD1) variants and response to naltrexone treatment for alcohol dependence: results from the VA Cooperative Study. Alcohol. Clin. Exp. Res. 31 (4), 555–563.

Gilman, J.M., Ramchandani, V.A., Davis, M.B., Bjork, J.M., Hommer, D.W., 2008. Why we like to drink: a functional magnetic resonance imaging study of the rewarding and anxiolytic effects of alcohol. J. Neurosci. 28, 4583–4591.

Grahame, N.J., Low, M.J., Cunningham, C.L., 1998. Intravenous self-administration of ethanol in beta-endorphin-deficient mice. Alcohol. Clin. Exp. Res. 22 (5), 1093–1098.

Huang, P., Chen, C., Mague, S.D., Blendy, J.A., Liu-Chen, L.Y., 2012. A common single nucleotide polymorphism A118G of the mu opioid receptor alters its N-glycosylation and protein stability. Biochem. J. 441, 379–386.

Johnson, S.W., North, R.A., 1992. Opioids excite dopamine neurons by hyperpolarization of local interneurons. J. Neurosci. 12, 483–488.

Kiianmaa, K., Hoffman, P.L., Tabakoff, B., 1983. Antagonism of the behavioral effects of ethanol by naltrexone in BALB/c, C57BL/6, and DBA/2 mice. Psychopharmacology (Berl.) 79, 291–294.

Kim, S.G., Kim, C.M., Choi, S.W., Jae, Y.M., Lee, H.G., Son, B.K., Kim, J.G., Choi, Y.S., Kim, H.O., Kim, S.Y., Oslin, D.W., 2009. A micro opioid receptor gene polymorphism (A118G) and naltrexone treatment response in adherent Korean alcohol-dependent patients. Psychopharmacology (Berl) 201, 611–618.

Koob, G.F., Volkow, N.D., 2010. Neurocircuitry of addiction. Neuropsychopharmacology 35, 217–238.

Krystal, J.H., Cramer, J.A., Krol, W.F., Kirk, G.F., Rosenheck, R.A., 2001. Naltrexone in the treatment of alcohol dependence. N. Engl. J. Med. 345, 1734–1739.

Mague, S.D., Isiegas, C., Huang, P., Liu-Chen, L.Y., Lerman, C., Blendy, J.A., 2009. Mouse model of OPRM1 (A118G) polymorphism has sex-specific effects on drug-mediated behavior. Proc. Natl. Acad. Sci. U.S.A. 106, 10847–10852.

Mahmoud, S., Thorsell, A., Sommer, W.H., Heilig, M., Holgate, J.K., Bartlett, S.E., Ruiz-Velasco, V., 2011. Pharmacological consequence of the A118G μ opioid receptor polymorphism on morphine- and fentanyl-mediated modulation of Ca^{2+} channels in humanized mouse sensory neurons. Anesthesiology 115, 1054–1062.

Marinelli, P.W., Kiianmaa, K., Gianoulakis, C., 2000. Opioid propeptide mRNA content and receptor density in the brains of AA and ANA rats. Life Sci. 66 (20), 1915–1927.

Monterosso, J.R., Flannery, B.A., Pettinati, H.M., et al., 2001. Predicting treatment response to naltrexone: the influence of craving and family history. Am. J. Addict. 10 (3), 258–268.

Myers, R.D., Robinson, D.E., 1999. Mu and D2 receptor antisense oligonucleotides injected in nucleus accumbens suppress high alcohol intake in genetic drinking HEP rats. Alcohol 18 (2–3), 225–233.

Myers, R.D., Borg, S., Mossberg, R., 1986. Antagonism by naltrexone of voluntary alcohol selection in the chronically drinking macaque monkey. Alcohol 3, 383–388.

O'Malley, S.S., Jaffe, A.J., Chang, G., Schottenfeld, R.S., Meyer, R.E., Rounsaville, B., 1992. Naltrexone and coping skills therapy for alcohol dependence: a controlled study. Arch. Gen. Psychiatry 49, 881–887.

O'Malley, S.S., Krishnan-Sarin, S., Farren, C., Sinha, R., Kreek, M.J., 2002. Naltrexone decreases craving and alcohol self-administration in alcohol-dependent subjects and activates the hypothalamo-pituitary-adrenocortical axis. Psychopharmacology (Berl.) 160, 19–29.

Oslin, D., Berrettini, W.H., Kranzler, H.R., Pettinati, H., Gelernter, J., Volpicelli, J.R., O'Brien, C.P., 2003. A functional polymorphism in the mu opioid receptor gene is associated with therapeutic response in alcohol-dependent patients treated with naltrexone. Neuropsychopharmacol 28, 1546–1552.

Oslin, D.W., Leong, S.H., Lynch, K.G., Berrettini, W.H., O'Brien, C.P., Gordon, A.J., Rustalis, M., 2015. A randomized clinical trial of naltrexone versus placebo for the treatment of alcohol dependence: investigating the pharmacogenetics of the mu-opioid receptor gene polymorphism, rs1799971. JAMA Psychiatry 72, 430–437.

Pettinati, H.M., O'Brien, C.P., Rabinowitz, A.R., Wortman, S.P., Oslin, D.W., Kampman, K.M., Dackis, C.A., 2006. The status of naltrexone in the treatment of alcohol dependence: specific effects on heavy drinking. J. Clin. Psychopharmacol. 26, 610–625.

Popp, R.L., Erickson, C.K., 1998. The effect of an acute ethanol exposure on the rat brain POMC opiopeptide system. Alcohol 16 (2), 139–148.

Ramchandani, V.A., Umhau, J., Pavon, F.J., Ruiz-Valasco, V., Margas, W., Sun, H., Damadzic, R., Eskay, R., Schoor, M., Thorsell, A., Schwandt, M.L., Sommer, W.H., George, D.T., Parsons, L.H., Herscovitch, P., Hommer, D., Heilig, M., 2011. A genetic determinant of the striatal dopamine response to alcohol in men. Mol. Psychiatry 16, 809–817.

Rasmussen, D.D., Bryant, C.A., Boldt, B.M., Colasurdo, E.A., Levin, N., Wilkinson, C.W., 1998. Acute alcohol effects on opiomelanocortinergic regulation. Alcohol. Clin. Exp. Res. 22 (4), 789–801.

Ray, L.A., Hutchison, K.E., 2004. A polymorphism of the mu-opioid receptor gene (OPRM1) and sensitivity to the effects of alcohol in humans. Alcohol. Clin. Exp. Res. 28 (12), 1789–1795.

Ray, L.A., Hutchison, K.E., 2007. Effects of naltrexone on alcohol sensitivity and genetic moderators of medication response: a double-blind placebo-controlled study. Arch. Gen. Psychiatry 64 (9), 1069–1077.

Ray, L.A., Miranda Jr., R., Tidey, J.W., McGeary, J.E., MacKillop, J., Gwaltney, C.J., Rohsenow, D.J., Swift, R.M., Monti, P.M., 2010. Polymorphisms of the mu-opioid receptor and dopamine D4 receptor genes and subjective responses to alcohol in the natural environment. J. Abnorm. Psychol. 119, 115–125.

Ray, L.A., Bujarski, S., Squeglia, L.M., Ashenhurst, J.R., Anton, R.F., 2014. Interactive effects of OPRM1 and DAT1 genetic variation on subjective responses to alcohol. Alcohol Alcohol. 49, 261–270.

Schacht, J.P., Anton, R.F., Voronin, K.E., Randall, P.K., Li, X., Henderson, S., Myrick, H., 2013. Interacting effects of naltrexone and OPRM1 and DAT1 variation on the neural response to alcohol cues. Neuropsychopharmacology 38, 414–422.

Setiawan, E., Pihl, R.O., Cox, S.M., Gianoulakis, C., Palmour, R.M., Benkelfat, C., Leyton, M., 2011. The effect of naltrexone on alcohol's stimulant properties and self-administration behavior in social drinkers: influence of gender and genotype. Alcohol. Clin. Exp. Res. 35, 1134–1141.

Sobell, L.C., Sobell, M.B., 1992. Timeline follow-back: a technique for assessing self-reported alcohol consumption. In: Litten, R., Allen, J. (Eds.), Measuring Alcohol Consumption. Humana Press Inc, Totowa, NJ, pp. 41–65.

Spanagel, R., 2009. Alcoholism: a systems approach from molecular physiology to addictive behavior. Physiol. Rev. 89, 649–705.

Spanagel, R., Herz, A., Shippenberg, T.S., 1992. Opposing tonically active endogenous opioid systems modulate the mesolimbic dopaminergic pathway. Proc. Natl. Acad. Sci. U.S.A. 89, 2046–2050.

Srisurapanont, M., Jarusuraisin, N., 2005. Naltrexone for the treatment of alcoholism: a meta-analysis of randomized controlled trials. Int. J. Neuropsychopharmacol. 8, 267–280.

Szeto, C.Y., Tang, N.L., Lee, D.T., Stadlin, A., 2001. Association between mu opioid receptor gene polymorphisms and Chinese heroin addicts. Neuroreport 12, 1103–1106.

Tan, E.C., Tan, C.H., Karupathivan, U., Yap, E.P., 2003. Mu opioid receptor gene polymorphisms and heroin dependence in Asian populations. Neuroreport 14, 569–572.

Tanda, G.L., Di Chiara, G., 1998. A dopamine mu(1) opioid link in the rat ventral tegmentum shared by palatable food (Fonzies) and non-psychostimulant drugs of abuse. Eur. J. Neurosci. 10, 1179–1187.

Vallender, E.J., Ruedi-Bettschen, D., Miller, G.M., Platt, D.M., 2010. A pharmacogenetic model of naltrexone-induced attenuation of alcohol consumption in rhesus monkeys. Drug Alcohol Depend. 109, 252–256.

Volpicelli, J.R., Davis, M.A., Olgin, J.E., 1986. Naltrexone blocks the post-shock increase of ethanol consumption. Life Sci. 38, 841–847.

Volpicelli, J.R., Alterman, A.I., Hayashida, M., O'Brien, C.P., 1992. Naltrexone in the treatment of alcohol dependence. Arch. Gen. Psychiatry 49, 876–880.

Volpicelli, J.R., Watson, N.T., King, A.C., Sherman, C.E., O'Brien, C.P., 1995. Effect of naltrexone on alcohol "high" in alcoholics. Am. J. Psychiatry 152 (4), 613–615.

Zhang, Y., Wang, D.X., Johnson, A.D., Papp, A.C., Sadee, W., 2005. Allelic expression imbalance of human mu opioid receptor (OPRM1) caused by variant A118G. J. Biol. Chem. 280, 32618–32624.

CHAPTER

Competing neurobehavioral decision systems theory of cocaine addiction: From mechanisms to therapeutic opportunities

14

Warren K. Bickel*[,1], Sarah E. Snider*, Amanda J. Quisenberry*, Jeffrey S. Stein*, Colleen A. Hanlon[†]

**Addiction Recovery Research Center, Virginia Tech Carilion Research Institute, Roanoke, VA, USA*

[†]Medical University of South Carolina, Charleston, SC, USA

[1]Corresponding author: Tel.: +1-540-526-2088, Fax: 540-985-3361, e-mail address: wkbickel@vtc.vt.edu

Abstract

Cocaine dependence is a difficult-to-treat, chronically relapsing disorder. Multiple scientific disciplines provide distinct perspectives on this disorder; however, connections between disciplines are rare. The competing neurobehavioral decision systems (CNDS) theory posits that choice results from the interaction between two decision systems (impulsive and executive) and that regulatory imbalance between systems can induce pathology, including addiction. Using this view, we integrate a diverse set of observations on cocaine dependence, including bias for immediacy, neural activity and structure, developmental time course, behavioral comorbidities, and the relationship between cocaine dependence and socioeconomic status. From the CNDS perspective, we discuss established and emerging behavioral, pharmacological, and neurological treatments and identify possible targets for future treatments. The ability of the CNDS theory to integrate diverse findings highlights its utility for understanding cocaine dependence and supports that dysregulation between the decision systems contributes to addiction.

Keywords

Cocaine dependence, Competing neurobehavioral decision systems, Impulsivity, Self-control, Executive function, Delay discounting, Dual systems, Transcranial magnetic stimulation

1 INTRODUCTION

Cocaine is a powerful psychoactive and addictive substance. Approximately 15% of cocaine users develop dependence within the first decade after initial use, with lifetime incidence of dependence estimated at 20% (Lopez-Quintero et al., 2011; Wagner and Anthony, 2002). In some racial minorities, these estimates are even higher (e.g., 35% lifetime incidence of dependence in African American users) (Lopez-Quintero et al., 2011). Cocaine dependence is difficult to treat and is recognized as a chronically relapsing disorder, in which affected individuals choose continued drug use despite negative consequences, and return to use after periods of abstinence. Understanding the processes that undergird these choices is an important undertaking for the science and treatment of this disorder.

A variety of scientific approaches have tried to understand and explain cocaine dependence. Some have focused on molecular variables, such as pharmacological action (Volkow et al., 1999); others have focused on demographics, including age, race, and socioeconomic status (SES) (Lopez-Quintero et al., 2011; Palamar et al., 2015). These multiple levels of analysis provide distinct perspectives on cocaine dependence, but connections across levels have been rare. A thorough understanding of these multilevel phenomena, in our view, will require a scientific theory or paradigm that not only can integrate observations across levels in a compelling way, but can also suggest novel hypotheses. As Henri Poincaré noted in his classic text, *Science and Hypothesis* (Poincaré, 1905), "Science is built up of facts, as a house is built of stones; but an accumulation of facts is no more a science than a heap of stones is a house (p. 157)."

The question we should ask is what would we want from such a theory that could set the extant facts in order? At the very least, any such theory should integrate the neuroscience of cocaine's effects on the brain, developmental processes associated with drug use initiation, the relationship of SES to cocaine use, and the high prevalence of certain comorbidities. Such a theory should also have the capacity to suggest novel treatments and perhaps reveal mechanisms underlying established treatments.

We have been involved with formulating a view, referred to as the competing neurobehavioral decision systems (CNDS) theory (Bickel and Yi, 2008; Bickel et al., 2007, 2012a) that has considerable integrative power. This view, consistent with a broad array of dual-systems theories, suggests that choices result from the interaction between the two decision systems and that those who are experiencing addiction suffer from imbalance or dysregulation between these two systems. In this chapter, we will examine cocaine dependence from the perspective of this theory. To accomplish this, we will first give a brief synopsis of this theoretical view and examine the evidence to support the dysregulation between the dual systems in individuals with cocaine dependence. Next, we will examine how this perspective provides insight on the relationship between cocaine dependence and developmental life course, SES, and comorbidities. Finally, we will examine the implications of the CNDS perspective for existing and emerging approaches to the treatment of cocaine dependence.

2 THE COMPETING NEUROBEHAVIORAL DECISION SYSTEMS THEORY

Dual-systems models of decision-making have been discussed since Descartes and have evolved to many variations and applications (Sanfey and Chang, 2008), particularly in the areas of self-control (Metcalfe and Mischel, 1999) and addiction (Bechara, 2005; Goldstein and Volkow, 2002, 2011; Jentsch and Taylor, 1999). In decision-making research, most models refer to the dual systems as System 1 and System 2. System 1 refers to unconscious and automatic processes, requiring little effort, while System 2 refers to conscious, controlled, and effortful processes (Evans, 2008; Evans and Stanovich, 2013).

The CNDS theory is a dual-systems model that accounts for self-control failure (Bickel et al., 2007, 2011a), has been directly applied to addiction (Bickel et al., 2011a; Sofis et al., 2014) and emphasizes the relative control between impulsive and executive decision systems. The impulsive system, comprised of the limbic and paralimbic brain regions, and executive system, comprised of the prefrontal and parietal cortices, are interdependent and compete for relative control during decision-making (see Bickel et al., 2012a for pictorial representations). Normal functioning results when the systems are in regulatory balance; however, when the two systems are not in regulatory balance, pathology may result (Bickel et al., 2015). Although worthwhile, systematic comparison of the CNDS theory and other dual-systems models is beyond the scope of this chapter, thus we reserve such comparisons for future discussions.

Importantly, delay discounting is a behavioral measure of self-control that designates the relative strength of the competing decision systems (Bickel et al., 2012b; McClure and Bickel, 2014). Delay discounting procedures measure future valuation by asking participants if they would prefer a smaller, immediate amount of a commodity or a larger, delayed amount. The immediate amount is titrated until a point of subjective equality (the indifference point) is determined. A hyperbolic function often best accounts for the fit of the indifference points across delays and is represented by the equation (Mazur, 1987),

$$V = A/(1 + kD),$$

where V is the subjective value of the reinforcer, A is the amount of the reinforcer, D is the delay to receipt of the reinforcer, and k is a free parameter that serves as an index of discounting (higher values of k indicate higher rates of discounting). Nicotine- (Bickel et al., 1999), alcohol- (Petry, 2001), cocaine- (Bickel et al., 2011b, 2014a; Heil et al., 2006), and heroin-dependent (Madden et al., 1997) individuals discount future rewards more than controls. Higher rates of discounting, then, reflect hyperactive control by the impulsive decision system, consistent with the bias for immediate reward evident in addiction (Bickel et al., 2011a).

The study of neuroeconomics, which combines psychology, economics, and neuroscience (Bickel et al., 2011a), has provided confirmatory neural evidence for the actions of the CNDS (described in the following sections). When participants complete delay-discounting procedures in an MRI scanner, relative activity between the executive and impulsive systems varies, dependent upon the choice being made. For example, choices for the immediate and delayed reinforcer result in greater activity in the impulsive and executive systems, respectively (McClure et al., 2004, 2007). Moreover, when the reinforcer is delayed for both choices, the limbic system shows no differential activation. Thus, activation of the impulsive decision system depends on the presence of an immediate reinforcer (McClure et al., 2004).

2.1 THE IMPULSIVE DECISION SYSTEM

The impulsive decision system, comparable to System 1, is embodied in the limbic (e.g., midbrain, amygdala, habenular commissure, and striatum) and paralimbic (e.g., insula and nucleus accumbens) brain regions (Bickel et al., 2007). Habit formation, emotional responding, and the acquisition of primary reinforcers to satisfy biological needs (Bickel et al., 2013) are controlled by the impulsive decision system.

As discussed above, imaging studies have confirmed that the impulsive decision system is involved in the choice for immediate reinforcers in delay discounting. Choice for immediate reinforcers (McClure et al., 2004) selectively activate the paralimbic cortex and parts of the limbic system, including ventral striatum, medial orbitofrontal cortex, medial prefrontal cortex, posterior cingulate cortex, and left posterior hippocampus (McClure et al., 2004, 2007).

2.2 THE EXECUTIVE DECISION SYSTEM

The second decision system of the CNDS, comparable to System 2, is embodied in the parietal lobes and portions of the prefrontal cortex, including the dorsolateral prefrontal cortex (Bickel et al., 2007). Some overlap of function in the decisions systems exists for several areas of the prefrontal cortex, including the orbitofrontal cortex. The cortical pathways of the executive decision system are responsible for planning, memory, attention, and future valuation (Bickel et al., 2013). Neuroeconomic evidence has demonstrated activation of the lateral prefrontal cortex and parietal lobe during decision-making in delay discounting for monetary and primary reinforcers (i.e., juice) regardless of delay, indicating the executive system is involved in all decisions (McClure et al., 2004, 2007). Moreover, greater activation in the executive system structures occurs during the more difficult choices requiring greater executive function.

2.3 THE COMPETING NEUROBEHAVIORAL DECISION SYSTEMS THEORY IN HEALTH AND ADDICTION

When regulatory balance is achieved between the impulsive and executive decision systems, an individual is considered self-controlled and is likely to have no dysfunction (Bickel et al., 2015). Conversely, hyperactive control by either the impulsive or executive decision system can lead to pathological behavior. Many combinations of relative strength of each system are possible (Bickel et al., 2013). Consider Fig. 1 that shows a continuum from low to high control by the impulsive decision system on the *y*-axis and on the *x*-axis, low to high executive control. The diagonal line represents regulatory balance between the two decision systems. Shaded regions represent high risk for engaging in negative health behaviors. The pathological decision-making strategies associated with these behaviors emerge when control by the impulsive decision system overpowers control by the executive decision system. For example, high impulsive system control coupled with low or medium executive control results in greater relative control by the impulsive decision system and can result in pathological decision-making (e.g., bias toward smaller, immediate over larger, delayed consequences) (Bickel et al., 2011a, 2013).

Imbalance of the CNDS is evident in many disease states where individuals have bias for immediate consequences over delayed, healthier choices. Hyperactive control of the impulsive decision system results in patterns of behavior consistent with obesity, legal and illicit substance use, and gambling problems. Regulatory imbalance can also be a result of hyperactive control by the executive system.

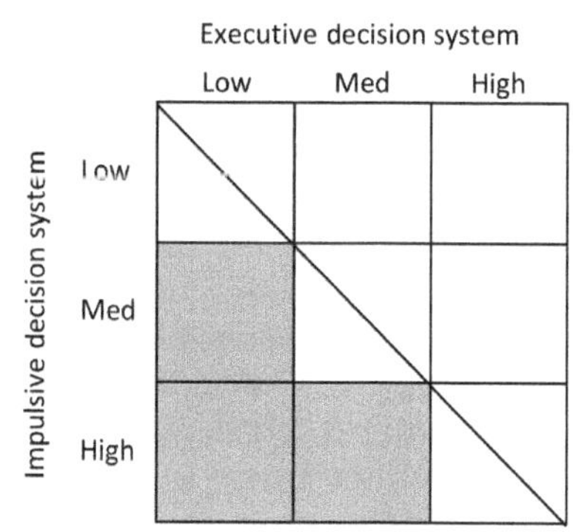

FIGURE 1

The relative control of the impulsive and executive decision systems, represented graphically. The diagonal line represents regulatory balance between the systems. The shaded regions indicate an imbalance between the two systems producing a bias for immediate over delayed rewards.

3 THE COMPETING NEUROBEHAVIORAL DECISIONS SYSTEMS THEORY AND COCAINE

3.1 NEURAL EVIDENCE OF THE IMBALANCE OF DECISION SYSTEMS

Imbalance of the CNDS contributes to excessive discounting and addiction behaviors (Bickel et al., 2012b). Addiction occurs when the executive system is weak and the hyperactive impulsive decision system drives choice (Bechara, 2005). Cocaine use, via neuronal plasticity, promotes a transition in regulation from the prefrontal cortices to the striatum leading to compulsive and habitual drug seeking (Everitt et al., 2008). Advances in imaging have provided us with tools to examine the neural evidence of this imbalance in cocaine addiction.

3.1.1 Hyperactivation of the impulsive system

The impulsive decision system is comprised of regions of the limbic system and related areas (McClure et al., 2004). One of these regions, the orbitofrontal cortex, is associated with: (1) the reinforcing aspects of cocaine, (2) immediate choice preferences, and (3) craving and cocaine salience (Lucantonio et al., 2012; McClure et al., 2004; Steinberg, 2007). Compared to healthy controls, cocaine addicts show increased activation in limbic regions (i.e., the amygdala, anterior cingulate cortex, and striatum) following exposure to cocaine cues (Childress et al., 1999; Garavan et al., 2000). These increases in activation suggest regions responsible for drug craving and hyperactivation in craving states. Hyperactivation of the impulsive system in cocaine users is also consistent with findings of acute withdrawal circuits becoming hypermetabolic during spontaneous craving (Kalivas and Volkow, 2005; Lucantonio et al., 2012). Moreover, hyperactivation of the medial orbitofrontal cortex and anterior cingulate cortex (structures with impulsive functions) occurs following acute methylphenidate administration in cocaine addicts (Wilcox et al., 2011), suggestive of system over-activation following repeated stimulant administration (akin to sensitization observed in animals) (Robinson and Berridge, 1993).

Interestingly, while hyperactivation and hypermetabolism of the limbic system occurs under certain conditions, cocaine-dependent participants show an overall reduction in aspects of the impulsive decision system compared to healthy controls. These reductions include decreased activation of the orbitofrontal cortex and cingulate gyrus (Volkow et al., 1993) and decreased gray matter volume of the amygdala (Makris et al., 2004) and the ventromedial, orbitofrontal, anterior cingulate, and anteroventral insular cortices (Franklin et al., 2002). Although these findings may seem counterintuitive from the viewpoint of CNDS (i.e., reduced function and structure of the impulsive decision system in cocaine addicts), cocaine may prime the limbic regions associated with cue salience and motivation, consistent with hyperactivation of the impulsive decision system following drug administration and contributes to increased craving and compulsive intake (Volkow et al., 2005). Consistently, acute methylphenidate administration may normalize limbic activation by working similarly to cocaine but with slower pharmacokinetics. That is, in cocaine addicts, acute methylphenidate increases activation in the anterior cingulate cortex during a cue-reactivity task (Goldstein et al., 2010), restores response levels to normal after fatigue in a Stroop task (Moeller et al., 2012), and increases resting-state functional connectivity in limbic regions, including the anterior cingulate cortex (Konova et al., 2013). Thus, hyperactivation of the impulsive decision system, as a consequence of cocaine priming the system, weakens relative control of the executive system and decreases self-control (Noel et al., 2013).

3.1.2 Hypoactivation of the executive system

In addition to hyperactivity of the impulsive system, drugs of abuse cause an interruption of the top-down processes required for self-control (Dalley et al., 2011). Neural evidence suggests that cocaine induces executive dysfunction. Although hyperactivation may occur in some instances, as mentioned above, overall reductions in signaling, glucose metabolism (Kalivas and Volkow, 2005), and structural volume (Franklin et al., 2002) in both the impulsive and executive systems are observed after cocaine use. Moreover, the degree of cocaine use is associated with both structural and functional deficits in the executive system (Beveridge et al., 2008).

3.2 DEVELOPMENTAL PROCESSES AND COCAINE ADDICTION

3.2.1 Differential development

Evidence of differential development between the CNDS explains impaired self-control in adolescents, as the two decision systems appear to differentially mature. During the first half of adolescence (i.e., ages 10–15), dopaminergic activity increases dramatically in brain areas associated with the impulsive decision system (Sisk and Zehr, 2005), including a dramatic dopamine and dendritic synaptic overexpression in the striatum (Andersen et al., 2000). Related to this overexpression, the nucleus accumbens, an area of the impulsive decision system responsible for the rewarding properties of stimuli, and orbitofrontal cortex display hyperactivation in children and adolescents compared to adults in resting state (Galvan et al., 2006) and when completing a monetary reward task (Ernst et al., 2005). Moreover, differential myelination between limbic and nonlimbic regions enhances activation in the impulsive decision system (Galvan et al., 2006). As adolescents mature, the overexpression and hyperactivation of the impulsive system begins to prune to model an inverted U-shaped function over time (Sisk and Zehr, 2005; Teicher et al., 1995). That is, after the overexpression peaks, extra connectivity begins to decline while the slower to mature executive decision system continues to develop.

Development of the executive decision system includes increases in parietal gray matter volume (Sisk and Zehr, 2005) along with dramatic dopamine and dendritic synaptic overexpression in the prefrontal cortex (Andersen et al., 2000). Gray matter density development and myelination in the frontal and parietal cortices continues into adulthood (Sowell et al., 2003), thus increasing relative control of the executive over the impulsive decision systems with age.

3.2.2 Related behaviors

The differential development of the two systems and inverted U-shaped curve of impulsive decision system development is evident in self-control. Paralleling the overexpression of dopamine and activation of the impulsive decision system, a drastic increase in risky behavior is present in adolescence. Specifically, self-reported sensation-seeking and risky sexual behavior increases drastically, peaks in early adolescence, and declines as self-regulatory behavior begins to mature (Baams et al., 2015; Steinberg, 2007; Steinberg et al., 2008).

Importantly, a longitudinal study modeled the imbalance of the CNDS and reported that high rates of delay discounting and poor working memory (both measures of weak executive control) predicted greater subsequent initiation of drug use (i.e., alcohol, marijuana, and tobacco) (Khurana et al., 2015). Using data from two large national surveys (Substance Abuse and Mental Health Services Administration, 2004, 2013), Fig. 2 highlights this increased vulnerability in adolescents by illustrating the percentage of adolescents who used cocaine in the last 30 days by age group. Note, the percentage of use rapidly peaks in adolescence and declines with increasing age.

3.3 SOCIOECONOMIC STATUS AND COCAINE ADDICTION

A widely demonstrated negative linear relationship exists between SES and illicit drug use, health problems, and mortality. This monotonic gradient, describing the relationship between SES and health status, extends from the lowest to the highest ends of the socioeconomic spectrum. As a result, this relationship cannot be entirely accounted for by poverty-induced deprivation or healthcare access (Adler and Stewart, 2010). This gradient represents the health disparity in prevalence of

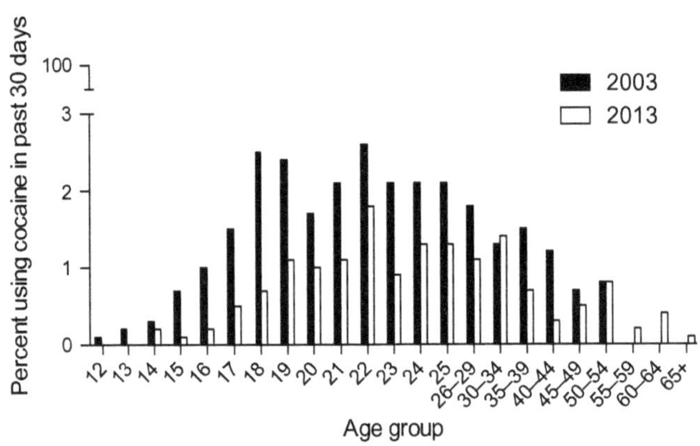

FIGURE 2

Percentage of cocaine use in the past month, by age group. The results from the 2003 and 2013 National Survey on Drug Use and Health Surveys are presented. Percentage of use increases with age, which then slowly dissipates over time.

negative health behaviors (e.g., drug use, risky sexual behavior, and obesity) such that a lower prevalence of disease states is observed in high-SES individuals and a higher prevalence is observed in lower SES individuals. The greater the income inequality within a society, the larger the health disparity (Banks et al., 2006). This trend is apparent within the United States and is representative of the general trend showing larger health disparities in countries with more income inequality (Wilkinson and Pickett, 2011). Rates of mental illness, obesity, and substance use are disease states strongly associated with SES inequalities (Pampel et al., 2010).

One measure included in SES, education level, contributes to the prevalence of past year cocaine use and exemplifies this general trend in health disparities. In 2012, 2.4% of people who did not graduate from high school, while only 1.1% of college graduates, used cocaine in the past year. This relationship between SES and cocaine use began in the 1990s when risk perception of using cocaine increased and therefore became less culturally acceptable. As a result, high-SES individuals were more likely to discontinue cocaine use while low SES individuals continued use (Miech, 2008). The increased prevalence of cocaine use among lower income individuals demonstrates the negative socioeconomic gradient present across a wide variety of negative health behaviors, including cigarette smoking (Hiscock et al., 2012), illicit drug use (Buka, 2002), and obesity (Baum and Ruhm, 2009).

The CNDS theory can be used as a conceptual framework to explain the discrepancy between the prevalence of negative health behaviors among individuals with varying SES (Bickel et al., 2014b). The experiences associated with low SES, including increased allostatic load and lack of resources (Haushofer and Fehr, 2014; Mani et al., 2013), disrupt the development and regulatory balance between the impulsive and executive decision systems (Bickel et al., 2014b; Noble et al., 2012). Exposure to these environmental circumstances facilitates a biased decision-making process favoring immediate over delayed, healthier consequences. This executive dysfunction results from hyperactivation of the impulsive decision system and results in continued choice for immediate rewards, which perpetuates the disparity in negative health behaviors, including cocaine use.

3.4 COMORBIDITIES WITH OTHER SUBSTANCE USE AND RISKY SEXUAL BEHAVIOR

The CNDS theory also provides a framework for understanding the relationship between comorbid disease states and cocaine use. Regulatory imbalance of the systems resulting from hyperactivity of the impulsive decision system may explain the presence of decision-making favoring immediate rewards (e.g., cocaine use and risky sexual behavior) (Chesson et al., 2006; Johnson and Bruner, 2012) over delayed, more healthy consequences. These decision-making processes are central to many disease states, which contributes to the incidence of comorbidity (Bickel and Mueller, 2009). Comorbid substance use, including tobacco (Budney et al., 1993; Burling et al., 1996), alcohol (Bierut et al., 2008), marijuana (Narvaez et al., 2014), and opiate use (Bierut et al., 2008), is common in cocaine users although few treatments intended for cocaine dependence take these comorbidities into account (Yoon et al., 2013).

4 THE CNDS AND COCAINE TREATMENT

The CNDS theory has been used previously to understand and categorize the effects of various delay-discounting manipulations (Koffarnus et al., 2013). Here, we apply a similar analysis to current and emerging treatments for cocaine dependence.

4.1 CONVENTIONAL TREATMENT FOR COCAINE DEPENDENCE

A number of therapies have been successfully used to treat cocaine dependence, among which cognitive behavioral therapy (CBT) has the largest evidence base (Carroll and Onken, 2005; Carroll et al., 2008; Maude-Griffin et al., 1998). However, behavioral measures of executive dysfunction (e.g., poor Stroop performance) consistently predict poor response to these treatments (Aharonovich et al., 2006; Bleiberg et al., 1994; Moeller et al., 2001; Simpson et al., 1999; Streeter et al., 2008; Worhunsky et al., 2013; Xu et al., 2010). Likewise, functional and structural neuroimaging data, such as diminished prefrontal cortex activation and white matter integrity, further implicate executive dysfunction in poor treatment response (Brewer et al., 2008; Moeller et al., 2005; Worhunsky et al., 2013; Xu et al., 2010).

Many of these conventional treatments, including CBT, require a complex repertoire of executive skills (e.g., coping strategies or the ability to recognize dynamic relapse cues and modify behavior accordingly), which are likely compromised in individuals demonstrating regulatory imbalance between decision systems. From the viewpoint of the CNDS, a more promising approach would be to precisely target areas of dysfunction to produce more uniformly efficacious treatment outcomes compared to conventional treatment strategies (Bickel et al., 2012b). In the sections that follow, we consider a number of treatments that may accomplish this goal.

4.2 TREATMENTS TO DECREASE CONTROL OF THE IMPULSIVE DECISION SYSTEM

4.2.1 Contingency management

One of the most reliable treatments for cocaine and other substance dependence in recent decades has been contingency management, a behavioral approach that arranges immediate delivery of monetary or other tangible reinforcers contingent on physiologically verified drug abstinence (Higgins et al., 1991, 1994) (for review and meta-analysis, see Lussier et al., 2006; Prendergast et al., 2006). This approach rapidly reduces cocaine use (Robles et al., 2000) and maintains abstinence over long periods of time (Poling et al., 2006; Rawson et al., 2002), even in the absence of continued treatment (Epstein et al., 2003; Higgins et al., 1995; Petry and Martin, 2002). Moreover, contingency management for cocaine use may be implemented successfully at relatively low cost (Petry and Martin, 2002; Petry et al., 2004) and may be paired with adjunctive therapies (e.g., CBT) (Epstein et al., 2003) to further improve treatment outcomes.

As discussed previously, substance use may be viewed as an intertemporal choice between immediate drug reinforcement and the temporally diffuse and distant outcomes associated with drug abstinence (e.g., sustained physical and mental health and attainment of occupational goals). Regulatory imbalance between decision systems may predispose individuals toward cocaine use by disproportionately weighting the value of immediate drug reinforcement. With this in mind, the provision of extrinsic, relatively immediate reinforcement for abstinence in contingency management therapies may supplant the naturalistic, delayed outcomes of abstinence (e.g., improved health and social function) that are otherwise insufficient to impact behavior in those suffering from regulatory imbalance. In addition, cessation of cocaine use during contingency management likely facilitates initial contact with these naturalistic outcomes, perhaps contributing to continued abstinence following treatment (Epstein et al., 2003; Higgins et al., 1995; Petry and Martin, 2002). Consistent with these mechanisms, a recent study examining contingency management for opioid abuse allowed participants to either redeem these earnings immediately at each laboratory visit or accumulate their earnings in an account over the course of the study (Bickel et al., 2010). Participants with the highest baseline rates of delay discounting more frequently redeemed their earnings immediately than participants with lower rates of delay discounting, demonstrating the selective importance of immediate outcomes for participants with regulatory imbalance. Future studies should be designed to determine whether a similar finding would be observed with contingency management for cocaine dependence.

4.2.2 Medications

Currently no approved medication exists for stimulant addiction (Brackins et al., 2011) and replacement therapies with stimulants for cocaine and methamphetamine addiction have produced equivocal results (Moeller et al., 2008). However, the possibility remains that replacement agonist therapy may be a viable avenue to decrease or buffer the hyperactivation of the impulsive decision system during or to prevent crave states.

For example, dexamphetamine and methylphenidate are long-acting stimulants, with similar mechanisms of action to cocaine (i.e., increases in extracellular dopamine) and have shown positive results in reducing behaviors related to cocaine addiction. In intravenous cocaine users, dexamphetamine reduced positive urine samples for cocaine, self-reported use, craving, and criminal activity (Shearer et al., 2003). Dexamphetamine maintenance also reduces choice preferences for immediate cocaine over money (Rush et al., 2009). Likewise, methylphenidate reduces reaction to cocaine cues and attenuates anterior cingulate cortex activation in cocaine-dependent individuals (see review Mariani and Levin, 2012) without impairing inhibitory control in a go/no-go task (Vansickel et al., 2008), offering a stimulant agonist medication without over activating the impulsive decision system.

Moreover, replacement therapies such as methylphenidate, dexamphetamine, and atomoxetine are pharmacologically safe for maintenance therapy (Grabowski et al., 1997; Rush et al., 2009; Stoops et al., 2008). Thus, the benefit of longer acting agonist medications for use as partial agonist therapies offers a potential avenue to buffer hyperactivation of the impulsive decision system in cocaine-dependent individuals.

4.2.3 Neurotherapeutic stimulation

Transcranial magnetic stimulation (TMS) is a noninvasive brain stimulation tool which enables us to selectively activate or inhibit populations of neurons by altering the frequency and placement of cortical stimulation. When stimulation is delivered repetitively, at frequencies known to induce long-term potentiation (LTP) or depression (LTD) of cortical activity, this technique is known as repetitive TMS (rTMS) (Fitzgerald et al., 2006; Hoogendam et al., 2010; Thickbroom, 2007; Ziemann et al., 2008). LTP of both behavioral and neural activity is possible by applying either a single high frequency (e.g., 10 Hz) or an intermittent theta burst frequency to the cortex. In contrast, transient LTD of behavioral and neural activity is possible by applying either a single low-frequency (e.g., 1 Hz) or continuous theta burst frequency to the cortex. rTMS is an FDA-approved treatment for depression and is the only noninvasive brain stimulation tool available for humans.

A growing body of substance dependence literature suggests that we may be able to directly dampen limbic circuitry or amplify executive control circuitry in substance-dependent individuals through rTMS. Consequently, rTMS has garnered significant attention as an innovative tool for treating substance dependence from both the National Institutes of Health and in the literature (Barr et al., 2011; Bellamoli et al., 2014; Gorelick et al., 2014; Wing et al., 2013). In context with the CNDS, several strategies could be used to develop treatments for substance dependence, including altering the relative control of the impulsive and executive decision systems.

The vulnerability to drug-related cues in treatment-seeking cocaine users is likely sustained by high functional activity in the impulsive decision system (Ersche et al., 2012; Moeller et al., 2010; Moreno-Lopez et al., 2012). Consequently, application of low-frequency TMS, for example, applying LTD-like stimulation to the limbic system may reduce sensitivity to cocaine and other substance cues. Given that the nucleus accumbens is one of the primary brain regions involved in craving (Robinson and Berridge, 1993) and the medial prefrontal cortex is that structure's primary cortical input, targeting the medial prefrontal cortex would be a method to modulate nucleus accumbens activity among substance-dependent populations. Recent work by Cho et al. (2015) demonstrated that LTP-like rTMS (i.e., 10 Hz) to the medial prefrontal cortex in a group of healthy, nondrug-using individuals was associated with a significant decrease in dopamine binding potential in the dorsal striatum, reflecting a release of dopamine in these areas. Although they did not find a significant change in dopamine binding in the nucleus accumbens, LTP-like stimulation of the medial prefrontal–striatal circuit increased delay discounting (a behavioral marker of executive dysfunction). This finding suggests that an LTD-like rTMS strategy over the medial prefrontal cortex would attenuate activity in this neural circuit and may reduce drug craving and impulsive decision system control. Prior data from our laboratory demonstrate that in cocaine users, continuous theta burst stimulation to the frontal lobe selectively decreases activation in the medial prefrontal cortex and nucleus accumbens (Hanlon et al., 2015). Given that craving for cocaine is associated with an increase in striatal dopamine, decreasing the sensitivity of this circuit through rTMS may be a valuable treatment strategy. Future research is required to determine whether stimulating this location is tolerable in substance-dependent populations because medial prefrontal cortex stimulation has not been widely pursued and is subjectively more painful than dorsolateral prefrontal cortex rTMS.

4.3 TREATMENTS TO INCREASE CONTROL OF THE EXECUTIVE DECISION SYSTEM

4.3.1 Neurocognitive training

Executive function deficits in chronic cocaine users are well established (Bolla et al., 2000). Specifically, compared to healthy controls, cocaine-dependent individuals demonstrate significant impairments of multiple measures of attention, visual and spatial memory, language and sensory perception functions (Jovanovski et al., 2005). This executive dysfunction is related to retention rates for relapse prevention therapy in cocaine users (Aharonovich et al., 2003, 2006). Because functional and regional overlap exists between executive function areas, including those involved in making delay-discounting decisions for the delayed reinforcer (Bickel et al., 2011c; Wesley and Bickel, 2014), training specific executive functions, such as working memory, may increase executive decision system control leading to program retention and a rebalance of the CNDS.

4.3.1.1 Working memory training

Of the impaired executive systems in cocaine addicts, working memory is an executive function mediated by the prefrontal cortex and is involved in goal-directed behavior (Miller and Cohen, 2001). Interestingly, following working memory training, healthy participants demonstrate increases in prefrontal and parietal region activation (Olesen et al., 2004). Consistent with the CNDS theory, more activation in these areas indicate increases in executive decision system functionality and is important because greater frontoparietal activity occurs when participants choose larger delayed rewards (McClure et al., 2004). In fact, we have demonstrated decreased delay discounting of monetary rewards following working memory training in cocaine addicts (Bickel et al., 2011c), thus providing support for this potential approach to increase executive system functionality. In addition to working memory training, a second potential treatment, episodic future thinking, shows beneficial executive neurocognitive improvement capabilities.

4.3.1.2 Episodic future thinking

Episodic future thinking is a form of prospection which involves mental simulation of future events (Atance and O'Neill, 2001). Neural evidence demonstrates that future thinking tasks activate frontal cortices (Okuda et al., 2003) associated with the executive decision system. Moreover, goal-directed simulations activate the prefrontal cortex and associated regions (Gerlach et al., 2011). Behaviorally, episodic future thinking decreases delay discounting, which is predicted by anterior cingulate cortex activation (Daniel et al., 2013; Peters and Buchel, 2010). Thus, given that poor performance of future thinking is associated with poor executive function (de Vito et al., 2012), repetition of either working memory training or episodic future thinking may increase control of the executive decision system, improve valuation of future rewards, and provide a valuable adjunct to cocaine cessation therapy.

4.3.2 Medications

Modafinil acts on several neurotransmitter systems including glutamate, GABA, and dopamine. Similar to the previously proposed agonist therapies to decrease control of the impulsive decision system, modafinil produces a similar mechanism of action to cocaine (i.e., increases in dopamine) and produces protracted mild stimulant properties to promote wakefulness. Modafinil reduces activity in the ventral tegmental area, an impulsive decision system brain region, and reduces self-reported craving in response to cocaine cues (Goudriaan et al., 2013), indicative of an attenuation of craving. Though modafinil has been investigated as an agonist replacement therapy (i.e., to buffer hyperactivation of the impulsive decision system) with mixed results (Dackis et al., 2012; Hart et al., 2008), modafinil's actions may be most beneficial by activating the executive decision system. Modafinil promotes enhanced activation of the frontoparietal regions and reduced activation of the ventromedial prefrontal cortex (Schmaal et al., 2014), both regions associated with the valuation of rewards. Behaviorally, modafinil increases several measures of working memory and attention in cocaine users (Kalechstein et al., 2013). Modafinil reduces delay discounting in alcohol-dependent participants compared to controls (Schmaal et al., 2014), and importantly, modafinil does not impair inhibitory control in a go/no-go task in cocaine-dependent individuals (Vansickel et al., 2008) offering another stimulant medication that increases executive function without overactivating the impulsive system.

Modafinil, alongside other medications, has been classified as a nootropic, or a cognitive enhancer. Nootropics are reported to increase working and visual memory, decision-making, and planning (Turner et al., 2004), indicating that pharmacological interventions can improve executive decision system function and regulatory balance of the CNDS. Interestingly, improving deficits in neurotransmitter systems with nicotine agonists, norepinephrine transporter inhibitors, or alpha-2 adrenergic agonists, coincide with some improved attention, response inhibition, and working memory (Sofuoglu, 2010). Evidence that these other systems modulate executive function warrants further investigation into nootropics enhancing the executive decision system to improve treatment outcomes. Moreover, the benefits of pharmacological treatments can provide a valuable adjunct therapy to behavioral interventions such as contingency management or working memory training, allowing for synergistic treatment.

4.3.3 Neurotherapeutic stimulation

Vulnerability to drug-related cues may be due to low functional activity in the executive decision system of substance-dependent individuals (Goldstein et al., 2004; Kubler et al., 2005; Moeller et al., 2010) suggesting that an LTP-like rTMS stimulation of the executive decision system (e.g., dorsolateral prefrontal cortex) might enable better resistance against drug cues. To date, the vast majority of rTMS studies in addiction have targeted the dorsolateral prefrontal cortex (Amiaz et al., 2009; Camprodon et al., 2007; Eichhammer et al., 2003; Herremans et al., 2012, 2013; Hoppner et al., 2011; Li et al., 2013; Mishra et al., 2010; Politi et al., 2008; Pripfl et al., 2014). While many of these studies demonstrated that LTP-like rTMS stimulation to the dorsolateral prefrontal cortex can result in a significant reduction of craving, the neurobiological mechanism is unclear. For example, in a comprehensive review on the efficacy of rTMS for smoking cessation, Wing et al. (2013) reported beneficial effects on tobacco craving following LTP-like rTMS on the dorsolateral prefrontal cortex.

Neurotherapeutic stimulation is a developing area of research for treatment of drug dependence. Future research needs to resolve two questions, which cortical location should be targeted in order to maximally affect the circuitry associated with regulatory balance between decision systems and what stimulation frequency should be used. Identification of a single "optimal" protocol for all individuals or all drug classes is not likely. For example, some individuals may benefit the most from a treatment strategy that amplifies the executive decision system (e.g., 10 Hz dorsolateral prefrontal cortex stimulation) while others may benefit most from a strategy that attenuates the impulsive decision system (e.g., 1 Hz medial prefrontal stimulation). Before moving forward with expensive and slow multisite clinical trials investigating the efficacy of rTMS as a viable treatment tool for addiction, exploration of these combinations of frequencies and cortical targets to maximize potential impact should be considered. TMS may provide a powerful new tool to use as an adjunct to behavioral and pharmacotherapeutic addiction treatment. Given that no FDA-approved pharmacotherapy for cocaine dependence exists, brain stimulation may be a particularly useful therapeutic technique.

5 CONCLUSION AND FUTURE DIRECTIONS

Integration of findings from multiple scientific disciplines and levels of analysis into a robust conceptual system will permit and suggest experiments, and perhaps lead to novel treatments for cocaine dependence. Scientific paradigms in the field of addiction have continuously evolved and have had at least four major paradigm shifts in the last hundred years (Bickel et al., 2013). The CNDS theory constitutes the most recent paradigm shift and is a valuable perspective for addiction research in two ways. First, it stipulates that a fundamental contributor to the addiction process is a dysregulation between the impulsive and executive decision systems. Second, it identifies those two decision systems as targets for interventions.

In this chapter, we have shown that numerous observations could be integrated when viewed from the perspective of the CNDS. Armed with that view, we connect observations regarding the immediacy bias evident in addiction, neural activity and structure, the developmental pattern associated with cocaine and other drug use vulnerabilities, the SES gradient of cocaine and other drug dependencies, and the presence of comorbidities. Such integration supports use of the CNDS theory to guide treatment strategies.

For treatment of cocaine dependence, our view is that treatments or interventions should be supported by a theoretical conceptualization. If the conceptualization of a disorder changes, that change should force a reevaluation of the treatment efficacy. The CNDS is a relatively new conceptualization and permits understanding of the efficacy of existing treatments (e.g., CBT), but also suggests novel approaches (e.g., rTMS) to either decrease activity in the impulsive decision system or increase activity in the executive decision system. Efficacy of these novel approaches will, in part, continue to test the CNDS and indicate the range of its relevance.

The CNDS, like many paradigmatic approaches, is an approximation of a more complete paradigm. The examination and use of the CNDS in the treatment of cocaine and other drug dependence disorders are not based on the ultimate value of the theory, but rather its proximal utility in making new discoveries and assisting those trapped by cocaine dependence. Whether the CNDS continues to provide new research insights that contribute to treatment or will instead give way to an even more robust perspective will await subsequent investigation. In either case, the continued exploration and elaboration of this integrated view contributes to the science of addiction, in general, and cocaine dependence in particular.

ACKNOWLEDGMENTS

The following grants contributed to the support of the authors during the development of this work: NIH grants U19CA157345, R01DA034755, R01AA021529, and R01DA036617.

REFERENCES

Adler, N.E., Stewart, J., 2010. Health disparities across the lifespan: meaning, methods, and mechanisms. Ann. N. Y. Acad. Sci. 1186, 5–23.

Aharonovich, E., Nunes, E., Hasin, D., 2003. Cognitive impairment, retention and abstinence among cocaine abusers in cognitive-behavioral treatment. Drug Alcohol Depend. 71 (2), 207–211.

Aharonovich, E., Hasin, D.S., Brooks, A.C., Liu, X., Bisaga, A., Nunes, E.V., 2006. Cognitive deficits predict low treatment retention in cocaine dependent patients. Drug Alcohol Depend. 81 (3), 313–322.

Amiaz, R., Levy, D., Vainiger, D., Grunhaus, L., Zangen, A., 2009. Repeated high-frequency transcranial magnetic stimulation over the dorsolateral prefrontal cortex reduces cigarette craving and consumption. Addiction 104 (4), 653–660.

Andersen, S.L., Thompson, A.T., Rutstein, M., Hostetter, J.C., Teicher, M.H., 2000. Dopamine receptor pruning in prefrontal cortex during the periadolescent period in rats. Synapse 37 (2), 167–169.

Atance, C.M., O'Neill, D.K., 2001. Episodic future thinking. Trends Cogn. Sci. 5 (12), 533.

Baams, L., Dubas, J.S., Overbeek, G., van Aken, M.A.G., 2015. Transitions in body and behavior: a meta-analytic study on the relationship between pubertal development and adolescent sexual behavior. J. Adolesc. Health 56 (6), 586–598.

Banks, J., Marmot, M., Oldfield, Z., Smith, J.P., 2006. Disease and disadvantage in the United States and in England. JAMA 295 (17), 2037–2045.

Barr, M.S., Farzan, F., Wing, V.C., George, T.P., Fitzgerald, P.B., Daskalakis, Z.J., 2011. Repetitive transcranial magnetic stimulation and drug addiction. Int. Rev. Psychiatry 23 (5), 454–466.

Baum II, C.L., Ruhm, C.J., 2009. Age, socioeconomic status and obesity growth. J. Health Econ. 28 (3), 635–648.

Bechara, A., 2005. Decision making, impulse control and loss of willpower to resist drugs: a neurocognitive perspective. Nat. Neurosci. 8 (11), 1458–1463.

Bellamoli, E., Manganotti, P., Schwartz, R.P., Rimondo, C., Gomma, M., Serpelloni, G., 2014. rTMS in the treatment of drug addiction: an update about human studies. Behav. Neurol. 2014, 815215.

Beveridge, T.J., Gill, K.E., Hanlon, C.A., Porrino, L.J., 2008. Parallel studies of cocaine-related neural and cognitive impairment in humans and monkeys. Philos. Trans. R. Soc. Lond. Ser. B Biol. Sci. 363 (1507), 3257–3266.

Bickel, W.K., Mueller, E.T., 2009. Toward the study of trans-disease processes: a novel approach with special reference to the study of co-morbidity. J. Dual Diagn. 5 (2), 131–138.

Bickel, W.K., Yi, R., 2008. Temporal discounting as a measure of executive function: Insights from the competing neuro-behavioral decision system hypothesis of addiction. In: Houser, D., McCabe, K. (Eds.), Neuroeconomics: Advances in Health Services Research. Vol. 20. Emerald Group Publishing, Bingley, UK, pp. 289–309.

Bickel, W.K., Odum, A.L., Madden, G.J., 1999. Impulsivity and cigarette smoking: delay discounting in current, never, and ex-smokers. Psychopharmacology (Berl) 146 (4), 447–454.

Bickel, W.K., Miller, M.L., Yi, R., Kowal, B.P., Lindquist, D.M., Pitcock, J.A., 2007. Behavioral and neuroeconomics of drug addiction: competing neural systems and temporal discounting processes. Drug Alcohol Depend. 90S, S85–S91.

Bickel, W.K., Jones, B.A., Landes, R.D., Christensen, D.R., Jackson, L., Mancino, M., 2010. Hypothetical intertemporal choice and real economic behavior: delay discounting predicts voucher redemptions during contingency-management procedures. Exp. Clin. Psychopharmacol. 18 (6), 546–552.

Bickel, W.K., Jarmolowicz, D.P., Mueller, E.T., Gatchalian, K.M., 2011a. The behavioral economics and neuroeconomics of reinforcer pathologies: implications for etiology and treatment of addiction. Curr. Psychiatry Rep. 13 (5), 406–415.

Bickel, W.K., Landes, R.D., Christensen, D.R., et al., 2011b. Single- and cross-commodity discounting among cocaine addicts: the commodity and its temporal location determine discounting rate. Psychopharmacology (Berl) 217 (2), 177–187.

Bickel, W.K., Yi, R., Landes, R.D., Hill, P.F., Baxter, C., 2011c. Remember the future: working memory training decreases delay discounting among stimulant addicts. Biol. Psychiatry 69 (3), 260–265.

Bickel, W.K., Jarmolowicz, D.P., Mueller, E.T., Gatchalian, K.M., McClure, S.M., 2012a. Are executive function and impulsivity antipodes? A conceptual reconstruction with special reference to addiction. Psychopharmacology (Berl) 221 (3), 361–387.

Bickel, W.K., Jarmolowicz, D.P., Mueller, E.T., Koffarnus, M.N., Gatchalian, K.M., 2012b. Excessive discounting of delayed reinforcers as a trans-disease process contributing to addiction and other disease-related vulnerabilities: emerging evidence. Pharmacol. Ther. 134 (3), 287–297.

Bickel, W.K., Mueller, E.T., Jarmolowicz, D.P., 2013. What is addiction? In: McCrady, B., Epstein, E. (Eds.), Addictions: A Comprehensive Guidebook, second ed. Oxford University Press, New York, pp. 3–16.

Bickel, W.K., Landes, R.D., Kurth-Nelson, Z., Redish, A.D., 2014a. A quantitative signature of self-control repair: rate-dependent effects of successful addiction treatment. Clin. Psychol. Sci. 2 (6), 685–695.

Bickel, W.K., Moody, L., Quisenberry, A.J., Ramey, C.T., Sheffer, C.E., 2014b. A competing neurobehavioral decision systems model of SES-related health and behavioral disparities. Prev. Med. 68, 37–43.

Bickel, W.K., Quisenberry, A.J., Moody, L., Wilson, A.G., 2015. Therapeutic opportunities for self-control repair in addiction and related disorders: change and the limits of change in trans-disease processes. Clin. Psychol. Sci. 3 (1), 140–153.

Bierut, L.J., Strickland, J.R., Thompson, J.R., Afful, S.E., Cottler, L.B., 2008. Drug use and dependence in cocaine dependent subjects, community-based individuals, and their siblings. Drug Alcohol Depend. 95, 14–22.

Bleiberg, J.L., Devlin, P., Croan, J., Briscoe, R., 1994. Relationship between treatment length and outcome in a therapeutic community. Int. J. Addict. 29 (6), 729–740.

Bolla, K.I., Funderburk, F.R., Cadet, J.L., 2000. Differential effects of cocaine and cocaine alcohol on neurocognitive performance. Neurology 54 (12), 2285–2292.

Brackins, T., Brahm, N.C., Kissack, J.C., 2011. Treatments for methamphetamine abuse: a literature review for the clinician. J. Pharm. Pract. 24 (6), 541–550.

Brewer, J.A., Worhunsky, P.D., Carroll, K.M., Rounsaville, B.J., Potenza, M.N., 2008. Pretreatment brain activation during stroop task is associated with outcomes in cocaine-dependent patients. Biol. Psychiatry 64 (11), 998–1004.

Budney, A.J., Higgins, S.T., Hughes, J.R., Bickel, W.K., 1993. Nicotine and caffeine use in cocaine-dependent individuals. J. Subst. Abus. Treat. 5 (2), 117–130.

Buka, S.L., 2002. Disparities in health status and substance use: ethnicity and socioeconomic factors. Public Health Rep. 117, S118–S125.

Burling, T.A., Salvio, M.A., Seidner, A.L., Ramsey, T.G., 1996. Cigarette smoking in alcohol and cocaine abusers. J. Subst. Abus. 8 (4), 445–452.

Camprodon, J.A., Martinez-Raga, J., Alonso-Alonso, M., Shih, M.C., Pascual-Leone, A., 2007. One session of high frequency repetitive transcranial magnetic stimulation (rtms) to the right prefrontal cortex transiently reduces cocaine craving. Drug Alcohol Depend. 86 (1), 91–94.

Carroll, K.M., Onken, L.S., 2005. Behavioral therapies for drug abuse. Am. J. Psychiatry 162 (8), 1452–1460.

Carroll, K.M., Ball, S.A., Martino, S., et al., 2008. Computer-assisted delivery of cognitive-behavioral therapy for addiction: a randomized trial of CBT4CBT. Am. J. Psychiatry 165 (7), 881–888.

Chesson, H.W., Leichliter, J.S., Zimet, G.D., Rosenthal, S.L., Bernstein, D.I., Fife, K.H., 2006. Discount rates and risky sexual behavior among teenagers and young adults. J. Risk Uncertain. 32 (3), 217–230.

Childress, A.R., Mozley, P.D., McElgin, W., Fitzgerald, J., Reivich, M., O'Brien, C.P., 1999. Limbic activation during cue-induced cocaine craving. Am. J. Psychiatr. 156 (1), 11–18.

Cho, S.S., Koshimori, Y., Aminian, K., et al., 2015. Investing in the future: stimulation of the medial prefrontal cortex reduces discounting of delayed rewards. Neuropsychopharmacology 40 (3), 546–553.

Dackis, C.A., Kampman, K.M., Lynch, K.G., et al., 2012. A double-blind, placebo-controlled trial of modafinil for cocaine dependence. J. Subst. Abus. Treat. 43 (3), 303–312.

Dalley, J.W., Everitt, B.J., Robbins, T.W., 2011. Impulsivity, compulsivity, and top-down cognitive control. Neuron 69 (4), 680–694.

Daniel, T.O., Stanton, C.M., Epstein, L.H., 2013. The future is now: reducing impulsivity and energy intake using episodic future thinking. Psychol. Sci. 24 (11), 2339–2342.

de Vito, S., Gamboz, N., Brandimonte, M.A., Barone, P., Amboni, M., Della, Sala S., 2012. Future thinking in Parkinson's disease: an executive function? Neuropsychologia 50 (7), 1494–1501.

Eichhammer, P., Johann, M., Kharraz, A., et al., 2003. High-frequency repetitive transcranial magnetic stimulation decreases cigarette smoking. J. Clin. Psychiatry 64 (8), 951–953.

Epstein, D.H., Hawkins, W.E., Covi, L., Umbricht, A., Preston, K.L., 2003. Cognitive-behavioral therapy plus contingency management for cocaine use: findings during treatment and across 12-month follow-up. Psychol. Addict. Behav. 17 (1), 73–82.

Ernst, M., Nelson, E.E., Jazbec, S., et al., 2005. Amygdala and nucleus accumbens in response to receipt and omission of gains in adults and adolescents. Neuroimage 25 (4), 1279–1291.

Ersche, K.D., Turton, A.J., Chamberlain, S.R., Muller, U., Bullmore, E.T., Robbins, T.W., 2012. Cognitive dysfunction and anxious-impulsive personality traits are endophenotypes for drug dependence. Am. J. Psychiatry 169 (9), 926–936.

Evans, J.S., 2008. Dual-processing accounts of reasoning, judgment, and social cognition. Annu. Rev. Psychol. 59, 255–278.

Evans, J.S., Stanovich, K.E., 2013. Dual-process theories of higher cognition-advancing the debate. Perspect. Psychol. Sci. 8 (3), 223–241.

Everitt, B.J., Belin, D., Economidou, D., Pelloux, Y., Dalley, J.W., Robbins, T.W., 2008. Neural mechanisms underlying the vulnerability to develop compulsive drug-seeking habits and addiction. Philos. Trans. R. Soc. B 363 (1507), 3125–3135.

Fitzgerald, P.B., Fountain, S., Daskalakis, Z.J., 2006. A comprehensive review of the effects of rTMS on motor cortical excitability and inhibition. Clin. Neurophysiol. 117 (12), 2584–2596.

Franklin, T.R., Acton, P.D., Maldjian, J.A., et al., 2002. Decreased gray matter concentration in the insular, orbitofrontal, cingulate, and temporal cortices of cocaine patients. Biol. Psychiatry 51 (2), 134–142.

Galvan, A., Hare, T.A., Parra, C.E., et al., 2006. Earlier development of the accumbens relative to orbitofrontal cortex might underlie risk-taking behavior in adolescents. J. Neurosci. 26 (25), 6885–6892.

Garavan, H., Pankiewicz, J., Bloom, A., et al., 2000. Cue-induced cocaine craving: neuroanatomical specificity for drug users and drug stimuli. Am. J. Psychiatr. 157 (11), 1789–1798. November 1, 2000.

Gerlach, K.D., Spreng, R.N., Gilmore, A.W., Schacter, D.L., 2011. Solving future problems: default network and executive activity associated with goal-directed mental simulations. Neuroimage 55 (4), 1816–1824.

Goldstein, R.Z., Volkow, N.D., 2002. Drug addiction and its underlying neurobiological basis: neuroimaging evidence for the involvement of the frontal cortex. Am. J. Psychiatr. 159 (10), 1642–1652.

Goldstein, R.Z., Volkow, N.D., 2011. Dysfunction of the prefrontal cortex in addiction: neuroimaging findings and clinical implications. Nat. Rev. Neurosci. 12 (11), 652–669.

Goldstein, R.Z., Leskovjan, A.C., Hoff, A.L., et al., 2004. Severity of neuropsychological impairment in cocaine and alcohol addiction: association with metabolism in the prefrontal cortex. Neuropsychologia 42 (11), 1447–1458.

Goldstein, R.Z., Woicik, P.A., Maloney, T., et al., 2010. Oral methylphenidate normalizes cingulate activity in cocaine addiction during a salient cognitive task. Proc. Natl. Acad. Sci. U. S. A. 107 (38), 16667–16672.

Gorelick, D.A., Zangen, A., George, M.S., 2014. Transcranial magnetic stimulation in the treatment of substance addiction. Ann. N. Y. Acad. Sci. 1327, 79–93.

Goudriaan, A.E., Veltman, D.J., van den Brink, W., Dom, G., Schmaal, L., 2013. Neurophysiological effects of modafinil on cue-exposure in cocaine dependence: a randomized placebo-controlled cross-over study using pharmacological fMRI. Addict. Behav. 38 (2), 1509–1517.

Grabowski, J., Roache, J.D., Schmitz, J.M., Rhoades, H., Creson, D., Korszun, A., 1997. Replacement medication for cocaine dependence: methylphenidate. J. Clin. Psychopharmacol. 17 (6), 485–488.

Hanlon, C.A., Dowdle, L.T., Austelle, C.W., et al., 2015. What goes up, can come down: novel brain stimulation paradigms may attenuate craving and craving-related neural circuitry in substance dependent individuals. Brain Res. In press.

Hart, C.L., Haney, M., Vosburg, S.K., Rubin, E., Foltin, R.W., 2008. Smoked cocaine self-administration is decreased by modafinil. Neuropsychopharmacology 33 (4), 761–768.

Haushofer, J., Fehr, E., 2014. On the psychology of poverty. Science 344 (6186), 862–867.

Heil, S.H., Johnson, M.W., Higgins, S.T., Bickel, W.K., 2006. Delay discounting in currently using and currently abstinent cocaine-dependent outpatients and non-drug-using matched controls. Addict. Behav. 31 (7), 1290–1294.

Herremans, S.C., Baeken, C., Vanderbruggen, N., et al., 2012. No influence of one right-sided prefrontal hf-rtms session on alcohol craving in recently detoxified alcohol-dependent patients: results of a naturalistic study. Drug Alcohol Depend. 120 (1–3), 209–213.

Herremans, S.C., Vanderhasselt, M.A., De Raedt, R., Baeken, C., 2013. Reduced intra-individual reaction time variability during a Go-NoGo task in detoxified alcohol-dependent patients after one right-sided dorsolateral prefrontal HF-rTMS session. Alcohol Alcohol. 48 (5), 552–557.

Higgins, S.T., Delaney, D.D., Budney, A.J., et al., 1991. A behavioral approach to achieving initial cocaine abstinence. Am. J. Psychiatr. 148 (9), 1218–1224.

Higgins, S.T., Budney, A.J., Bickel, W.K., Foerg, F.E., Donham, R., Badger, G.J., 1994. Incentives improve outcome in outpatient behavioral treatment of cocaine dependence. Arch. Gen. Psychiatry 51 (7), 568–576.

Higgins, S.T., Budney, A.J., Bickel, W.K., Badger, G., Foerg, F., Ogden, D., 1995. Outpatient behavioral treatment for cocaine dependence: one-year outcome. Exp. Clin. Psychopharmacol. 3 (2), 205–212.

Hiscock, R., Bauld, L., Amos, A., Fidler, J.A., Munafo, M., 2012. Socioeconomic status and smoking: a review. Ann. N. Y. Acad. Sci. 1248, 107–123.

Hoogendam, J.M., Ramakers, G.M., Di Lazzaro, V., 2010. Physiology of repetitive transcranial magnetic stimulation of the human brain. Brain Stimul. 3 (2), 95–118.

Hoppner, J., Broese, T., Wendler, L., Berger, C., Thome, J., 2011. Repetitive transcranial magnetic stimulation (rTMS) for treatment of alcohol dependence. World J. Biol. Psychiatry 12 (Suppl. 1), 57–62.

Jentsch, J.D., Taylor, J.R., 1999. Impulsivity resulting from frontostriatal dysfunction in drug abuse: implications for the control of behavior by reward-related stimuli. Psychopharmacology (Berl) 146 (4), 373–390.

Johnson, M.W., Bruner, N.R., 2012. The sexual discounting task: HIV risk behavior and the discounting of delayed sexual rewards in cocaine dependence. Drug Alcohol Depend. 123 (1–3), 15–21.

Jovanovski, D., Erb, S., Zakzanis, K.K., 2005. Neurocognitive deficits in cocaine users: a quantitative review of the evidence. J. Clin. Exp. Neuropsychol. 27 (2), 189–204.

Kalechstein, A.D., Mahoney 3rd, J.J., Yoon, J.H., Bennett, R., De la Garza 2nd, R., 2013. Modafinil, but not escitalopram, improves working memory and sustained attention in long-term, high-dose cocaine users. Neuropharmacology 64, 472–478.

Kalivas, P.W., Volkow, N.D., 2005. The neural basis of addiction: a pathology of motivation and choice. Am. J. Psychiatry 162 (8), 1403–1413.

Khurana, A., Romer, D., Betancourt, L.M., Brodsky, N.L., Giannetta, J.M., Hurt, H., 2015. Experimentation versus progression in adolescent drug use: a test of an emerging neurobehavioral imbalance model. Dev. Psychopathol. 27, 901–913.

Koffarnus, M.N., Jarmolowicz, D.P., Mueller, E.T., Bickel, W.K., 2013. Changing delay discounting in the light of the competing neurobehavioral decision systems theory: a review. J. Exp. Anal. Behav. 99 (1), 32–57.

Konova, A.B., Moeller, S.J., Tomasi, D., Volkow, N.D., Goldstein, R.Z., 2013. Effects of methylphenidate on resting-state functional connectivity of the mesocorticolimbic dopamine pathways in cocaine addiction. JAMA Psychiatry 70 (8), 857–868.

Kubler, A., Murphy, K., Garavan, H., 2005. Cocaine dependence and attention switching within and between verbal and visuospatial working memory. Eur. J. Neurosci. 21 (7), 1984–1992.

Li, X., Hartwell, K.J., Owens, M., et al., 2013. Repetitive transcranial magnetic stimulation of the dorsolateral prefrontal cortex reduces nicotine cue craving. Biol. Psychiatry 73 (8), 714–720.

Lopez-Quintero, C., Perez de los Cobos, J., Hasin, D.S., et al., 2011. Probability and predictors of transition from first use to dependence on nicotine, alcohol, cannabis, and cocaine: results of the national epidemiologic survey on alcohol and related conditions (NESARC). Drug Alcohol Depend. 115 (1–2), 120–130.

Lucantonio, F., Stalnaker, T.A., Shaham, Y., Niv, Y., Schoenbaum, G., 2012. The impact of orbitofrontal dysfunction on cocaine addiction. Nat. Neurosci. 15 (3), 358–366.

Lussier, J.P., Heil, S.H., Mongeon, J.A., Badger, G.J., Higgins, S.T., 2006. A meta-analysis of voucher-based reinforcement therapy for substance use disorders. Addiction 101 (2), 192–203.

Madden, G.J., Petry, N.M., Badger, G.J., Bickel, W.K., 1997. Impulsive and self-control choices in opioid-dependent patients and non-drug-using control participants: drug and monetary rewards. Exp. Clin. Psychopharmacol. 5 (3), 256–262.

Makris, N., Gasic, G.P., Seidman, L.J., et al., 2004. Decreased absolute amygdala volume in cocaine addicts. Neuron 44 (4), 729–740.

Mani, A., Mullainathan, S., Shafir, E., Zhao, J., 2013. Poverty impedes cognitive function. Science 341 (6149), 976–980.

Mariani, J.J., Levin, F.R., 2012. Psychostimulant treatment of cocaine dependence. Psychiatr. Clin. North Am. 35 (2), 425–439.

Maude-Griffin, P.M., Hohenstein, J.M., Humfleet, G.L., Reilly, P.M., Tusel, D.J., Hall, S.M., 1998. Superior efficacy of cognitive-behavioral therapy for urban crack cocaine abusers: main and matching effects. J. Consult. Clin. Psychol. 66 (5), 832–837.

Mazur, J.E., 1987. An adjusting procedure for studying delayed reinforcement. In: Commons, M.L., Mazur, J.E., Nevin, J.A., Rachlin, H. (Eds.), Quantitative Analyses of Behavior, vol. 5. Erlbaum, Hillsdale, NJ, pp. 55–73.

McClure, S.M., Bickel, W.K., 2014. A dual-systems perspective on addiction: contributions from neuroimaging and cognitive training. Ann. N. Y. Acad. Sci. 1327, 62–78.

McClure, S.M., Laibson, D.I., Loewenstein, G., Cohen, J.D., 2004. Separate neural systems value immediate and delayed monetary rewards. Science 306 (5695), 503–507. October 15, 2004.

McClure, S.M., Ericson, K.M., Laibson, D.I., Loewenstein, G., Cohen, J.D., 2007. Time discounting for primary rewards. J. Neurosci. 27 (21), 5796–5804.

Metcalfe, J., Mischel, W., 1999. A hot/cold-system analysis of delay of gratification: dynamics of willpower. Psychol. Rev. 106 (1), 3–19.

Miech, R., 2008. The formation of a socioeconomic health disparity: the case of cocaine use during the 1980s and 1990s. J. Health Soc. Behav. 49 (3), 352–366.

Miller, E.K., Cohen, J.D., 2001. An integrative theory of prefrontal cortex function. Annu. Rev. Neurosci. 24 (1), 167–202.

Mishra, B.R., Nizamie, S.H., Das, B., Praharaj, S.K., 2010. Efficacy of repetitive transcranial magnetic stimulation in alcohol dependence: a sham-controlled study. Addiction 105 (1), 49–55.

Moeller, F.G., Dougherty, D.M., Barratt, E.S., Schmitz, J.M., Swann, A.C., Grabowski, J., 2001. The impact of impulsivity on cocaine use and retention in treatment. J. Subst. Abus. Treat. 21 (4), 193–198.

Moeller, F.G., Hasan, K.M., Steinberg, J.L., et al., 2005. Reduced anterior corpus callosum white matter integrity is related to increased impulsivity and reduced discriminability in cocaine-dependent subjects: diffusion tensor imaging. Neuropsychopharmacology 30 (3), 610–617.

Moeller, F.G., Schmitz, J.M., Herin, D., Kjome, K.L., 2008. Use of stimulants to treat cocaine and methamphetamine abuse. Curr. Psychiatry Rep. 10 (5), 385–391.

Moeller, F.G., Steinberg, J.L., Schmitz, J.M., et al., 2010. Working memory fMRI activation in cocaine-dependent subjects: association with treatment response. Psychiatry Res. 181 (3), 174–182.

Moeller, S.J., Tomasi, D., Honorio, J., Volkow, N.D., Goldstein, R.Z., 2012. Dopaminergic involvement during mental fatigue in health and cocaine addiction. Transl. Psychiatry 2, e176.

Moreno-Lopez, L., Catena, A., Fernandez-Serrano, M.J., et al., 2012. Trait impulsivity and prefrontal gray matter reductions in cocaine dependent individuals. Drug Alcohol Depend. 125 (3), 208–214.

Narvaez, J.C., Jansen, K., Pinheiro, R.T., et al., 2014. Psychiatric and substance-use comorbidities associated with lifetime crack cocaine use in young adults in the general population. Compr. Psychiatry 55 (6), 1369–1376.

Noble, K.G., Houston, S.M., Kan, E., Sowell, E.R., 2012. Neural correlates of socioeconomic status in the developing human brain. Dev. Sci. 15 (4), 516–527.

Noel, X., Brevers, D., Bechara, A., 2013. A neurocognitive approach to understanding the neurobiology of addiction. Curr. Opin. Neurobiol. 23 (4), 632–638.

Okuda, J., Fujii, T., Ohtake, H., et al., 2003. Thinking of the future and the past: the roles of the frontal pole and the medial temporal lobes. Neuroimage 19 (4), 1369–1380.

Olesen, P.J., Westerberg, H., Klingberg, T., 2004. Increased prefrontal and parietal activity after training of working memory. Nat. Neurosci. 7 (1), 75–79.

Palamar, J.J., Davies, S., Ompad, D.C., Cleland, C.M., Weitzman, M., 2015. Powder cocaine and crack use in the United States: an examination of risk for arrest and socioeconomic disparities in use. Drug Alcohol Depend. 149, 108–116.

Pampel, F.C., Krueger, P.M., Denney, J.T., 2010. Socioeconomic disparities in health behaviors. Annu. Rev. Sociol. 36, 349–370.

Peters, J., Buchel, C., 2010. Episodic future thinking reduces reward delay discounting through an enhancement of prefrontal-mediotemporal interactions. Neuron 66 (1), 138–148.

Petry, N.M., 2001. Delay discounting of money and alcohol in actively using alcoholics, currently abstinent alcoholics, and controls. Psychopharmacology (Berl) 154 (3), 243–250.

Petry, N.M., Martin, B., 2002. Low-cost contingency management for treating cocaine- and opioid-abusing methadone patients. J. Consult. Clin. Psychol. 70 (2), 398–405.

Petry, N.M., Tedford, J., Austin, M., Nich, C., Carroll, K.M., Rounsaville, B.J., 2004. Prize reinforcement contingency management for treating cocaine users: how low can we go, and with whom? Addiction 99 (3), 349–360.

Poincaré, H., 1905. Science and Hypothesis. The Walter Scott Publishing Company, Ltd., New York.

Poling, J., Oliveto, A., Petry, N., et al., 2006. Six-month trial of bupropion with contingency management for cocaine dependence in a methadone-maintained population. Arch. Gen. Psychiatry 63 (2), 219–228.

Politi, E., Fauci, E., Santoro, A., Smeraldi, E., 2008. Daily sessions of transcranial magnetic stimulation to the left prefrontal cortex gradually reduce cocaine craving. Am. J. Addict. 17 (4), 345–346.

Prendergast, M., Podus, D., Finney, J., Greenwell, L., Roll, J., 2006. Contingency management for treatment of substance use disorders: a meta-analysis. Addiction 101 (11), 1546–1560.

Pripfl, J., Tomova, L., Riecansky, I., Lamm, C., 2014. Transcranial magnetic stimulation of the left dorsolateral prefrontal cortex decreases cue-induced nicotine craving and EEG delta power. Brain Stimul. 7 (2), 226–233.

Rawson, R.A., Huber, A., McCann, M., et al., 2002. A comparison of contingency management and cognitive-behavioral approaches during methadone maintenance treatment for cocaine dependence. Arch. Gen. Psychiatry 59 (9), 817–824.

Robinson, T.E., Berridge, K.C., 1993. The neural basis of drug craving: an incentive-sensitization theory of addiction. Brain Res. Rev. 18 (3), 247–291.
Robles, E., Silverman, K., Preston, K.L., et al., 2000. The brief abstinence test: voucher-based reinforcement of cocaine abstinence. Drug Alcohol Depend. 58 (1–2), 205–212.
Rush, C.R., Stoops, W.W., Hays, L.R., 2009. Cocaine effects during D-amphetamine maintenance: a human laboratory analysis of safety, tolerability and efficacy. Drug Alcohol Depend. 99 (1–3), 261–271.
Sanfey, A.G., Chang, L.J., 2008. Multiple systems in decision making. Ann. N. Y. Acad. Sci. 1128, 53–62.
Schmaal, L., Goudriaan, A.E., Joos, L., et al., 2014. Neural substrates of impulsive decision making modulated by modafinil in alcohol-dependent patients. Psychol. Med. 44 (13), 2787–2798.
Shearer, J., Wodak, A., van Beek, I., Mattick, R.P., Lewis, J., 2003. Pilot randomized double blind placebo-controlled study of dexamphetamine for cocaine dependence. Addiction 98 (8), 1137–1141.
Simpson, D.D., Joe, G.W., Fletcher, B.W., Hubbard, R.L., Anglin, M.D., 1999. A national evaluation of treatment outcomes for cocaine dependence. Arch. Gen. Psychiatry 56 (6), 507–514.
Sisk, C.L., Zehr, J.L., 2005. Pubertal hormones organize the adolescent brain and behavior. Front. Neuroendocrinol. 26 (3–4), 163–174.
Sofis, M.J., Jarmolowicz, D.P., Martin, L.E., 2014. Competing neurobehavioral decision systems and the neuroeconomics of craving in opioid addiction. Neurosci. Neuroecon. 3, 87–89.
Sofuoglu, M., 2010. Cognitive enhancement as a pharmacotherapy target for stimulant addiction. Addiction 105 (1), 38–48.
Sowell, E.R., Peterson, B.S., Thompson, P.M., Welcome, S.E., Henkenius, A.L., Toga, A.W., 2003. Mapping cortical change across the human life span. Nat. Neurosci. 6 (3), 309–315.
Steinberg, L., 2007. Risk taking in adolescence: new perspectives from brain and behavioral science. Curr. Dir. Psychol. Sci. 16 (2), 55–59.
Steinberg, L., Albert, D., Cauffman, E., Banich, M., Graham, S., 2008. Age differences in sensation seeking and impulsivity as indexed by behavior and self-report: evidence for a dual systems model. Dev. Psychol. 44 (6), 1764–1778.
Stoops, W.W., Blackburn, J.W., Hudson, D.A., Hays, L.R., Rush, C.R., 2008. Safety, tolerability and subject-rated effects of acute intranasal cocaine administration during atomoxetine maintenance. Drug Alcohol Depend. 92 (1–3), 282–285.
Streeter, C.C., Terhune, D.B., Whitfield, T.H., et al., 2008. Performance on the stroop predicts treatment compliance in cocaine-dependent individuals. Neuropsychopharmacology 33 (4), 827–836.
Substance Abuse and Mental Health Services Administration, 2004. Results from the 2003 National Survey on Drug Use and Health: National Findings. Office of Applied Studies, Rockville, MD. SMA 04 3964.
Substance Abuse and Mental Health Services Administration, 2013. Results from the 2013 National Survey on Drug Use and Health: Summary of National Findings. Office of Applied Studies, Rockville, MD. SMA 14-4863.
Teicher, M.H., Andersen, S.L., Hostetter Jr., J.C., 1995. Evidence for dopamine receptor pruning between adolescence and adulthood in striatum but not nucleus accumbens. Brain Res. Dev. Brain Res. 89 (2), 167–172.

Thickbroom, G.W., 2007. Transcranial magnetic stimulation and synaptic plasticity: experimental framework and human models. Exp. Brain Res. 180 (4), 583–593.

Turner, D.C., Clark, L., Dowson, J., Robbins, T.W., Sahakian, B.J., 2004. Modafinil improves cognition and response inhibition in adult attention-deficit/hyperactivity disorder. Biol. Psychiatry 55 (10), 1031–1040.

Vansickel, A.R., Fillmorex, M.T., Hays, L.R., Rush, C.R., 2008. Effects of potential agonist-replacement therapies for stimulant dependence on inhibitory control in cocaine abusers. Am. J. Drug Alcohol Abuse 34 (3), 293–305.

Volkow, N.D., Fowler, J.S., Wang, G.J., et al., 1993. Decreased dopamine D2 receptor availability is associated with reduced frontal metabolism in cocaine abusers. Synapse 14 (2), 169–177.

Volkow, N.D., Wang, G.J., Fowler, J.S., et al., 1999. Prediction of reinforcing responses to psychostimulants in humans by brain dopamine D2 receptor levels. Am. J. Psychiatry 156 (9), 1440–1443.

Volkow, N.D., Wang, G.J., Ma, Y., et al., 2005. Activation of orbital and medial prefrontal cortex by methylphenidate in cocaine-addicted subjects but not in controls: relevance to addiction. J. Neurosci. 25 (15), 3932–3939.

Wagner, F.A., Anthony, J.C., 2002. From first drug use to drug dependence; developmental periods of risk for dependence upon marijuana, cocaine, and alcohol. Neuropsychopharmacology 26, 479–488.

Wesley, M.J., Bickel, W.K., 2014. Remember the future II: meta-analyses and functional overlap of working memory and delay discounting. Biol. Psychiatry 75 (6), 435–448.

Wilcox, C.E., Teshiba, T.M., Merideth, F., Ling, J.Y., Mayer, A.R., 2011. Enhanced cue reactivity and fronto-striatal functional connectivity in cocaine use disorders. Drug Alcohol Depend. 1115 (1), 137–144.

Wilkinson, R., Pickett, K., 2011. Greater equality: the hidden key to better health and higher scores. Am. Educ. 35 (1), 5–9.

Wing, V.C., Barr, M.S., Wass, C.E., et al., 2013. Brain stimulation methods to treat tobacco addiction. Brain Stimul. 6 (3), 221–230.

Worhunsky, P.D., Stevens, M.C., Carroll, K.M., et al., 2013. Functional brain networks associated with cognitive control, cocaine dependence, and treatment outcome. Psychol. Addict. Behav. 27 (2), 477–488.

Xu, J., DeVito, E.E., Worhunsky, P.D., Carroll, K.M., Rounsaville, B.J., Potenza, M.N., 2010. White matter integrity is associated with treatment outcome measures in cocaine dependence. Neuropsychopharmacology 35 (7), 1541–1549.

Yoon, J.H., Newton, T.F., Haile, C.N., et al., 2013. Effects of d-cycloserine on cue-induced craving and cigarette smoking among concurrent cocaine- and nicotine-dependent volunteers. Addict. Behav. 38, 1518–1526.

Ziemann, U., Paulus, W., Nitsche, M.A., et al., 2008. Consensus: motor cortex plasticity protocols. Brain Stimul. 1 (3), 164–182.

CHAPTER

Clinical neuroscience of amphetamine-type stimulants: From basic science to treatment development

15

Kelly E. Courtney, Lara A. Ray[1]

Department of Psychology, University of California, Los Angeles, CA, USA

[1]Corresponding author: Tel.: +1-310-794-5383; Fax: +1-310-206-5895, e-mail address: lararay@psych.ucla.edu

Abstract

Abuse of amphetamine-type stimulants (ATS) poses a significant public health concern with known neurotoxic and neurocognitive effects to the user. In this chapter, we seek to integrate the latest research on ATS, particularly methamphetamine, by covering areas of pharmacology, neurocognitive effects, and the treatment of ATS use disorders with the goal of advancing the clinical neuroscience of ATS and highlighting avenues for future research.

Keywords

Amphetamine, Stimulants, Methamphetamine, Addiction, Clinical neuroscience, Treatment, ATS use disorders

1 INTRODUCTION

Amphetamine-type stimulants (ATS), including amphetamine, dextroamphetamine (D-amphetamine), methamphetamine, and amphetamine-like drugs such as methylphenidate, have a long history of use in the United States (U.S.) and continue to pose a significant public health concern in the U.S. and worldwide. Synthetic amphetamine was first popularized in the U.S. in the 1930s as an over-the-counter nasal decongestant and was used to reduce fatigue and suppress appetite during World War II. In the 1950s and 1960s, amphetamine was commonly prescribed as a medication for depression and obesity, with approximately 31 million prescriptions filled in the U.S. in 1967 (Anglin et al., 2000). Shortly thereafter, legislation was passed in

Progress in Brain Research, Volume 223, ISSN 0079-6123, http://dx.doi.org/10.1016/bs.pbr.2015.07.010

attempt to restrict the availability of amphetamine, and medicinal use began to decline (Gonzales et al., 2010); however, this reclassification of amphetamine to a more restrictive schedule led to a surge in illicit manufacturing and use of methamphetamine. Furthermore, the relatively recent increase in attention deficit hyperactivity disorder diagnoses has been accompanied by a resurgence of prescriptions for stimulant medications with diversion of these medications a growing concern for the nation (Rabiner, 2013). Despite multiple legislative attempts to limit public access, illicit ATS use remains highly prevalent.

Currently, ATS are the second most commonly used class of illicit drugs worldwide (UNODC, World Drug Report 2012); approximately 0.7% of the global population (33.8 million people) aged 15–64 years old reported using an ATS in 2010 (UNODC, World Drug Report 2013). In the U.S., estimates from 2013 suggest over 21.7 million people ages 12 years and older (8.3% of total responders) have used ATS for nonmedical purposes in their lifetimes, 3.5 million people (1.3%) reported past year use, and approximately 1.4 million (0.5%) of those identified as past month users. Further, 12 million (4.7%) of the individuals surveyed reported lifetime use of methamphetamine specifically, with approximately 440,000 (0.2%) of those identified as past month users (Substance Abuse and Mental Health Services Administration (SAMHSA), 2013a). Importantly, these estimates appear to be growing both in terms of supply and demand (UNODC, 2013).

Subsequently, the prevalence of ATS use disorders is also on the rise. In 2012, 535,000 (0.2%) individuals were estimated to meet the *Diagnostic and Statistical Manual of Mental Disorders* (4th ed., DSM–IV; American Psychiatric Association, 1994) criteria of ATS abuse or dependence, a significant increase from the 329,000 (0.1%) in 2011 (SAMHSA, 2013a). This increase was especially pronounced among individuals aged 18–25 years, with 0.5% meeting criteria in 2012, up from 0.3% in 2011. Furthermore, primary methamphetamine/amphetamine treatment admissions were more likely than all drug treatment admissions combined to receive long-term rehabilitation/residential treatment (16% vs. 7%) (SAMHSA, 2013b), suggestive of the exceedingly high costs associated with the treatment of ATS use disorders and underscoring the need for more efficacious, cost effective, and easily deliverable treatments.

Developing a greater understanding of the clinical neuroscience underlying the consequences of ATS use is an important step toward the development of more efficacious treatments for ATS use disorders. This is especially important with respect to medications development given the lack of any current FDA-approved medications for ATS dependence. Significant advances in preclinical and clinical research have begun to identify the neurochemical pathways affected by ATS use and highlight potential targets for intervention. Thus, knowledge of the pharmacological and neurological adaptations associated with ATS use could lead to the development of more efficacious medications and further inform psychosocial interventions for ATS use disorders (Table 1).

Given that methamphetamine is the most frequently used ATS worldwide (UNODC, 2013), and that studies of neurodegeneration, neurocognitive functioning,

Table 1 Chapter Highlights

- Amphetamine-type stimulants (ATS) are the second most commonly used class of illicit drugs worldwide and the prevalence of ATS use disorders are on the rise
- ATS have pervasive, and potentially long-lasting effects on the dopaminergic, noradrenergic, serotonergic, and opioidergic neurotransmitter systems throughout the brain and can result in detrimental effects to cognitive processes in heavy users
- More efficacious treatment options, such as FDA-approved pharmacotherapies, are greatly needed for ATS use disorders
- Promising medications currently under study for the treatment of ATS use disorders include oxytocin, bupropion, mirtazapine, topiramate, modafinil, and naltrexone
- The integration of basic neuroscience and treatment development research could improve clinical outcomes in ATS use disorders by facilitating targeted treatment approaches

and treatment most commonly target methamphetamine using populations, the majority of this chapter presents the current understanding of the clinical neuroscience behind methamphetamine use and associated disorders, expanded to ATS more broadly where applicable.

2 PHARMACOLOGY AND NEUROTOXICITY

As with most ATS, methamphetamine stimulates the release, and partially blocks the reuptake, of newly synthesized catecholamines in the CNS (Cho and Melega, 2002). Due to its structural similarity, methamphetamine interacts with the dopamine transporter (DAT), noradrenaline transporter (NET), serotonin transporter (SERT), and vesicular monoamine transporter-2 (VMAT-2) and reverses their endogenous function, thereby redistributing monoamines from storage vesicles into the cytosol. This process results in the release of dopamine, noradrenaline, and serotonin into the synapse, which then stimulate postsynaptic monoamine receptors (Cruickshank and Dyer, 2009). Methamphetamine also attenuates the metabolism of monoamines by inhibiting monoamine oxidase (Sulzer et al., 2005), further enabling the buildup of excess monoamines in the synapse.

The monoamines released due to the presence of ATS act on the major noradrenergic, serotonergic, and dopaminergic pathways of the brain. The medial basal forebrain, the hippocampus, as well as the prefrontal cortex (PFC) represent noradrenergic regions of interest for ATS effects, with various affected functions related to arousal, memory consolidation, and cognitive processing, respectively (Berridge and Waterhouse, 2003). Affected serotonergic neurons are dispersed throughout the brain, regulating diverse functions such as respiration, pain perception, sexual drive, reward, and higher-order cognitive processing (Hornung, 2003). In the case of dopamine, methamphetamine activates the mesolimbic, mesocortical circuit, and the nigrostriatal pathways, which have been related to the euphoric effects observed immediately after the ingestion of the drug (Homer et al., 2008).

Although no differences in striatal dopamine release between amphetamine and methamphetamine are observed (Melega et al., 1995), amphetamine is thought to result in a slightly greater dopamine release in the PFC, which may be responsible for the subtle differences between these drugs on behavioral tolerance and working memory measures (Shoblock et al., 2003a,b).

Repeated exposure to moderate to high levels of methamphetamine has been related to neurotoxic effects on the dopaminergic and serotonergic systems, leading to potentially irreversible loss of nerve terminals and/or neuron cell bodies (Cho and Melega, 2002). Preclinical evidence suggests that D-amphetamine, even when administered at commonly prescribed therapeutic doses, also results in toxicity to brain dopaminergic axon terminals (Ricaurte et al., 2005). Although the precise mechanisms remain unclear, the culmination of evidence suggests that the high level of cytoplasmic dopamine released as a result of ATS use leads to the accumulation of reactive oxygen species and severe oxidative stress on the neuron (Berman et al., 2008). Furthermore, frequent use of methamphetamine has been associated with reductions in striatal D_2-receptor availability (Groman et al., 2012; Volkow et al., 2001a), VMAT-2 density (Johanson et al., 2006), SERT density (Sekine et al., 2006), and DAT site density (McCann et al., 1998; Villemagne et al., 1998; Volkow et al., 2001b,c), with some markers (i.e., DAT density) showing improvement following prolonged (greater than 12 months) abstinence (Volkow et al., 2001b). Reduced markers of neuronal integrity and increased markers of glial content are also observed in chronic methamphetamine abusers, possibly indicating the proliferation of glial cells following neural damage (Chang et al., 2007; Ernst et al., 2000).

The potentiation of dopaminergic neurotransmission within the mesocorticolimbic circuit is thought to underlie the reinforcing properties of drugs of abuse, although evidence is accumulating on a converging role of the endogenous opioid systems in the establishment of reinforcement (Boutrel, 2008). In terms of neuroanatomy, endogenous opioid receptors are widely distributed throughout the CNS, with differential distributions per opioid receptor type. Importantly, opioid receptors and peptides are highly expressed in brain areas involved in reward and motivation, such as the ventral tegmental area (VTA) and nucleus accumbens (NAcc) (Mansour et al., 1995). Administration of classical exogenous opioids facilitates dopamine release in the mesolimbic reward system by activating μ- and δ-opioid receptors in the NAcc (Hirose et al., 2005; Murakawa et al., 2004) and by decreasing GABA-inhibition via μ- and κ-opioid receptors, which are mainly located on GABA interneurons in the VTA (Bonci and Williams, 1997; Shoji et al., 1999). Many nonopioid drugs of abuse, including ATS, are also known to interact with the endogenous opioid system (for a review, see Trigo et al., 2010), and this interaction may mediate some of the rewarding properties associated with acute ATS use (Boutrel, 2008). For example, acute amphetamine administration has been linked with increased β-endorphin levels in the NAcc (Olive et al., 2001), increased striatonigral dynorphin-like immunoreactivity (Bustamante et al., 2002; Hanson et al., 1988), and changes in the endogenous opioid mRNA expression in the striatum (Hurd and Herkenham, 1992; Smith and

McGinty, 1994; Wang and McGinty, 1995). Further, preclinical data suggest that the endogenous opioid system is involved in the induction and expression of methamphetamine-induced behavioral (locomotor) sensitization (Chiu et al., 2006), analogous to compulsive drug-seeking behavior in humans (i.e., drug craving; Itzhak and Ali, 2002), through its modulatory actions of the mesolimbic dopamine system (Ford et al., 2006).

In summary, methamphetamine and other ATS have pervasive and potentially long-lasting effects not only on the dopaminergic system but also on noradrenergic, serotonergic, and opioidergic neurotransmitter systems throughout the brain. It is through the culmination of these complex neurochemical modulations that significant behavioral and cognitive changes result.

3 NEUROCOGNITIVE EFFECTS

Many ATS are used therapeutically to improve attention and cognition; however, a review of the literature suggests dosage, and route of administration is a key determinant of the cognitive effects of these drugs (see Wood et al., 2014). Wood et al. (2014) argue that cognitive effects of ATS, including prescription medications such as D-amphetamine and methylphenidate, follow an inverted U dose–response curve, such that high doses result in detrimental effects on cognitive processing in domains such as learning and memory. In fact, a recent study of frequent recreational users of D-amphetamine observed impairments in performance on executive functioning and memory consolidation tasks, in addition to a trend toward reduced striatal DAT site binding and a blunted hemodynamic response to methylphenidate challenge, when compared to healthy controls (Schouw et al., 2013).

Chronic methamphetamine use, more specifically, has been associated with alterations across a broad spectrum of neurocognitive processes, although differentiating preexisting deficits from methamphetamine-induced cognitive deficits poses significant challenges (Dean et al., 2013), and concerns regarding the interpretation of these discrepancies and their clinical significance have been raised (Hart et al., 2012). The culmination of evidence acquired through various methodologies (e.g., preclinical, cross-sectional human, and brain imaging studies), however, supports the assertion that methamphetamine abuse does indeed cause cognitive decline in at least some individuals (i.e., individuals at the age of early-to-middle adulthood), and that individual difference factors such as education level and genotype further moderate this relationship (Dean et al., 2013). Cognitive domains including episodic memory, complex information processing speed, executive functions (e.g., response inhibition, novel problem solving), and psychomotor functions appear to be most affected in individuals with methamphetamine use disorders, with smaller, yet significant, effects also observed on measures of attention/working memory, language, and visuoconstruction (Scott et al., 2007).

A number of these cognitive discrepancies and other behavioral changes associated with methamphetamine abuse have been related to methamphetamine-induced alterations in neurotransmission, such as memory deficits and impaired psychomotor coordination associated with reduced DAT site density (Volkow et al., 2001c), and increased aggression associated with reduced SERT density (Sekine et al., 2006). Further, preclinical evidence suggests D_2-specific alterations of the dopaminergic system may subserve some of the disturbances in learning observed with repeated methamphetamine use. Specifically, using a reversal learning task and PET in a preclinical sample of vervet monkeys given a chronic, escalating-dose regimen of methamphetamine revealed associations between the change in response to positive feedback and individual differences in the change in dopamine D_2-like receptor availability in the striatum, assessed pre- and postmethamphetamine regimen (Groman et al., 2012).

Functional neuroimaging procedures have begun to identify region-specific alterations in glucose metabolism and blood-oxygen-level-dependent measures of brain activation associated with these potentially affected cognitive processes. For example, glucose metabolism in the anterior and middle cingulate gyrus and the insula was negatively correlated with error rates on an auditory vigilance task indexing attentional processing in recently abstinent (4–7 days) methamphetamine abusers (London et al., 2005). Evidence also suggests frontal and insular involvement in learning and cognitive control changes associated with methamphetamine abuse. On a color-word Stroop task administered during functional magnetic resonance imaging, methamphetamine abusers display reduced reaction time (RT) adjustments and reduced PFC activity following conflict (i.e., incongruent) trials (Salo et al., 2009, 2013), and reduced RT, increased error rate, and reduced activation of the right inferior frontal gyrus (IFG), supplementary motor cortex/anterior cingulate gyrus, and the anterior insular cortex during the incongruent condition (Nestor et al., 2011).

Region-specific alterations in brain activation have also been observed on decision-making tasks in methamphetamine abusers. Methamphetamine abusers displayed reduced activation in the right IFG and the left medial frontal gyrus during a two-choice prediction task (where only 50% of the responses are reinforced with a correct response outcome), and a decrease in dorsolateral PFC (dlPFC) and right orbitofrontal cortex (OFC) activity in the active compared to control conditions, as opposed to the increase of activation in these areas observed in the healthy controls (Paulus et al., 2002). In a follow-up study using the same task, recently abstinent (average 25 days) individuals with methamphetamine dependence displayed reduced activation of the OFC, dlPFC, anterior cingulate cortex (ACC), and parietal cortex irrespective of the outcome, and attenuation of specific "success-related" patterns of brain activation as compared to healthy controls (Paulus et al., 2003). Furthermore, the degree of activation in the right middle frontal gyrus, middle temporal gyrus, and posterior cingulate during the two-choice prediction task in early remission (3–4 weeks abstinent) was predictive of relapse during a 1-year follow-up (Paulus et al., 2005).

On a temporal discounting task indexing reward-related decision-making, contrasting "hard choices," where roughly equivalent preference is obtained for the immediate and delayed reward choices, and "easy choices," in which the choices differ dramatically in value and preference, revealed less activation in the precuneus, right caudate nucleus, ACC, and dlPFC in recently abstinent (2–8 weeks) individuals with methamphetamine dependence (Hoffman et al., 2008), and less activation of the left dlPFC and right intraparietal sulcus in active methamphetamine abusers (Monterosso et al., 2007), as compared to healthy controls. Furthermore, methamphetamine-dependent individuals undergoing treatment display disrupted risk-related processing, a component of decision-making, on the Risky Gains Task in both the ACC and insula (Gowin et al., 2013).

In summary, ATS abuse is associated with specific task-related behavioral and neural processing differences across a number of cognitive domains, which appear to be moderated by dose, route of administration, and other individual difference variables. Importantly, evidence is accumulating to suggest some of these differences are associated with altered dopaminergic processing (Groman et al., 2012) and clinically meaningful outcomes (Paulus et al., 2005), suggestive of a functional role for these cognitive differences in the development and maintenance of methamphetamine addiction.

4 TREATMENT

At present, few effective options exist for individuals seeking treatment for ATS use disorders, and to date, these options have been limited to psychosocial interventions. A systematic review of cognitive and behavioral treatments as applied specifically to methamphetamine use disorders concluded that good clinical outcomes are achieved with cognitive behavioral treatment (CBT; with and without motivational interviewing [MI]) and contingency management (CM) therapies involving the systematic use of reinforcement (Lee and Rawson, 2008). A number of caveats must be considered when interpreting these conclusions, however, such as the durability of treatment effects (especially with respect to CM programs). Furthermore, the effectiveness of psychosocial interventions is compromised by poor rates of treatment induction and retention (Shearer, 2007), and methamphetamine-related cognitive deficits in executive functioning, particularly those related to inhibitory control, have been hypothesized to potentially render heavily cognitive-based treatments ineffective (Baicy and London, 2007).

Given these important caveats of psychosocial interventions, and the heavy focus on the neurobiology of methamphetamine dependence, attention has shifted to the development of efficacious pharmacotherapies for methamphetamine addiction (NIDA, 2005). At present, no medication is approved by the U.S. Food and Drug Administration (FDA) for use in ATS use disorders. Numerous classes of medications are currently under study for methamphetamine use disorders, primarily in small clinical trials (for a recent focused review, see Brensilver et al., 2013). Some of the most promising medications include bupropion, mirtazapine, topiramate, modafinil, and naltrexone.

Bupropion, commonly prescribed as an antidepressant or smoking-cessation agent, is known to affect several biological targets. Widely described as a dopamine and norepinephrine reuptake inhibitor (Stahl et al., 2004), bupropion also acts as a noncompetitive antagonist of several neuronal nicotinic acetylcholine (nACh) receptors (Slemmer et al., 2000). Clinical use of bupropion has been associated with reduced use of methamphetamine among baseline light, but not heavy, methamphetamine users (identified in a *post hoc* analysis; Shoptaw et al., 2008); however, its precise mechanism of action remains unclear.

In a 12-week trial, mirtazapine, a noradrenergic and specific serotonergic antidepressant, combined with CBT/MI counseling has also been associated with significant reductions in methamphetamine use (percent positive urines at the week 12 visit) in a sample of methamphetamine-dependent men who have sex with men (Colfax et al., 2011). The clinical efficacy of this agent may be related to its ability to enhance of the release of norepinephrine and 5-HT1A-mediated serotonergic transmission (Anttila and Leinonen, 2001).

Topiramate, a sulfamate fructopyranose derivative and anticonvulsant, has been associated with reductions in methamphetamine use in large multisite clinical trial; however, no effects on total abstinence (negative urines during 6–12-week follow-up) were observed (Elkashef et al., 2012). Further analysis of this data identified a small subgroup of patients who exhibited consistent reductions of use or achieved abstinence during follow-up which were associated with topiramate treatment. This subgroup consisted of individuals who were more likely to have discontinued methamphetamine use (i.e., have a negative last urine) during the week prior to randomization (Ma et al., 2013) suggesting that topiramate may function best for relapse prevention. GABAergic modulation may be one possible mechanism underlying the potential efficacy of topiramate for methamphetamine treatment. Topiramate is known to facilitate GABAergic function via enhancement of inhibitory $GABA_A$-mediated currents at nonbenzodiazepine sites on the $GABA_A$ receptor (White et al., 2000). Topiramate also antagonizes glutaminergic activity through an effect at kainate/alpha-amino-3-hydroxy-5-methylisoxazole-4-propionic acid receptors (Gryder and Rogawski, 2003). Through these processes, topiramate is thought to modulate cortico-mesolimbic dopaminergic activity (Johnson, 2004), potentially stabilizing this activity and subsequently helping to prevent relapse or reduce methamphetamine use.

Cognitive-enhancing medications such as modafinil, an analeptic drug with known cognitive-enhancing properties, have garnered recent attention given the known cognitive deficits associated with chronic methamphetamine use (e.g., Ghahremani et al., 2011). Modafinil combined with CBT was associated with reduced methamphetamine use within a small sample of HIV+ gay men dependent on methamphetamine (McElhiney et al., 2009), although recent trials have not found strong support for a direct effect of modafinil on abstinence outcomes (e.g., Anderson et al., 2012; Heinzerling et al., 2010). The mechanism of action of modafinil is complex, involving multiple neurotransmitter systems, but its potential effects on ATS use may be related to inhibition of catecholamine transporters (Madras et al., 2006; Volkow et al., 2009), thereby increasing extracellular dopamine and norepinephrine levels.

Lastly, naltrexone, an opioid antagonist with greatest affinity for the μ- and κ-opioid receptors in humans (Emmerson et al., 1994; Toll et al., 1998), has been associated with reduced amphetamine use and greater abstinence rates in a sample of amphetamine-dependent individuals (Jayaram-Lindstrom et al., 2008). Further, amphetamine dependent patients with high levels of naltrexone (≥2 ng/ml) in their blood were 2.27 times more likely to be abstinent than patients with low naltrexone blood levels (<2 ng/ml; Grant et al., 2010). Naltrexone-related reductions of cue-induced craving and subjective responses to methamphetamine administration have recently been observed in nontreatment seeking individuals with methamphetamine use disorders (Ray et al., 2015), advancing naltrexone as a potential treatment for methamphetamine addiction as well. The blockage of ATS-induced dopamine release in the mesolimbic dopamine system has been proposed as the neural mechanism underlying naltrexone's effects on craving and subjective reward (Ashenhurst et al., 2012; Benjamin et al., 1993; Jayaram-Lindstrom et al., 2004; Lee et al., 2005; Naleid et al., 2005; Widdowson and Holman, 1992) which may underlie the observed attenuation of ATS use.

In summary, the clinically limiting caveats of psychosocial treatments have engendered a strong interest in medication development for the treatment of ATS use disorders. A number of medications are currently under study in clinical research for the treatment of ATS use disorders, many with promising preliminary results. Preclinical research is also continuously advancing novel pharmacological agents that may progress to human trials for ATS use disorders. For example, oxytocin, a mammalian neuropeptide, has shown promise in reducing responding for intravenous methamphetamine in rodent models (Carson et al., 2010; Cox et al., 2013), which may one day translate to improved clinical outcomes associated with oxytocin treatment in humans. Treatment development for ATS use disorders has been a challenging enterprise, yet consistent with the addiction field broadly (Litten et al., 2012), efforts to refocus the field toward medications with novel therapeutic targets (e.g., the opioidergic system, cognitive enhancement) hold considerable promise for these complex disorders.

5 CONCLUSION AND FUTURE DIRECTIONS

Illicit ATS use continues to be highly prevalent despite numerous attempts to limit public access to the drugs and their precursors. Methamphetamine in particular is the most frequently used ATS worldwide and has the highest abuse potential, yet the diversion of stimulant medications is also a growing concern. Through actions on the brain's major dopaminergic, noradrenergic, serotonergic, and opioidergic pathways, repeated use of ATS (especially methamphetamine) is associated with significant neurotoxic effects and neurocognitive deficits, with only a few of such effects known to remediate following sustained abstinence. Thus, early identification of problematic ATS use and effective treatment implementation is critical to successful outcomes.

Advances in the identification of the neural pathways affected by ATS use have begun to highlight potential targets for intervention. The development of efficacious pharmacologic interventions is most promising in this regard, particularly given the profound neurochemical alterations associated with ATS use. Medications that act on the dopaminergic, GABAergic, and serotonergic systems have shown promise in reducing ATS use in clinical samples, and increasing evidence for the opioidergic system's role in the development of ATS use disorders has advanced pharmacologic agents targeting this pathway as plausible treatments.

By integrating basic neuroscience into treatment development research, one may elucidate how psychosocial and pharmacological interventions function to reduce ATS use and for whom specific interventions may be most efficacious. For example, the most effective medications may function via novel mechanisms such as enhancing the effectiveness of existent psychosocial interventions (e.g., via decreasing cognitive impairment) and by targeting intermediate phenotypes of addiction (e.g., relapse prevention/craving) (NIDA, 2005). Further, current research suggests that clinical outcomes may be improved by tailoring interventions to differences in patient presentation (e.g., heaviness of use, age of user, cognitive capability), some of which effects may be driven by individual differences in dopaminergic processing. Clinical neuroscience research is well positioned to address these questions and ultimately provide relief to thousands of individuals currently struggling to overcome their addiction to these stimulating and reinforcing drugs.

REFERENCES

American Psychiatric Association, 1994. Diagnostic and Statistical Manual of Mental Disorders. American Psychiatric Association, Washington, DC.

Anderson, A.L., Li, S.H., Biswas, K., Mcsherry, F., Holmes, T., Iturriaga, E., Kahn, R., Chiang, N., Beresford, T., Campbell, J., Haning, W., Mawhinney, J., Mccann, M., Rawson, R., Stock, C., Weis, D., Yu, E., Elkashef, A.M., 2012. Modafinil for the treatment of methamphetamine dependence. Drug Alcohol Depend. 120, 135–141.

Anglin, M.D., Burke, C., Perrochet, B., Stamper, E., Dawud-Noursi, S., 2000. History of the methamphetamine problem. J. Psychoactive Drugs 32, 137–141.

Anttila, S.A., Leinonen, E.V., 2001. A review of the pharmacological and clinical profile of mirtazapine. CNS Drug Rev. 7, 249–264.

Ashenhurst, J.R., Bujarski, S., Ray, L.A., 2012. Delta and kappa opioid receptor polymorphisms influence the effects of naltrexone on subjective responses to alcohol. Pharmacol. Biochem. Behav. 103, 253–259.

Baicy, K., London, E.D., 2007. Corticolimbic dysregulation and chronic methamphetamine abuse. Addiction 102 (Suppl. 1), 5–15.

Benjamin, D., Grant, E.R., Pohorecky, L.A., 1993. Naltrexone reverses ethanol-induced dopamine release in the nucleus accumbens in awake, freely moving rats. Brain Res. 621, 137–140.

Berman, S., O'neill, J., Fears, S., Bartzokis, G., London, E.D., 2008. Abuse of amphetamines and structural abnormalities in the brain. Ann. N. Y. Acad. Sci. 1141, 195–220.

Berridge, C.W., Waterhouse, B.D., 2003. The locus coeruleus-noradrenergic system: modulation of behavioral state and state-dependent cognitive processes. Brain Res. Brain Res. Rev. 42, 33–84.

Bonci, A., Williams, J.T., 1997. Increased probability of GABA release during withdrawal from morphine. J. Neurosci. 17, 796–803.

Boutrel, B., 2008. A neuropeptide-centric view of psychostimulant addiction. Br. J. Pharmacol. 154, 343–357.

Brensilver, M., Heinzerling, K.G., Shoptaw, S., 2013. Pharmacotherapy of amphetamine-type stimulant dependence: an update. Drug Alcohol Rev. 32, 449–460.

Bustamante, D., You, Z.B., Castel, M.N., Johansson, S., Goiny, M., Terenius, L., Hokfelt, T., Herrera-Marschitz, M., 2002. Effect of single and repeated methamphetamine treatment on neurotransmitter release in substantia nigra and neostriatum of the rat. J. Neurochem. 83, 645–654.

Carson, D.S., Cornish, J.L., Guastella, A.J., Hunt, G.E., Mcgregor, I.S., 2010. Oxytocin decreases methamphetamine self-administration, methamphetamine hyperactivity, and relapse to methamphetamine-seeking behaviour in rats. Neuropharmacology 58, 38–43.

Chang, L., Alicata, D., Ernst, T., Volkow, N., 2007. Structural and metabolic brain changes in the striatum associated with methamphetamine abuse. Addiction 102 (Suppl. 1), 16–32.

Chiu, C.T., Ma, T., Ho, I.K., 2006. Methamphetamine-induced behavioral sensitization in mice: alterations in mu-opioid receptor. J. Biomed. Sci. 13, 797–811.

Cho, A.K., Melega, W.P., 2002. Patterns of methamphetamine abuse and their consequences. J. Addict. Dis. 21, 21–34.

Colfax, G.N., Santos, G.M., Das, M., Santos, D.M., Matheson, T., Gasper, J., Shoptaw, S., Vittinghoff, E., 2011. Mirtazapine to reduce methamphetamine use: a randomized controlled trial. Arch. Gen. Psychiatry 68, 1168–1175.

Cox, B.M., Young, A.B., See, R.E., Reichel, C.M., 2013. Sex differences in methamphetamine seeking in rats: impact of oxytocin. Psychoneuroendocrinology 38, 2343–2353.

Cruickshank, C.C., Dyer, K.R., 2009. A review of the clinical pharmacology of methamphetamine. Addiction 104, 1085–1099.

Dean, A.C., Groman, S.M., Morales, A.M., London, E.D., 2013. An evaluation of the evidence that methamphetamine abuse causes cognitive decline in humans. Neuropsychopharmacology 38, 259–274.

Elkashef, A., Kahn, R., Yu, E., Iturriaga, E., Li, S.H., Anderson, A., Chiang, N., Ait-Daoud, N., Weiss, D., Mcsherry, F., Serpi, T., Rawson, R., Hrymoc, M., Weis, D., Mccann, M., Pham, T., Stock, C., Dickinson, R., Campbell, J., Gorodetzky, C., Haning, W., Carlton, B., Mawhinney, J., Li, M.D., Johnson, B.A., 2012. Topiramate for the treatment of methamphetamine addiction: a multi-center placebo-controlled trial. Addiction 107, 1297–1306.

Emmerson, P.J., Liu, M.R., Woods, J.H., Medzihradsky, F., 1994. Binding affinity and selectivity of opioids at mu, delta and kappa receptors in monkey brain membranes. J. Pharmacol. Exp. Ther. 271, 1630–1637.

Ernst, T., Chang, L., Leonido-Yee, M., Speck, O., 2000. Evidence for long-term neurotoxicity associated with methamphetamine abuse: a 1H MRS study. Neurology 54, 1344–1349.

Ford, C.P., Mark, G.P., Williams, J.T., 2006. Properties and opioid inhibition of mesolimbic dopamine neurons vary according to target location. J. Neurosci. 26, 2788–2797.

Ghahremani, D.G., Tabibnia, G., Monterosso, J., Hellemann, G., Poldrack, R.A., London, E.D., 2011. Effect of modafinil on learning and task-related brain activity in methamphetamine-dependent and healthy individuals. Neuropsychopharmacology 36, 950–959.

Gonzales, R., Mooney, L., Rawson, R.A., 2010. The methamphetamine problem in the United States. Annu. Rev. Public Health 31, 385–398.

Gowin, J.L., Stewart, J.L., May, A.C., Ball, T.M., Wittmann, M., Tapert, S.F., Paulus, M.P., 2013. Altered cingulate and insular cortex activation during risk-taking in methamphetamine dependence: losses lose impact. Addiction 109, 237–247.

Grant, J.E., Odlaug, B.L., Kim, S.W., 2010. A double-blind, placebo-controlled study of N-acetyl cysteine plus naltrexone for methamphetamine dependence. Eur. Neuropsychopharmacol. 20, 823–828.

Groman, S.M., Lee, B., Seu, E., James, A.S., Feiler, K., Mandelkern, M.A., London, E.D., Jentsch, J.D., 2012. Dysregulation of D(2)-mediated dopamine transmission in monkeys after chronic escalating methamphetamine exposure. J. Neurosci. 32, 5843–5852.

Gryder, D.S., Rogawski, M.A., 2003. Selective antagonism of GluR5 kainate-receptor-mediated synaptic currents by topiramate in rat basolateral amygdala neurons. J. Neurosci. 23, 7069–7074.

Hanson, G.R., Merchant, K.M., Letter, A.A., Bush, L., Gibb, J.W., 1988. Characterization of methamphetamine effects on the striatal-nigral dynorphin system. Eur. J. Pharmacol. 155, 11–18.

Hart, C.L., Marvin, C.B., Silver, R., Smith, E.E., 2012. Is cognitive functioning impaired in methamphetamine users? A critical review. Neuropsychopharmacology 37, 586–608.

Heinzerling, K.G., Swanson, A.N., Kim, S., Cederblom, L., Moe, A., Ling, W., Shoptaw, S., 2010. Randomized, double-blind, placebo-controlled trial of modafinil for the treatment of methamphetamine dependence. Drug Alcohol Depend. 109, 20–29.

Hirose, N., Murakawa, K., Takada, K., Oi, Y., Suzuki, T., Nagase, H., Cools, A.R., Koshikawa, N., 2005. Interactions among mu- and delta-opioid receptors, especially putative delta1- and delta2-opioid receptors, promote dopamine release in the nucleus accumbens. Neuroscience 135, 213–225.

Hoffman, W.F., Schwartz, D.L., Huckans, M.S., Mcfarland, B.H., Meiri, G., Stevens, A.A., Mitchell, S.H., 2008. Cortical activation during delay discounting in abstinent methamphetamine dependent individuals. Psychopharmacology 201, 183–193.

Homer, B.D., Solomon, T.M., Moeller, R.W., Mascia, A., Deraleau, L., Halkitis, P.N., 2008. Methamphetamine abuse and impairment of social functioning: a review of the underlying neurophysiological causes and behavioral implications. Psychol. Bull. 134, 301–310.

Hornung, J.P., 2003. The human raphe nuclei and the serotonergic system. J. Chem. Neuroanat. 26, 331–343.

Hurd, Y.L., Herkenham, M., 1992. Influence of a single injection of cocaine, amphetamine or GBR 12909 on mRNA expression of striatal neuropeptides. Brain Res. Mol. Brain Res. 16, 97–104.

Itzhak, Y., Ali, S.F., 2002. Behavioral consequences of methamphetamine-induced neurotoxicity in mice: relevance to the psychopathology of methamphetamine addiction. Ann. N. Y. Acad. Sci. 965, 127–135.

Jayaram-Lindstrom, N., Wennberg, P., Hurd, Y.L., Franck, J., 2004. Effects of naltrexone on the subjective response to amphetamine in healthy volunteers. J. Clin. Psychopharmacol. 24, 665–669.

Jayaram-Lindstrom, N., Hammarberg, A., Beck, O., Franck, J., 2008. Naltrexone for the treatment of amphetamine dependence: a randomized, placebo-controlled trial. Am. J. Psychiatry 165, 1442–1448.

Johanson, C.E., Frey, K.A., Lundahl, L.H., Keenan, P., Lockhart, N., Roll, J., Galloway, G.P., Koeppe, R.A., Kilbourn, M.R., Robbins, T., Schuster, C.R., 2006. Cognitive function and nigrostriatal markers in abstinent methamphetamine abusers. Psychopharmacology 185, 327–338.

Johnson, B.A., 2004. Topiramate-induced neuromodulation of cortico-mesolimbic dopamine function: a new vista for the treatment of comorbid alcohol and nicotine dependence? Addict. Behav. 29, 1465–1479.

Lee, N.K., Rawson, R.A., 2008. A systematic review of cognitive and behavioural therapies for methamphetamine dependence. Drug Alcohol Rev. 27, 309–317.

Lee, Y.K., Park, S.W., Kim, Y.K., Kim, D.J., Jeong, J., Myrick, H., Kim, Y.H., 2005. Effects of naltrexone on the ethanol-induced changes in the rat central dopaminergic system. Alcohol Alcohol. 40, 297–301.

Litten, R.Z., Egli, M., Heilig, M., Cui, C., Fertig, J.B., Ryan, M.L., Falk, D.E., Moss, H., Huebner, R., Noronha, A., 2012. Medications development to treat alcohol dependence: a vision for the next decade. Addict. Biol. 17, 513–527.

London, E.D., Berman, S.M., Voytek, B., Simon, S.L., Mandelkern, M.A., Monterosso, J., Thompson, P.M., Brody, A.L., Geaga, J.A., Hong, M.S., Hayashi, K.M., Rawson, R.A., Ling, W., 2005. Cerebral metabolic dysfunction and impaired vigilance in recently abstinent methamphetamine abusers. Biol Psychiatry 58, 770–778.

Ma, J.Z., Johnson, B.A., Yu, E., Weiss, D., Mcsherry, F., Saadvandi, J., Iturriaga, E., Ait-Daoud, N., Rawson, R.A., Hrymoc, M., Campbell, J., Gorodetzky, C., Haning, W., Carlton, B., Mawhinney, J., Weis, D., Mccann, M., Pham, T., Stock, C., Dickinson, R., Elkashef, A., Li, M.D., 2013. Fine-grain analysis of the treatment effect of topiramate on methamphetamine addiction with latent variable analysis. Drug Alcohol Depend. 130, 45–51.

Madras, B.K., Xie, Z., Lin, Z., Jassen, A., Panas, H., Lynch, L., Johnson, R., Livni, E., Spencer, T.J., Bonab, A.A., Miller, G.M., Fischman, A.J., 2006. Modafinil occupies dopamine and norepinephrine transporters in vivo and modulates the transporters and trace amine activity in vitro. J. Pharmacol. Exp. Ther. 319, 561–569.

Mansour, A., Fox, C.A., Akil, H., Watson, S.J., 1995. Opioid-receptor mRNA expression in the rat CNS: anatomical and functional implications. Trends Neurosci. 18, 22–29.

Mccann, U.D., Wong, D.F., Yokoi, F., Villemagne, V., Dannals, R.F., Ricaurte, G.A., 1998. Reduced striatal dopamine transporter density in abstinent methamphetamine and methcathinone users: evidence from positron emission tomography studies with [11C]WIN-35,428. J. Neurosci. 18, 8417–8422.

Mcelhiney, M.C., Rabkin, J.G., Rabkin, R., Nunes, E.V., 2009. Provigil (modafinil) plus cognitive behavioral therapy for methamphetamine use in HIV+ gay men: a pilot study. Am. J. Drug Alcohol Abuse 35, 34–37.

Melega, W.P., Williams, A.E., Schmitz, D.A., Distefano, E.W., Cho, A.K., 1995. Pharmacokinetic and pharmacodynamic analysis of the actions of D-amphetamine and D-methamphetamine on the dopamine terminal. J. Pharmacol. Exp. Ther. 274, 90–96.

Monterosso, J.R., Ainslie, G., Xu, J., Cordova, X., Domier, C.P., London, E.D., 2007. Frontoparietal cortical activity of methamphetamine-dependent and comparison subjects performing a delay discounting task. Hum. Brain Mapp. 28, 383–393.

Murakawa, K., Hirose, N., Takada, K., Suzuki, T., Nagase, H., Cools, A.R., Koshikawa, N., 2004. Deltorphin II enhances extracellular levels of dopamine in the nucleus accumbens via opioid receptor-independent mechanisms. Eur. J. Pharmacol. 491, 31–36.

Naleid, A.M., Grace, M.K., Cummings, D.E., Levine, A.S., 2005. Ghrelin induces feeding in the mesolimbic reward pathway between the ventral tegmental area and the nucleus accumbens. Peptides 26, 2274–2279.

National Institute on Drug Abuse (Nida), 2005. Medications Development Research for Treatment of Amphetamine and Methamphetamine Addiction. U.S. Department of Health and Human Services, National Institutes of Health, Bethesda, MD.

Nestor, L.J., Ghahremani, D.G., Monterosso, J., London, E.D., 2011. Prefrontal hypoactivation during cognitive control in early abstinent methamphetamine-dependent subjects. Psychiatry Res. 194, 287–295.

Olive, M.F., Koenig, H.N., Nannini, M.A., Hodge, C.W., 2001. Stimulation of endorphin neurotransmission in the nucleus accumbens by ethanol, cocaine, and amphetamine. J. Neurosci. 21, RC184.

Paulus, M.P., Hozack, N.E., Zauscher, B.E., Frank, L., Brown, G.G., Braff, D.L., Schuckit, M.A., 2002. Behavioral and functional neuroimaging evidence for prefrontal dysfunction in methamphetamine-dependent subjects. Neuropsychopharmacology 26, 53–63.

Paulus, M.P., Hozack, N., Frank, L., Brown, G.G., Schuckit, M.A., 2003. Decision making by methamphetamine-dependent subjects is associated with error-rate-independent decrease in prefrontal and parietal activation. Biol. Psychiatry 53, 65–74.

Paulus, M.P., Tapert, S.F., Schuckit, M.A., 2005. Neural activation patterns of methamphetamine-dependent subjects during decision making predict relapse. Arch. Gen. Psychiatry 62, 761–768.

Rabiner, D.L., 2013. Stimulant prescription cautions: addressing misuse, diversion and malingering. Curr. Psychiatry Rep. 15, 375.

Ray, L.A., Bujarski, S., Courtney, K.E., Moallem, N.R., Lunny, K., Roche, D., Leventhal, A.M., Shoptaw, S., Heinzerling, K., London, E.D., Miotto, K., 2015. The Effects of Naltrexone on Subjective Response to Methamphetamine in a Clinical Sample: a Double-Blind, Placebo-Controlled Laboratory Study. Neuropsychopharmacology, epub ahead of print.

Ricaurte, G.A., Mechan, A.O., Yuan, J., Hatzidimitriou, G., Xie, T., Mayne, A.H., Mccann, U.D., 2005. Amphetamine treatment similar to that used in the treatment of adult attention-deficit/hyperactivity disorder damages dopaminergic nerve endings in the striatum of adult nonhuman primates. J. Pharmacol. Exp. Ther. 315, 91–98.

Salo, R., Ursu, S., Buonocore, M.H., Leamon, M.H., Carter, C., 2009. Impaired prefrontal cortical function and disrupted adaptive cognitive control in methamphetamine abusers: a functional magnetic resonance imaging study. Biol. Psychiatry 65, 706–709.

Salo, R., Fassbender, C., Buonocore, M.H., Ursu, S., 2013. Behavioral regulation in methamphetamine abusers: an fMRI study. Psychiatry Res. 211, 234–238.

Schouw, M.L., Caan, M.W., Geurts, H.M., Schmand, B., Booij, J., Nederveen, A.J., Reneman, L., 2013. Monoaminergic dysfunction in recreational users of dexamphetamine. Eur. Neuropsychopharmacol. 23, 1491–1502.

Scott, J.C., Woods, S.P., Matt, G.E., Meyer, R.A., Heaton, R.K., Atkinson, J.H., Grant, I., 2007. Neurocognitive effects of methamphetamine: a critical review and meta-analysis. Neuropsychol. Rev. 17, 275–297.

Sekine, Y., Ouchi, Y., Takei, N., Yoshikawa, E., Nakamura, K., Futatsubashi, M., Okada, H., Minabe, Y., Suzuki, K., Iwata, Y., Tsuchiya, K.J., Tsukada, H., Iyo, M., Mori, N., 2006. Brain serotonin transporter density and aggression in abstinent methamphetamine abusers. Arch. Gen. Psychiatry 63, 90–100.

Shearer, J., 2007. Psychosocial approaches to psychostimulant dependence: a systematic review. J. Subst. Abus. Treat. 32, 41–52.

Shoblock, J.R., Maisonneuve, I.M., Glick, S.D., 2003a. Differences between d-methamphetamine and d-amphetamine in rats: working memory, tolerance, and extinction. Psychopharmacology 170, 150–156.

Shoblock, J.R., Sullivan, E.B., Maisonneuve, I.M., Glick, S.D., 2003b. Neurochemical and behavioral differences between d-methamphetamine and d-amphetamine in rats. Psychopharmacology 165, 359–369.

Shoji, Y., Delfs, J., Williams, J.T., 1999. Presynaptic inhibition of GABA(B)-mediated synaptic potentials in the ventral tegmental area during morphine withdrawal. J. Neurosci. 19, 2347–2355.

Shoptaw, S., Heinzerling, K.G., Rotheram-Fuller, E., Steward, T., Wang, J., Swanson, A.N., De La Garza, R., Newton, T., Ling, W., 2008. Randomized, placebo-controlled trial of bupropion for the treatment of methamphetamine dependence. Drug Alcohol Depend. 96, 222–232.

Slemmer, J.E., Martin, B.R., Damaj, M.I., 2000. Bupropion is a nicotinic antagonist. J. Pharmacol. Exp. Ther. 295, 321–327.

Smith, A.J., Mcginty, J.F., 1994. Acute amphetamine or methamphetamine alters opioid peptide mRNA expression in rat striatum. Brain Res. Mol. Brain Res. 21, 359–362.

Stahl, S.M., Pradko, J.F., Haight, B.R., Modell, J.G., Rockett, C.B., Learned-Coughlin, S., 2004. A review of the neuropharmacology of bupropion, a dual norepinephrine and dopamine reuptake inhibitor. Prim Care Companion J. Clin. Psychiatry 6, 159–166.

Substance Abuse and Mental Health Services Administration (SAMHSA), 2013. Results from the 2012 National Survey on Drug Use and Health: Summary of National Findings. Substance Abuse and Mental Health Services Administration, Rockville, MD. NSDUH Series H-46, HHS Publication No. (SMA) 13–4795.

Substance Abuse and Mental Health Services Administration (SAMHSA), Center for Behavioral Health Statistics and Quality, 2013b. Treatment Episode Data Set (TEDS): 2001–2011. National Admissions to Substance Abuse Treatment Services. Substance Abuse and Mental Health Services Administration, Rockville, MD.

Sulzer, D., Sonders, M.S., Poulsen, N.W., Galli, A., 2005. Mechanisms of neurotransmitter release by amphetamines: a review. Prog. Neurobiol. 75, 406–433.

Toll, L., Berzetei-Gurske, I.P., Polgar, W.E., Brandt, S.R., Adapa, I.D., Rodriguez, L., Schwartz, R.W., Haggart, D., O'brien, A., White, A., Kennedy, J.M., Craymer, K., Farrington, L., Auh, J.S., 1998. Standard binding and functional assays related to medications development division testing for potential cocaine and opiate narcotic treatment medications. NIDA Res. Monogr. 178, 440–466.

Trigo, J.M., Martin-García, E., Berrendero, F., Robledo, P., Maldonado, R., 2010. The endogenous opioid system: a common substrate in drug addiction. Drug Alcohol Depend. 108, 183–194.

United Nations Office on Drugs and Crime (UNODC), 2012. World Drug Report 2012. United Nations publication, Sales No. E.12.XI.1. United Nations Office on Drugs and Crime, Vienna.

United Nations Office on Drugs and Crime (UNODC), 2013. World Drug Report 2013. United Nations publication, Sales No. E.13.XI.6. United Nations Office on Drugs and Crime, Vienna.

Villemagne, V., Yuan, J., Wong, D.F., Dannals, R.F., Hatzidimitriou, G., Mathews, W.B., Ravert, H.T., Musachio, J., Mccann, U.D., Ricaurte, G.A., 1998. Brain dopamine neurotoxicity in baboons treated with doses of methamphetamine comparable to those recreationally abused by humans: evidence from [11C]WIN-35,428 positron emission tomography studies and direct in vitro determinations. J. Neurosci. 18, 419–427.

Volkow, N.D., Chang, L., Wang, G.J., Fowler, J.S., Ding, Y.S., Sedler, M., Logan, J., Franceschi, D., Gatley, J., Hitzemann, R., Gifford, A., Wong, C., Pappas, N., 2001a.

Low level of brain dopamine D2 receptors in methamphetamine abusers: association with metabolism in the orbitofrontal cortex. Am. J. Psychiatry 158, 2015–2021.

Volkow, N.D., Chang, L., Wang, G.J., Fowler, J.S., Franceschi, D., Sedler, M., Gatley, S.J., Miller, E., Hitzemann, R., Ding, Y.S., Logan, J., 2001b. Loss of dopamine transporters in methamphetamine abusers recovers with protracted abstinence. J. Neurosci. 21, 9414–9418.

Volkow, N.D., Chang, L., Wang, G.J., Fowler, J.S., Leonido-Yee, M., Franceschi, D., Sedler, M.J., Gatley, S.J., Hitzemann, R., Ding, Y.S., Logan, J., Wong, C., Miller, E.N., 2001c. Association of dopamine transporter reduction with psychomotor impairment in methamphetamine abusers. Am. J. Psychiatry 158, 377–382.

Volkow, N.D., Fowler, J.S., Logan, J., Alexoff, D., Zhu, W., Telang, F., Wang, G.J., Jayne, M., Hooker, J.M., Wong, C., Hubbard, B., Carter, P., Warner, D., King, P., Shea, C., Xu, Y., Muench, L., Apelskog-Torres, K., 2009. Effects of modafinil on dopamine and dopamine transporters in the male human brain: clinical implications. J. Am. Med. Assoc. 301, 1148–1154.

Wang, J.Q., Mcginty, J.F., 1995. Dose-dependent alteration in zif/268 and preprodynorphin mRNA expression induced by amphetamine or methamphetamine in rat forebrain. J. Pharmacol. Exp. Ther. 273, 909–917.

White, H.S., Brown, S.D., Woodhead, J.H., Skeen, G.A., Wolf, H.H., 2000. Topiramate modulates GABA-evoked currents in murine cortical neurons by a nonbenzodiazepine mechanism. Epilepsia 41 (Suppl. 1), S17–S20.

Widdowson, P.S., Holman, R.B., 1992. Ethanol-induced increase in endogenous dopamine release may involve endogenous opiates. J. Neurochem. 59, 157–163.

Wood, S., Sage, J.R., Shuman, T., Anagnostaras, S.G., 2014. Psychostimulants and cognition: a continuum of behavioral and cognitive activation. Pharmacol. Rev. 66, 193–221.

CHAPTER

Behavioral addictions in addiction medicine: from mechanisms to practical considerations

16

Barbara C. Banz*, Sarah W. Yip*, Yvonne H.C. Yau[†,‡], Marc N. Potenza*[,§,¶,1]

Department of Psychiatry, Yale University School of Medicine, New Haven, CT, USA

[†]*Department of Neurology and Neurosurgery, Montreal Neurological Institute, McGill University, Montréal, QC, Canada*

[‡]*Montreal Neurological Institute, 3801 Rue University, Montréal, QC, Canada*

[§]*Department of Neurobiology, Child Study Center, and CASA Columbia, Yale University School of Medicine, New Haven, CT, USA*

[¶]*Connecticut Mental Health Center, Yale University School of Medicine, New Haven, CT, USA*

[1]*Corresponding author: Tel.: (203) 974-7356; Fax: (203) 974-7366, e-mail address: marc.potenza@yale.edu*

Abstract

Recent progress has been made in our understanding of nonsubstance or "behavioral" addictions, although these conditions and their most appropriate classification remain debated and the knowledge basis for understanding the pathophysiology of and treatments for these conditions includes important gaps. Recent developments include the classification of gambling disorder as a "Substance-Related and Addictive Disorder" in the 5th edition of the Diagnostic and Statistical Manual of Mental Disorders (DSM-5) and proposed diagnostic criteria for Internet Gaming Disorder in Section 3 of DSM-5. This chapter reviews current neuroscientific understandings of behavioral addictions and the potential of neurobiological data to assist in the development of improved policy, prevention, and treatment efforts.

Keywords

Gambling disorder, Internet, Gaming, Addiction, Neuroscience, Treatment

Progress in Brain Research, Volume 223, ISSN 0079-6123, http://dx.doi.org/10.1016/bs.pbr.2015.08.003

1 INTRODUCTION

While debated, the concept of nonsubstance or "behavioral" addictions has gained traction as evidenced by the recent classification of gambling disorder as a "Substance-Related and Addictive Disorder" in the *Diagnostic and Statistical Manual of Mental Disorders* (5th edition) (DSM-5) and definition in Section 3 of DSM-5 of diagnostic criteria for Internet gaming disorder (IGD) (American Psychiatric Association, 2013; Petry and O'Brien, 2013). These classification and inclusion efforts have been informed by neuroscientific data. A current challenge exists in translating a neuroscientific understanding of these disorders into more effective treatments. Behaviors that may involve excessive or problematic engagement include gambling, Internet use, and gaming. The following chapter reviews the current neurobiological understanding of, and discusses treatment implications with respect to, excessive and interfering patterns of gambling, Internet use, and gaming.

2 GAMBLING DISORDER

The reclassification of gambling disorder in the DSM-5 was based upon evidence of clinical, neurobiological, and other similarities between substance-use and gambling disorders (Potenza, 2006). Due to the recent classification and renaming of "pathological gambling" (PG) in DSM-IV-TR to "gambling disorder" in DSM-5 (American Psychiatric Association, 2000, 2013; Potenza, 2014), this condition will be referred to as gambling disorder in this chapter despite a majority of data emanating from studies of PG.

3 NEURAL FEATURES OF GAMBLING DISORDER

Phenomenological similarities between substance-use and gambling disorders have been observed, leading to inclusionary criteria addressing tolerance, withdrawal, and interference in major areas of life functioning for these conditions. Recently, there have been various other reviews of neural function in gambling disorder (Leeman and Potenza, 2012, 2013; Meng et al., 2014). The current review will describe recent findings related to processes which may be beneficial for advancing treatment of this disorder.

3.1 NEUROCOGNITIVE FACETS

Neurocognitive measures allow for evaluation of possible dysfunction in a variety of cognitive facets and offer insight in to potential underlying neural regions of importance in behavioral addictions (Potenza, 2014). The evaluation of patterns of dysfunction allows for comparisons to healthy comparison subjects, across substance-use disorders, and various other populations of interest which allow for a more in-depth understanding of similarities and differences between these groups (Choi et al., 2014; Leeman and Potenza, 2012; Noël et al., 2013; Yan et al., 2014). Importantly, evaluation of neurocognitive function in PG through neurocognitive tasks has provided insight into the maintenance of this disorder (for review, see Brevers et al., 2013;

van Holst et al., 2010). Together, these data inform potential approaches to the identification of those at risk and the development of more effective treatments.

3.2 ELECTROPHYSIOLOGY

Electrophysiological studies involving electroencephalogram (EEG) data and tasks designed to elicit event-related potentials (ERPs) offer insight into neural function linked to sensory or cognitive processing. To date, these methods have not been extensively used within individuals with PG, with existing studies frequently using gambling tasks, as described below.

Feedback-related negativity (FRN), an ERP component elicited through feedback related to subject performance, has been evaluated. Healthy comparison subjects and those with PG presented with similar FRN amplitudes in win and loss conditions; however, in PG subjects, an additional FRN occurred earlier with latency and amplitude correlated with severity of PG (Oberg et al., 2011). In PG, blunted P3 amplitude and EEG power in theta-band activity were also found in response to high-risk scenarios (Oberg et al., 2011). More recently, Lole and colleagues (2015) found attenuated FRN and feedback-related positivity in response to losses and wins with no difference in P3b amplitude in response to large and small rewards in PG. These data suggest varied sensitivity to risk, reward, and loss in PG which can be evaluated through EEG.

During simulated blackjack, reward resulted in more positive reactivity in PG compared to healthy comparison subjects during a window after the FRN (between 270 and 320 ms); a difference in positivity was found within PG subjects between responses to rewards and losses, with no differences in healthy comparison subjects (Hewig et al., 2010). However, during varying loss conditions, PG subjects did not show differences in reactivity during conditions of near or full losses during this same window of activity, unlike healthy comparison subjects (Kreussel et al., 2013). When comparing occasional gamblers and PG subjects during a blackjack task, reactivity in these two groups differed in both low- and high-risk conditions during risk assessment, and PG subjects presented with greater negativity during reward processing (Miedl et al., 2014). Together, these studies differences in the electrophysiological brain correlates of reward/loss processing and suggest a need for additional study of cue-related craving effects on risk assessment, loss, and reward processing in PG.

3.3 FUNCTIONAL MAGNETIC RESONANCE IMAGING

Functional magnetic resonance imaging (fMRI) tasks offer insight into the neural circuitry associated with different neuropsychological processes that may be targeted in the treatment of gambling disorder. These tasks permit the evaluation of the neural underpinnings of cognitive processes, such as decision-making or processing of monetary rewards and losses (reviewed in Potenza, 2014). Data from fMRI studies implicate similar brain regions in both substance and behavioral addictions (Leeman and Potenza, 2013; Potenza, 2013). For example, several studies of monetary-reward processing have identified blunted activation of the ventral striatum (VS) during reward anticipation in gambling disorder (Balodis et al, 2012; Choi et al., 2012), resonating with findings in alcohol-, tobacco-, and cannabis-use disorders (reviewed in

Balodis and Potenza, 2015). Below, we will highlight findings from fMRI recent studies not covered in recent reviews.

In a recent study examining decision-making when varying risk and ambiguity, healthy control participants but not those with PG showed greater striatal, insular, and prefrontal cortical activations during decision-making under risk as compared to ambiguity, and individuals with PG as compared to those without showed greater striatal activation during betting as compared to "safe" choices (Brevers et al., 2015). Using a different task, Miedl and colleagues demonstrated that processing delayed rewards during decision-making involves widespread, bilateral activation in PG subjects compared to left side activation of healthy controls. In addition, indifferent compared to sure decisions elicited greater widespread activation in PG versus control subjects, where sure decisions for PG subjects only elicited activity in the inferior parietal and superior temporal areas, and greater activity in the cingulate gyrus, insula, and medial frontal gyrus in healthy control subjects (Miedl et al., 2015). Together, findings suggest a complex relationship between striatal activation and gambling disorder. It is likely that between-study differences in task design (e.g., delay-discounting vs. risky choice tasks) may relate to differences in findings.

Reward type may also impact neural reactivity. Monetary and erotic rewards have been used in order to assess possible differences between reward types. VS activation during reward anticipation for erotic stimuli was lower compared to monetary rewards in PG subjects, and this activation was correlated to subjective ratings for erotic but not monetary rewards in PG subjects (Sescousse et al., 2013). Posterior orbitofrontal cortical activation was greater during reward outcome for PG subjects for monetary gains (Sescousse et al., 2013). Differences based on reward type related to stress systems as VS activation and cortisol levels were correlated in PG subjects in response to monetary cues (Li et al., 2014).

fMRI tasks may model aspects of electronic-gambling machines including "near-miss" events that occur when symbols on two of three reels match. These tasks may relate more closely to specific gambling behaviors compared to other decision-making tasks. During near-miss events, both cocaine-dependent and PG subjects showed greater reactivity in ventrocortical and mesolimbic areas compared to those without either diagnosis, with PG subjects having greatest reactivity (Worhunsky et al., 2014). During a similar task, near-miss events elicited activity in the insula and right inferior frontal gyrus, and increased theta-band oscillations in the right orbitofrontal cortex (OFC) and insula (Dymond et al., 2014). These data suggest task content is also relevant to activity when comparing across addictions. Therefore, future studies should attempt to incorporate tasks which may be more ecologically valid.

Resting-state fMRI studies may offer valuable insight into functional brain connectivity at rest. In PG subjects, the supplementary motor area and paracingulate cortex show reduced connectivity at rest (Tschernegg et al., 2014). The right caudate appears more involved and the hippocampus less involved in information integration in PG versus control subjects (Tschernegg et al., 2014). These data suggest differences in PG and non-PG groups in networks involved in self-regulation and reward processing.

3.4 STRUCTURAL MRI

Structural MRI allows for volumetric comparison of tissue structures across different diagnostic groups. Using this method, comparisons between PG, substance-addicted and nonaddicted groups may improve understanding of the neural structural underpinnings of various addictions. Recently, smaller left hippocampal and right amygdalar volumes were found in PG versus control subjects (Rahman et al., 2014). Regional volumes were related to behavioral inhibition scores grouping PG subjects (Rahman et al., 2014). Additionally, problem-gambling subjects displayed similar gray matter volumes as did subjects with alcohol use disorder: lower volumes in left superior frontal cortex, bilateral precentral cortex, right insula, left thalamus, bilateral superior parietal cortex, and right supramarginal cortex (van Holst et al., 2012).

3.5 DIFFUSION TENSOR IMAGING

Diffusion tensor imaging (DTI) may assess white matter integrity. To date, two studies have used DTI to study PG. In PG versus control subjects, lower fractional anisotropy (FA) was present in the right and left genu of the corpus callosum, a pattern also seen in substance abuse (Yip et al., 2011). Lower FA in the corpus callosum was also seen in an independent study which reported widespread lower FA in PG (Jousta et al., 2011). Together, these studies suggest microstructural deficits present in PG which appear not to be accounted for by neurotoxic effects of substances.

3.6 NEUROCHEMISTRY

Preclinical, ligand-based imaging, and molecular genetic research methods may all be used to inform understanding of the role of different neurotransmitters systems in PG (reviewed in Potenza, 2013). In this section, we will focus on recent findings from human ligand-based and genetic studies, with an emphasis on dopamine and serotonin (5-HT). For a discussion of findings from pharmacological treatment studies conducted in gambling disorder, see Section 5.2.

The role of dopamine in PG remains poorly understood. Dopamine has been implicated in substance addictions and reward processing, among other behaviors. Positron emission tomography (PET) permits study of neurochemical and metabolic measures. As reviewed in Potenza (2013), data from PET studies using [11C]-raclopride suggest that individual differences in dopamine release and $D_{2/3}$ receptor availability are related to individual differences in clinical features of PG, such as positive urgency (Clark et al., 2012) and task performance (Linnet et al., 2012) or subjective experiences during gambling tasks (Jousta et al., 2012).

There have been two recent PET studies using the D_3-preferring radioligand $[^{11}C]$-(+)-propyl-hexahydro-naphtho-oxazin (Boileau et al., 2013, 2014). While no significant differences in D_3 receptor availability were found between individuals with and without PG (Boileau et al., 2013, 2014), significant decreases in receptor binding subsequent to amphetamine challenge were observed among individuals with PG (Boileau et al., 2014).

Problem-gambling severity has been linked to the serotonin system (reviewed in Leeman and Potenza, 2013; Potenza, 2013). A positive association between serotonin 1B receptor availability and problem-gambling severity (Potenza, 2013) and a genetic link between serotonin 2A receptors and PG have been reported; see genetic studies, below (Wilson et al., 2013).

The relationship between dopamine and serotonin and their complementary roles in value-adaptation and loss-chasing may be relevant to PG (Campbell-Meiklejohn et al., 2011). Other neurotransmitter systems (e.g., opioid, cannabinoid, glutamate) require consideration in PG.

4 GENETICS

Polymorphisms in genes encoding for dopamine-related moieties, including *DRD1* Ddel, *DRD2* Taq I A, and *DRD4* (exon III), have been reported, although negative results have also been reported (reviewed in Leeman and Potenza, 2013). Studies also suggest genes coding for the serotonin transporter, *5HTTLPR*, and MAO enzymes (e.g., *MAO-A*, *MAO-B*) may contribute to PG (reviewed in Leeman and Potenza, 2013). An association between the C/C genotype of the 5-HT-2A receptor gene and PG has also been reported recently (Wilson et al., 2013). Below, we review findings from recent genetic studies, focusing on findings not covered in recent reviews (Gyollai et al., 2014; Leeman and Potenza, 2013; Yau et al., 2014).

While candidate gene studies often fail to replicate, genome-wide association studies (GWASs) have arguably shown more consistency, although few have been performed in PG (one reported to date). GWAS data collected in a community-based Australian twin sample identified six single-nucleotide polymorphisms (SNPs) as related to disordered gambling, four of which the authors interpreted as "theoretically relevant," although none reached genome-wide significance (Lind et al., 2012). Identified SNPs included rs8064100, located downstream of metallothionein 1X (*MT1X*), previously implicated in alcohol and drug dependence (Lind et al., 2012).

Genetic contributions have been found to relate to the age of onset of gambling but not age of onset of drinking in males; however, in females, genetic contributions link to onsets of both behaviors (Richmond-Rakerd et al., 2014). Conversely, twin studies suggest nondisordered gambling engagement is equally influenced by family environmental factors in men and women (Slutske and Richmond-Rakerd, 2014). Recently, shared environmental and genetic factors have been linked to the age of gambling initiation predicting gambling behavior later in life (Slutske et al., 2014). Taken together, these findings suggest a complicated relationship between genetic contributions, environmental influences, sex, and gambling behaviors and disorders.

Twin studies offer valuable insight into relative genetic and environmental contributions to PG and co-occurring disorders. The Vietnam Era Twin Registry has demonstrated that PG shares genetic factors with both obsessive-compulsive classes

(Scherrer et al., 2015) and drug addictions (Xian et al., 2014), with environmental contributions less important for the overlap between PG and stimulant dependence. These findings suggest the need to identify specific genetic factors involved in these relationships and determine the extent to which these may represent appropriate targets for behavioral or pharmacological interventions in subgroups of individuals with PG.

5 TREATMENT OF GAMBLING DISORDER

Treatment of PG involves both pharmacological and nonpharmacological methods, as reviewed elsewhere (Yip and Potenza, 2014). Though various treatments exist, individuals with PG are often unlikely to seek treatment, and those that do have high dropout rates, highlighting the need for enhanced efforts to get individuals into treatment settings and maintain their attendance (Rash and Petry, 2014).

5.1 BEHAVIORAL TREATMENT

Two recent manuscripts have reviewed nonpharmacological therapies for PG (Cowlishaw et al., 2012; Rash and Petry, 2014). Cognitive behavioral therapy (CBT) has arguably the greatest support, with one CBT treatment adapted from that demonstrating efficacy in the treatment of substance-use disorders and another focused on targeting irrational cognitions. As compared to CBT for substance addictions, CBT for PG may also target financial problems and financial management. Other approaches that have shown efficacy in the treatment of PG involve imaginal desensitization, motivational enhancement, and brief interventions like those that show efficacy in the treatment of substance-use disorders. Mindfulness-based approaches for PG have also been proposed and are beginning to be investigated. Active ingredients of these therapies and biological mechanisms relating to their efficacies have been proposed, and further research is needed to investigate how these treatments work and for whom they might work best (Potenza et al., 2013).

5.2 PHARMACOLOGICAL TREATMENT

Serotonin selective reuptake inhibitors, dopaminergic agents, mood stabilizers, glutamatergic agents, opioid-receptor antagonists, and other drugs have been investigated in the treatment of PG (reviewed in Bullock and Potenza, 2012; Yip and Potenza, 2014). Arguably, the most consistent results are from studies of opioid-receptor antagonists (naltrexone, nalmefene), with four randomized clinical trials (RCTs) showing varying degrees of efficacy. However, the effect size of opioid-receptor antagonists in the treatment of PG may be modest (Bartley and Bloch, 2013), although the medication may be most helpful for specific subgroups (those with familial histories of alcoholism or strong gambling urges). Treatment algorithms have been proposed (Bullock and Potenza, 2012) largely based on co-occurring disorders (e.g., lithium in the treatment of individuals with PG and co-occurring bipolar-spectrum disorders) and willingness to take pharmacotherapies. For individuals less willing to consider pharmacotherapies, *n*-acetyl cysteine, a dietary supplement with glutamatergic properties, may represent an important therapeutic possibility. Recently, in a placebo-controlled RCT of *n*-acetyl cysteine in individuals with PG and nicotine dependence in which all participants received a behavioral therapy for gambling involving CBT, motivational, and imaginal-desensitization components, active *n*-acetyl cysteine was superior to placebo with respect to reducing smoking during treatment and reducing gambling behaviors at follow-up (Grant et al., 2014). These findings raise the intriguing possibility that *n*-acetyl cysteine may lead to greater durability of gambling-related behavioral therapies, perhaps augmenting a "sleeper" effect that has been described for CBT in the treatment of substance-use disorders. However, this and other aspects of how behavioral and pharmacological therapies may be used conjointly in the treatment of PG warrant additional investigation.

6 PROBLEMATIC INTERNET USE AND IGD

The extent to which Internet use may be considered the focus of a disorder has been debated, with some contending that the Internet may represent a vehicle for other behaviors (e.g., gambling) that constitute the true diagnostic focus (Petry and O'Brien, 2013). Additionally, if considered a disorder, debate exists regarding the extent to which problematic Internet use (PIU) may represent an addiction or not. Given these debates, we will use the term PIU in this review although other terms (e.g., Internet addiction) have been used in the literature. PIU may be conceptualized as involving the excessive or poorly controlled urges and behaviors relating to Internet use that lead to subjective distress and/or interference in major areas of life functioning. It is a heterogeneous construct that may include a multitude features relating to sexual, social networking, and gaming behaviors. Currently, IGD is included in Section 3 of the DSM-5 as this was determined to need more research, despite being considered the most well studied and interfering type of Internet use at the time of DSM-5 deliberations (Petry and O'Brien, 2013).

7 NEURAL FEATURES OF PIU

7.1 NEUROCOGNITIVE FACETS

Few neurocognitive studies have been used to assess PIU and IGD; however, differences in impulsivity have been noted. High urgency, a facet of impulsivity, has been shown to relate to PIU (Billeux et al., 2011). Higher self-reported impulsivity was also found in IGD and alcohol use disorder (AUD) compared to healthy controls and GD (Choi et al., 2014). Interestingly, IGD, AUD, and healthy controls displayed less compulsivity than did GD on a set-shifting task (Choi et al., 2014). Additional neurocognitive differences have been observed (see following sections). Together, these data suggest differences between IGD and healthy populations as well as between IGD and other behavioral addictions, and these may be important considerations for treatment development (Dong and Potenza, 2014).

7.2 ELECTROPHYSIOLOGY

Resting-state EEG studies are used to evaluate intrinsic neural activity which is not elicited through a task. In a PIU population, lower absolute beta-band and greater absolute gamma-band activities related to disorder severity and impulsivity measures (Choi et al., 2013). In those with comorbid depression, increased theta and decreased alpha-band power was found compared to nondepressed individuals with PIU where nondepressed individuals presented with decreased delta- and beta-band power compared to depressed individuals (Lee et al., 2014). These resting-state data suggest these neurobiological differences may be markers for PIU.

Studies evaluating ERP activity, such as a Go/No-Go or Stroop task, may reflect facets related to impulsivity and error processing. In a Go/No-Go task, N2 amplitude was lower in PIU and P3 amplitude was larger with a longer latency, and decreased activity in conflict detection was observed compared to healthy comparison subjects (Dong et al., 2010). Relative to healthy comparison subjects, an incorrect response during the No-Go condition elicited a decreased amplitude of error-related negativity in individuals with IGD which was associated with impulsivity assessed via task performance and self-report (Littel et al., 2012). Similarly, during Stroop performance, decreased medial frontal negativity, greater reaction times, and response errors were found during incongruent trials in the PIU group to the healthy comparison subjects (Dong et al., 2011). Together, these studies suggest differences in impulsivity which may be evaluated through ERP tasks, and which may relate to deficits in conflict processing and have implications for treatment development (Dong and Potenza, 2014).

7.3 FUNCTIONAL MRI

Reward circuitry, cognitive control, cue reactivity, and craving fMRI tasks have implicated some similar brain regions in PIU and substance-use disorders (reviewed in van Rooij and Prause, 2014; Yau and Potenza, 2015; Yau et al., 2012). In particular, mesolimbic and cortical regions relating to reward/motivation and behavioral control may contribute to the pathophysiology of PIU. A recent meta-analysis found subjects with IGD showed abnormal activation of the medial frontal, medial temporal, and cingulate gyrus regions in response to a range of cognitive tasks (Meng et al., 2014).

Studies of functional connectivity may provide insight into how brain regions interact in circuits. Resting-state MRI has been used to investigate task-independent functional connectivity between regions of the mesolimbic system. Graph-theoretical approaches have identified connectivity in limbic regions including the amygdala, the insula, and dorsolateral prefrontal cortex (dlPFC) correlated with features of PIU and IGD (Ko et al., 2015; Wang et al., 2015). Similarly, less functional connectivity in executive-control networks was observed among individuals with IGD compared to control subjects (Dong et al., 2015). These brain networks have been implicated in substance addictions and may contribute to the development and maintenance of addictive behaviors. Altered connectivity between regions of the default-mode network has also been observed among individuals with PIU, and degree of connectivity was related to PIU severity (Wee et al., 2014).

7.4 STRUCTURAL MRI

Several studies have also investigated structural abnormalities that may relate to PIU. Reduced regional gray matter volume has been observed in key nodes of executive-control networks (e.g., frontoinsular cortex, anterior cingulate cortex, dlPFC, and posterior parietal cortex) in association with greater PIU severity (Li et al., 2015). Furthermore, these differences correlated with performance on the Stroop task and may, therefore, reflect reduced inhibitory control and cognitive efficiency. Decreased cortical thickness (Hong et al., 2013) has also been observed in regions of the executive-control network among individuals with PIU.

In light of the recent DSM categorization changes relating to conditions with addictive potential, comparing PIU and substance-abusing populations may provide insight into the most appropriate categorization of PIU. Commonalities and differences exist between the two groups and a better understanding of this can help improve diagnosis (Yau et al., 2012). For example, among both IGD and alcohol-dependent populations, negative functional connectivity between the dlPFC and the OFC and positive connectivity between the dlPFC and the ACC have been observed (Han et al., 2015). However, subjects with IGD showed negative connectivity between the dlPFC and regions of the temporal lobe, and the striatum, whereas alcohol-dependent subjects had positive connectivity. Kim and colleagues (2015) recently found increased regional homogeneity in the posterior cingulate cortex in both IGD and alcohol-dependent individuals; however, regional homogeneity appeared selectively reduced in the superior temporal gyrus in IGD.

7.5 DIFFUSION TENSOR IMAGING

DTI studies have suggested poorer white matter integrity in PIU, although not consistently (Yau et al., 2012). In one study of adolescents, lower FA was observed in the OFC, corpus callosum, cingulum, inferior fronto-occipital fasciculus, corona radiata, and internal and external capsules, with FA measures in the genu of the corpus collusum inversely correlating with measures of anxiety and those in the external capsule relating inversely correlating with PIU severity (Lin et al., 2012). A separate study found lower white matter density in the inferior frontal gyrus, insula, amygdala, and anterior cingulate in IGD subjects relatively to control subjects (Lin et al., 2015).

7.6 NEUROCHEMISTRY

Several neurotransmitter systems may contribute to PIU. Of these, the dopaminergic system has arguably received the most research attention. Years of problematic gaming was negatively correlated with D_2-like receptor availability in the striatum and IGD subjects showed decreased glucose metabolism in the OFC, insula, and limbic regions (Tian et al, 2014). In a separate study, dopamine transporter expression in the striatum was significantly lower in individuals with PIU compared to healthy control subjects (Hou et al., 2012).

Beyond the dopaminergic system, other neurochemical systems have shown differences which may underlie PIU. Recently, lower levels of *N*-acetyl aspartate (NAA) and cystolic, choline-containing compound (Cho) levels were observed in the medial temporal cortices of IGD patients. Additionally, lower levels of NAA and Cho in the right frontal cortex were also seen, and the NAA was inversely related to Young Internet Addiction Scale scores and perseverative responses during the Wisconsin Card Sorting Task (Han et al., 2014). Further study of the underlying neurochemical differences among individuals with PIU is needed in order to guide treatment development efforts.

8 GENETICS

Genetic mechanisms underlying PIU are poorly understood and currently only preliminary studies exists. As with disordered gambling, relationships between PIU and the Taq1A1 allele of the *DRD2* gene (Han et al., 2007) and homozygosity of the short allelic variant of the *5-HTTLPR* gene (Lee et al., 2008) have been associated with PIU. However, these studies warrant replication and verification in larger samples, and other approaches (twin and GWAS studies) warrant undertaking.

9 TREATMENT

Few interventions have been systematically tested for PIU or IGD. Treatment strategies may be particularly important for adolescents and young adults given high prevalence estimates among these age groups (Spada, 2014).

9.1 BEHAVIORAL TREATMENT

Behavioral treatments have yet to be systematically studied in the context of PIU. Preliminary studies of CBT adapted for PIU (and more specifically in certain studies of IGD and problematic Internet pornography viewing) has demonstrated preliminary positive benefits (Twohig and Crosby, 2010; Wölfling et al., 2014; Young, 2013). Such CBT approaches aim to help individuals examine emotional motives and identify problematic cognition that may prompt them to engage excessively in online activities. As with disordered gambling, individuals are encouraged to explore alternative ways to satisfy those needs (e.g., developing other recreational pursuits) and to correct maladaptive thinking patterns. Family therapies involving increased socialization efforts have also been examined in PIU (Liu et al., 2015), with preliminary reports of possible treatment-related changes in striatal responsivity (Han et al., 2012). Other behavioral interventions (e.g., solution-focused brief therapy, mindfulness-based interventions) may also be helpful given the proposed addiction model of PIU but their efficacy has yet to be studied directly.

9.2 PHARMACOLOGICAL TREATMENT

Psychostimulants, opioid-receptor antagonists (e.g., naltrexone), antiepileptics, antipsychotics (e.g., olanzapine), antidepressants (e.g., bupropion), and glutamate-receptor antagonists have explored in treatments for PIU (reviewed in Camardese et al., 2012; Przepiorka et al., 2014; Winkler et al., 2013). Successful treatment may link changes in specific domains (e.g., depression, cognitive flexibility), and as in PG, comorbidity may be important to consider (Przepiorka et al., 2014). Although many pharmacotherapies have been explored in the treatment of PIU, placebo-controlled RCTs of significant size and duration are largely lacking, and as is the case with PG, no medications have approval by the U.S. Food and Drug Administration (FDA) with an indication for treating PIU or IGD.

10 FUTURE DIRECTIONS

While much research has been conducted recently on PG, PIU, and IGD, many research gaps exist, with arguably greater gaps and controversies for Internet-related behaviors and disorders. For example, research criteria for IGD have only recently been proposed, and these do not cover other forms of Internet use (e.g., social networking, pornography viewing) that might be problematic for individuals (Rehbein and Mößle, 2013). Additionally, the absence of agreed-upon criteria for PIU has led to marked variations in reported prevalence estimates and assessment of public health impacts (Petry and O'Brien, 2013). While treatment development efforts for PG have led to the availability of efficacious behavioral therapies for PG, such studies are at earlier stages for PIU. For PG, PIU, and IGD, no pharmacotherapy trials have led to the availability of medications with FDA indications for the disorders. With respect to prevention efforts, healthy levels of engagement in gambling, gaming, and Internet use remain discussed and/or debated. In these efforts, potentially vulnerable individuals should be considered, especially youth. Other factors should also be considered; for example, most studies of PG, PIU, and IGD have focused on males, and sex differences warrant consideration, particularly for specific Internet-related behaviors like social networking. Additionally, health disparities related to race/ethnicity should also be considered. In this context, neurobiological findings should be used to inform advancement of policy, prevention, and treatment efforts relating to behavioral addictions.

ACKNOWLEDGMENTS

This was supported by P20 DA027844, T32 AA015496, T32 DA007238, CASA Columbia, and the National Center for Responsible Gaming. The content of the manuscript reflects the views of the authors and not necessarily the funding agencies. The funding agencies did not have input into the content of the manuscript.

Disclosures: The authors report that they have no financial conflicts of interest with respect to the content of this manuscript. Dr. Potenza has received financial support or compensation for the following: Dr. Potenza has consulted for and advised Somaxon, Boehringer Ingelheim, Lundbeck, Ironwood, Shire, INSYS, and RiverMend Health; has received research support from the National Institutes of Health, Veteran's Administration, Mohegan Sun Casino, the National Center for Responsible Gaming, and Forest Laboratories, Ortho-McNeil, Oy-Control/Biotie, Glaxo-SmithKline, Pfizer, and Psyadon pharmaceuticals; has participated in surveys, mailings, or telephone consultations related to drug addiction, impulse control disorders, or other health topics; has consulted for gambling entities, law offices, and the federal public defender's office in issues related to impulse control disorders; provides clinical care in the Connecticut Department of Mental Health and Addiction Services Problem Gambling Services Program; has performed grant reviews for the National Institutes of Health and other agencies; has edited or guest-edited journal sections; has given academic lectures in grand rounds, CME events, and other clinical or scientific venues; and has generated books or book chapters for publishers of mental health texts.

REFERENCES

American Psychiatric Association, 2000. Diagnostic and Statistical Manual of Mental Disorders (DSM-IV-R). American Psychiatric Press, Inc., Washington, DC.

American Psychiatric Association, 2013. Diagnostic and Statistical Manual of Mental Disorders (DSM-5). American Psychiatric Press, Inc., Washington, DC.

Balodis, I.M., Potenza, M.N., 2015. Anticipatory reward processing in addicted populations: a focus on the monetary incentive delay task. Biol. Psychiatry 77 (5), 434–444.

Balodis, I.M., Kober, H., Worhunsky, P.D., Stevens, M.C., Pearlson, G.D., Potenza, M.N., 2012. Diminished frontostriatal activity during processing of monetary rewards and losses in pathological gambling. Biol. Psychiatry 71 (8), 749–757.

Bartley, C.A., Bloch, M.H., 2013. Meta-analysis: pharmacological treatment of pathological gambling. Expert. Rev. Neurother. 13 (8), 887–894.

Billeux, J., Chanal, J., Khazaal, Y., Rochat, L., Gay, P., Zullino, D., Van der Linden, M., 2011. Psychological predictors of problematic involvement in massively multiplayer online role-playing games: illustration in a sample of male cybercafé players. Psychopathology 44, 165–171.

Boileau, I., Payer, D., Chugani, B., Lobo, D., Behzadi, A., Rusjan, P., Houle, S., Wilson, A., Warsh, J., Kish, S., Zack, M., 2013. The D2/3 dopamine receptor in pathological gambling: a positron emission tomography study with [11C]-(+)-propyl-hexahydro-naphtho-oxazin and [11C]raclopride. Addiction 108 (5), 953–963.

Boileau, I., Payer, D., Chugani, B., Lobo, D.S., Houle, S., Wilson, A.A., Warsh, J., Kish, S., Zack, M., 2014. *In vivo* evidence for greater amphetamine-induced dopamine release in pathological gambling: a positron emission tomography study with [(11)C-(+)-PHNO. Mol. Psychiatry 19 (12), 1305–1313.

Brevers, D., Cleeremans, A., Goudriaan, A.E., Bechara, A., Kornreich, C., Verbanck, P., Noël, X., 2013. Decision making under ambiguity but not under risk is related to problem gambling severity. Psychiatry Res. 200, 568–574.

Brevers, D., Bechara, A., Hermoye, L., Divano, L., Kornreich, C., Verbanck, P., Noel, X., 2015. Comfort for uncertainty in pathological gamblers: a fMRI study. Behav. Brain Res. 278C, 262–270.

Bullock, S., Potenza, M.N., 2012. Neuropsychology and treatment. Psychopharmacology 1, 67–85.

Camardese, G., De Risio, L., Di Nicola, M., Pizi, G., Janiri, L., 2012. A role for pharmacotherapy in the treatment of "Internet addiction". Clin. Neuropharmacol. 35 (6), 283–289.

Campbell-Meiklejohn, D., Wakeley, J., Herbert, V., Cook, J., Scollo, P., Ray, M.K., Selvaraj, S., Passingham, R.E., Cowen, P., Rogers, R.D., 2011. Serotonin and dopamine play complimentary roles in gambling to recover losses. Neuropsychopharmacology 36, 402–410.

Choi, J.S., Shin, Y.C., Jung, W.H., Jang, J.H., Kang, D.H., Choi, C.H., Choi, S.W., Lee, J.Y., Hwang, J.Y., Kwon, J.S., 2012. Altered brain activity during reward anticipation in pathological gambling and obsessive-compulsive disorder. PLoS ONE 7 (9), e45938.

Choi, J.S., Park, S.M., Lee, J., Hwang, J.Y., Jung, H.Y., Choi, S.W., Kim, D.J., Oh, S., Lee, J.Y., 2013. Resting-state beta and gamma activity in Internet addiction. Int. J. Psychophysiol. 89, 328–333.

Choi, S.W., Kim, H.S., Kim, G.Y., Jeon, Y., Park, S.M., Lee, J.Y., Jung, H.Y., Sohn, B.K., Choi, J.S., Kim, D.J., 2014. Similarities and differences among Internet gaming disorder, gambling disorder and alcohol use disorder: a focus on impulsivity and compulsivity. J. Behav. Addict. 3 (4), 246–253.

Clark, L., Stokes, P.R., Wu, K., Michalczuk, R., Benecke, A., Watson, B.J., Egerton, A., Piccini, P., Nutt, D.J., Bowden-Jones, H., Lingford-Hughes, A.R., 2012. Striatal dopamine D_2/D_3 receptor binding in pathological gambling is correlated with mood-related impulsivity. NeuroImage 63, 40–46.

Cowlishaw, S., Merkouris, S., Dowling, N., Anderson, C., Jackson, A., Thomas, S., 2012. Psychological therapies for pathological and problem gambling. Cochrane Database Syst. Rev. 11. 1–91. http://dx.doi.org/10.1002/14651858.CD008937.pub2.

Dong, G., Potenza, M.N., 2014. A cognitive-behavioral model of Internet gaming disorder: theoretical underpinnings and clinical implications. J. Psychiatr. Res. 58, 7–11.

Dong, G., Lu, Q., Zhou, H., Zhao, X., 2010. Impulse inhibition in people with Internet addiction disorder: electrophysiological evidence from a Go/NoGo study. Neurosci. Lett. 485, 138–142.

Dong, G., Zhou, H., Zhao, X., 2011. Male Internet addicts show impaired executive control ability: evidence from a color-word Stroop task. Neurosci. Lett. 499, 114–118.

Dong, G., Lin, X., Potenza, M.N., 2015. Decreased functional connectivity in an executive control network is related to impaired executive function in Internet gaming disorder. Prog. Neuropsychopharmacol. Biol. Psychiatry 57, 76–85.

Dymond, S., Lawrence, N.S., Dunkley, B.T., Yuen, K.S.L., Hinton, E.C., Dixon, M.R., Cox, W.M., Hoon, A.E., Munnelly, A., Muthukumaraswamy, S.D., Singh, K.D., 2014. Almost winning: induced MEG theta power in insula and orbitofrontal cortex increases during gambling near-misses and is associated with BOLD signal and gambling severity. NeuroImage 91, 210–219.

Grant, J.E., Odlaug, B.L., Chamberlain, S.R., Potenza, M.N., Schreiber, L.R.N., Donahue, C.B., Kim, S.W., 2014. A randomized, placebo-controlled trial of N-acetyl cysteine plus imaginal desensitization for nicotine-dependent pathological gamblers. J. Clin. Psychiatry 75, 39–45.

Gyollai, Á., Griffiths, M.D., Barta, C., Vereczkei, A., Urbán, R., Kun, B., Kökönyei, G., Székely, A., Sasvári-Székely, M., Blum, K., Demetrovics, Z., 2014. The genetics of problem and pathological gambling: a systematic review. Curr. Pharm. Des. 20, 3993–3999.

Han, D.H., Lee, Y.S., Yang, K.C., Kim, E.Y., Lyoo, I.K., Renshaw, P.F., 2007. Dopamine genes and reward dependence in adolescents with excessive internet video game play. J. Addict. Med. 1 (3), 133–138.

Han, D.H., Kim, S.M., Lee, Y.S., Renshaw, P.F., 2012. The effect of family therapy on the changes in the severity of on-line game play and brain activity in adolescents with on-line game addiction. Psychiatry Res. 20 (22), 126–131.

Han, D.H., Lee, Y.S., Shi, X., Renshaw, P.F., 2014. Proton magnetic resonance spectroscopy (MRS) in on-line game addiction. J. Psychiatr. Res. 58, 63–68.

Han, J.W., Han, D.H., Bolo, N., Kim, B.N., Renshaw, P.F., 2015. Differences in functional connectivity between alcohol dependence and internet gaming disorder. Addict. Behav. 41, 12–19.

Hewig, J., Kretschmer, N., Trippe, R.H., Hecht, H., Coles, M.G.H., Holroyd, C.B., Miltner, W.H.R., 2010. Hypersensitivity to reward in problem gamblers. Biol. Psychiatry 67, 781–783.

Hong, S.B., Kim, J.W., Choi, E.J., Kim, H.H., Suh, J.E., Kim, C.D., Klauser, P., Whittle, S., Yucel, M., Pantelis, C., Yi, S.H., 2013. Reduced orbitofrontal cortical thickness in male adolescents with internet addiction. Behav. Brain Funct. 9 (11), 9081–9089.

Hou, H., Jia, S., Hu, S., Fan, R., Sun, W., Sun, T., Zhang, H., 2012. Reduced striatal dopamine transporters in people with internet addiction disorder. J. Biomed. Biotechnol. 2012, 854524.

Jousta, J., Saunavaara, J., Parkkola, R., Niemelä, S., Kaasinen, V., 2011. Extensive abnormality of brain white matter integrity in pathological gambling. Psychiatry Res. 194, 340–346.

Jousta, J., Johansson, J., Niemelä, S., Ollikainen, A., Hirvonen, M.K., Piepponen, P., Arponen, E., Alho, H., Voon, V., Rinne, J.O., Hietala, J., Kaasinen, V., 2012. Mesolimbic dopamine release is linked to symptom severity in pathological gambling. NeuroImage 60, 1992–1999.

Kim, H., Kim, Y.K., Gwak, A.R., Lim, J.A., Lee, J.Y., Jung, H.Y., Sohn, B.K., Choi, S.W., Kim, D.J., Choi, J.S., 2015. Resting-state regional homogeneity as a biological marker for patients with Internet gaming disorder: a comparison with patients with alcohol use disorder and healthy controls. Prog. Neuropsychopharmacol. Biol. Psychiatry 14 (60), 104–111.

Ko, C.H., Hsieh, T.J., Wang, P.W., Lin, W.C., Yen, C.F., Chen, C.S., Yen, J.Y., 2015. Altered gray matter density and disrupted functional connectivity of the amygdala in adults with Internet gaming disorder. Prog. Neuropsychopharmacol. Biol. Psychiatry 57, 185–192.

Kreussel, L., Hewig, J., Kretschmer, N., Hecht, H., Coles, M.G.H., Miltner, W.H.R., 2013. How bad was it? Differences in the time course of sensitivity to the magnitude of loss in problem gamblers and controls. Behav. Brain Res. 247, 140–145.

Lee, Y.S., Han, D., Yang, K.C., Daniels, M.A., Na, C., Kee, B.S., Renshaw, P.F., 2008. Depression like characteristics of 5HTTLPR polymorphism and temperament in excessive internet users. J. Affect. Disord. 1–2, 165–169.

Lee, J., Hwang, J.Y., Park, S.M., Jung, H.Y., Choi, S.W., Kim, D.J., Lee, J.Y., Choi, J.S., 2014. Differential resting-state EEG patterns associated with comorbid depression in Internet addiction. Prog. Neuropsychopharmacol. Biol. Psychiatry 50, 21–26.

Leeman, R.F., Potenza, M.N., 2012. Similarities and differences between pathological gambling and substance use disorders: a focus on impulsivity and compulsivity. Psychopharmacology 219 (2), 469–490.

Leeman, R.F., Potenza, M.N., 2013. A targeted review of the neurobiology and genetics of behavioral addictions: an emerging area of research. Can. J. Psychiatry 58 (5), 260–273.

Li, Y., Sescousse, G., Dreher, J.C., 2014. Endogenous cortisol levels are associated with an imbalanced striatal sensitivity to monetary versus non-monetary cues in pathological gamblers. Front. Behav. Neurosci. 8, 83.

Li, W., Li, Y., Yang, W., Zhang, Q., Wei, D., Li, W., Hitchman, G., Qiu, J., 2015. Brain structure and functional connectivity associated with individual difference in Internet tendency in healthy young adults. Neuropsychologia 70, 134–144.

Lin, F., Zhou, Y., Du, Y., Qin, L., Zhao, Z., Xu, J., Lei, H., 2012. Abnormal white matter integrity in adolescents with internet addiction disorder: a tract-based spatial statistics study. PLoS One 7 (1), e30253.

Lin, X., Dong, G., Wang, Q., Du, X., 2015. Abnormal gray matter and white matter volume in 'Internet gaming addicts'. Addict. Behav. 40, 137–143.

Lind, P.A., Zhu, G., Montgomery, G.W., Madden, P.A., Heath, A.C., Martin, N.G., Slutske, W.S., 2012. Genome-wide association study of a quantitative disordered gambling trait. Addict. Biol. 18 (3), 511–522.

Linnet, J., Mouridsen, K., Peterson, E., Møller, A., Doudet, D.J., Gjedde, A., 2012. Striatal dopamine release codes uncertainty in pathological gambling. Psychiatry Res. 204, 55–60.

Littel, M., van den Berg, I., Luijten, M., van Rooij, A.J., Keemink, L., Franken, I.H.A., 2012. Error processing and response inhibition in excessive computer game players: an event-related potential study. Addict. Biol. 17, 934–947.

Liu, Q.X., Fang, X.Y., Yan, N., Zhou, Z.K., Yuan, X.J., Lan, J., Liu, C.Y., 2015. Multi-family group therapy for adolescent Internet addiction: exploring the underlying mechanisms. Addict. Behav. 42, 1–8.

Lole, L., Gonsalvez, C.J., Barry, R.J., 2015. Reward and punishment hyposensitivity in problem gamblers: a study of event-related potentials using a principal components analysis. Clin. Neurophysiol. 126 (7), 1295–1309.

Meng, Y.J., Deng, W., Wang, H.Y., Guo, W.J., Li, T., Lam, C., Lin, X., 2014. Reward pathway dysfunction in gambling disorder: a meta-analysis of functional magnetic resonance imaging studies. Behav. Brain Res. 275, 243–251.

Miedl, S.F., Fehr, T., Herrmann, M., Meyer, G., 2014. Risk assessment and reward processing in problem gambling investigated by event-related potentials a fMRI-constrained source analysis. BMC Psychiatry 14, 229–240.

Miedl, S.F., Wiswede, D., Marco-Pollares, J., Ye, Z., Fehr, T., Herrmann, M., Munte, T.F., 2015. The neural basis of impulsive discounting in pathological gamblers. Brain Imaging Behav. epub ahead of print. http://link.springer.com/article/10.1007%2Fs11682-015-9352-1 (accessed 28.09.15.).

Noël, X., Brevers, D., Bechara, A., 2013. A neurocognitive approach to understanding the neurobiology of addiction. Curr. Opin. Neurobiol. 23 (4), 632–638.

Oberg, S.A.K., Christie, G.J., Tata, M.S., 2011. Problem gamblers exhibit reward hypersensitivity in medial frontal cortex during gambling. Neuropsychologia 49, 3768–3775.

Petry, N.M., O'Brien, C.P., 2013. Internet gaming disorder and the DSM-5. Addiction 108 (7), 1186–1187.

Potenza, M.N., 2006. Should addictive disorders include non-substance-related conditions? Addiction 101, 142–151.

Potenza, M.N., 2013. Neurobiology of gambling behaviors. Curr. Opin. Neurobiol. 23 (4), 660–667.

Potenza, M.N., 2014. Non-substance addictive behaviors in the context of DSM-5. Addict. Behav. 39 (1), 1–2.

Potenza, M.N., Balodis, I.M., Franco, C.A., Bullock, S., Xu, J., Grant, J.E., 2013. Neurobiological considerations in understanding behavioral treatments for pathological gambling. Psychol. Addict. Behav. 27, 380–392.

Przepiorka, A.M., Blachnio, A., Miziak, B., Czuczwar, S.J., 2014. Clinical approaches to treatment of Internet addiction. Pharmacol. Rep. 66 (2), 187–191.

Rahman, A.S., Xu, J., Potenza, M.N., 2014. Hippocampal and amygdalar volumetric differences in pathological gambling: a preliminary study of the associations with the behavioral inhibition system. Neuropsychopharmacology 39, 738–745.

Rash, C.J., Petry, N.M., 2014. Psychological treatments for gambling disorder. Psychol. Res. Behav. Manag. 7, 285–295.

Rehbein, F., Mößle, T., 2013. Video game and Internet addiction: is there a need for differentiation? SUCHT 59 (3), 129–142.

Richmond-Rakerd, L.S., Slutske, W.S., Heath, A.C., Martin, N.G., 2014. Genetic and environmental influences on the ages of drinking and gambling initiation: evidence for distinct aetiologies and sex differences. Addiction 109 (2), 323–331.

Scherrer, J.F., Xian, H., Slutske, W.S., Eisen, S.A., Potenza, M.N., 2015. Associations between obsessive-compulsive classes and pathological gambling in a national cohort of male twins. JAMA Psychiatry 72 (4), 342–349.

Sescousse, G., Barbalat, G., Domenech, P., Dreher, J.C., 2013. Imbalance in the sensitivity to different types of rewards in pathological gambling. Brain 136 (6), 2527–2538.

Slutske, W.S., Richmond-Rakerd, L.S., 2014. A closer look at the evidence for sex differences in the genetic and environmental influences on gambling in the National Longitudinal Study of Adolescent health: from disordered to ordered gambling. Addiction 109 (1), 120–127.

Slutske, W.S., Deutsch, A.R., Richmond-Rakerd, L.S., Chernyavskiy, P., Statham, D.J., Martin, N.G., 2014. Test of a potential causal influence of earlier age of gambling initiation on gambling involvement and disorder: a multilevel discordant twin design. Psychol. Addict. Behav. 28 (4), 1177–1189.

Spada, M.M., 2014. An overview of problematic Internet use. Addict. Behav. 39, 3–6.

Tian, M., Chen, Q., Zhang, Y., Du, F., Hou, H., Chao, F., Zhang, H., 2014. PET imaging reveals brain functional changes in internet gaming disorder. Nucl. Med. Mol. Imaging 41 (7), 1388–1397.

Tschernegg, M., Crone, J.S., Eigenberger, T., Schwartenbeck, P., Fauth-Bühler, M., Lemènager, T., Mann, K., Thon, N., Wurst, F.M., Kronbichler, M., 2014. Abnormalities of functional brain networks in pathological gambling: a graph-theoretical approach. Front. Hum. Neurosci. 7, 625. http://journal.frontiersin.org/article/10.3389/fnhum.2013.00625/abstract; (accessed 28.09.15.).

Twohig, M.P., Crosby, J.M., 2010. Acceptance and commitment therapy as a treatment for problematic Internet pornography viewing. Behav. Ther. 41, 285–295.

Van Holst, R.J., vna den Brink, W., Veltman, D.J., Goudriaan, A.E., 2010. Why gamblers fail to win: a review of cognitive and neuroimaging findings in pathological gambling. Neurosci. Biobehav. Rev. 34, 87–107.

Van Holst, R.J., de Ruiter, M.B., van den Brink, W., Veltman, D.J., Goudriaan, A.E., 2012. A voxel-based morphometry study comparing problem gamblers, alcohol abusers, and healthy controls. Drug Alcohol Depend. 124, 142–148.

Van Rooij, A.J., Prause, N., 2014. A critical review of "Internet addiction" criteria with suggestions for the future. J. Behav. Addict. 3 (4), 203–213.

Wang, C.W., Ho, R.T., Chan, C.L., Tse, S., 2015. Exploring personality characteristics of Chinese adolescents with internet-related addictive behaviors: trait differences for gaming addiction and social networking addiction. Addict. Behav. 42, 32–35.

Wee, C.Y., Zhao, Z., Yap, P.T., Wu, G., Shi, F., Price, T., Du, Y., Xu, J., Zhou, Y., Shen, D., 2014. Disrupted brain functional network in internet addiction disorder: a resting-state functional magnetic resonance imaging study. PLoS One 9 (9), e107306.

Wilson, D., da Silva Lobo, D.S., Tavares, H., Gentil, V., Vallada, H., 2013. Family-based association analysis of serotonin genes in pathological gambling disorder: evidence of vulnerability risk in the 5HT-2A receptor gene. J. Mol. Neurosci. 49, 550–553.

Winkler, A., Dörsing, B., Rief, W., Shen, Y., Glombiewski, J.A., 2013. Treatment of internet addiction: a meta analysis. Clin. Psychol. Rev. 33, 317–329.

Wölfling, K., Beutel, M.E., Dreier, M., Muller, K.W., 2014. Treatment outcomes in patients with internet addiction: a clinical pilot study on the effects of a cognitive-behavioral therapy program. Biomed. Res. Int. 2014, 425924. http://www.hindawi.com/journals/bmri/2014/425924/(accessed 28.09.15.).

Worhunsky, P.D., Malison, R.T., Rogers, R.D., Potenza, M.N., 2014. Altered neural correlates of reward and loss processing during simulated slot-machine fMRI in a pathological gambling and cocaine dependence. Drug Alcohol Depend. 145, 77–86.

Xian, H., Giddens, J.L., Scherrer, J.F., Eisen, S.A., Potenza, M.N., 2014. Environmental factors selectively impact co-occurrence of problem/pathological gambling with specific drug-use disorders in male twins. Addiction 109 (4), 635–644.

Yan, W.S., Li, Y.H., Xiao, L., Zhu, N., Bechara, A., Sui, N., 2014. Working memory and affective decision-making in addiction: a neurocognitive comparison between heroin addicts, pathological gamblers and healthy controls. Drug Alcohol Depend. 134, 194–200.

Yau, Y.H.C., Crowley, M.J., Mayes, L.C., Potenza, M.N., 2012. Are internet use and video-game playing addictive behaviors? Biological, clinical and public health implications for youths and adults. Minerva Psichiatr. 53, 153–170.

Yau, Y.H.C., Yip, S.W., Potenza, M.N., 2014. Principles of Addiction Medicine, fifth ed. Lippincott Williams & Wilkins, Philadelphia, PA.

Yau, Y.H.C., Potenza, M.N., 2015. Gambling disorder and other behavioral addictions: recognition and treatment. Harv. Rev. Psychiatr. 23 (2), 134–146.

Yip, S.W., Potenza, M., 2014. Treatment of gambling disorders. Curr. Treat. Options Neurol. 1 (2), 189–203.

Yip, S.W., Lacadie, C., Xu, J., Worhunsky, P.D., Fulbright, R.K., Constable, R.T., Potenza, M.N., 2011. Reduced genual corpus callosal white matter integrity in pathological gambling and its relationship to alcohol abuse or dependence. World J. Biol. Psychiatry 14, 129–138.

Young, K.S., 2013. Treatment outcomes using CBT-IA with Internet-addicted patients. J. Behav. Addict. 2 (4), 209–215.

CHAPTER

Neural systems implicated in obesity as an addictive disorder: from biological to behavioral mechanisms

17

Erica M. Schulte*, Sonja Yokum[†], Marc N. Potenza[‡,§,¶], Ashley N. Gearhardt*[,1]

**Department of Psychology, University of Michigan, Ann Arbor, MI, USA*

[†]Oregon Research Institute, Eugene, OR, USA

[‡]Department of Psychiatry, Yale University School of Medicine, New Haven, CT, USA

[§]Department of Neurobiology, Child Study Center, Yale University School of Medicine, New Haven, CT, USA

[¶]CASAColumbia, Yale University School of Medicine, New Haven, CT, USA

[1]Corresponding author: Tel.: (734)647-3920; Fax: (734)763-7744, e-mail address: agearhar@umich.edu

Abstract

Contributing factors to obesity have been identified, yet prevention and treatment efforts have had limited long-term success. It has recently been suggested that some individuals may experience an addictive-like response to certain foods, such as losing control over consumption and continued consumption despite negative consequences. In support, shared biological and behavioral features seem to exist between "food addiction" and traditional substance-use disorders. "Food addiction" may be another important contributor to obesity. The current chapter reviews existing literature regarding neural systems implicated similarly in obesity and addiction, discusses unique considerations for addictive-like eating, and proposes directions for future research regarding "food addiction" as an emerging construct for addiction medicine.

Keywords

Obesity, Addiction, Substance dependence, Food addiction, Reward

1 INTRODUCTION

As obesity rates continue to rise, increased attention has been given to mechanisms associated with overeating behaviors. It has been proposed that an addictive-like

Progress in Brain Research, Volume 223, ISSN 0079-6123, http://dx.doi.org/10.1016/bs.pbr.2015.07.011

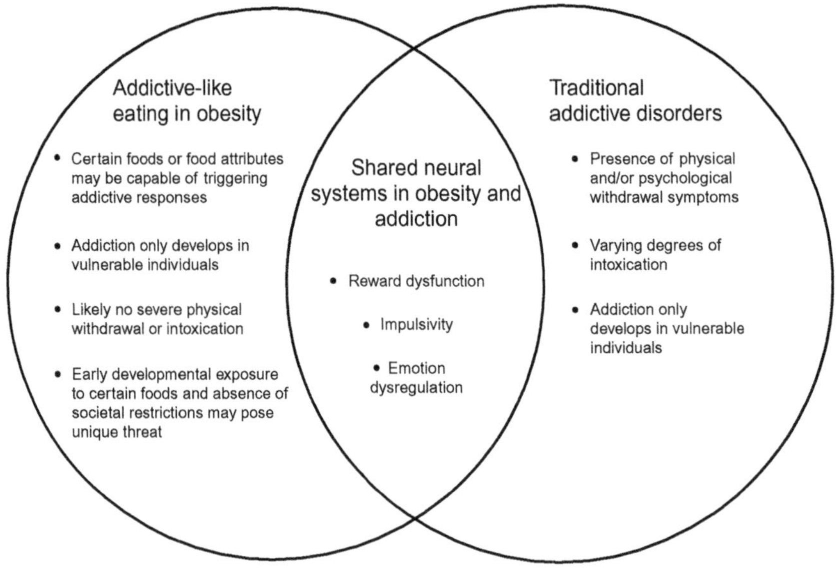

FIGURE 1

Shared neural systems in obesity and addiction and unique features for addictive-like eating and traditional addictive disorders.

process may underlie problematic eating for some individuals (Gearhardt et al., 2009b), although this point has been debated (Ziauddeen et al., 2012). Significant behavioral overlap exists between obesity and addictive disorders (particularly for some groups like those with Binge Eating Disorder; Gearhardt et al., 2011b), such as a loss of control over consumption and continued consumption despite negative consequences (Gearhardt et al., 2009a). The following chapter will review literature regarding shared neural systems in traditional addictive disorders and obesity (see Fig. 1), discuss differentiating factors of addictive-like eating, and offer essential next steps in neuroimaging research for "food addiction."

2 SHARED NEURAL SYSTEMS: REWARD DYSFUNCTION

Dopamine (DA) is a main catecholamine neurotransmitter implicated in reinforcement- and reward-related processes, such as motivation and craving. Food and drugs of abuse both increase DA signaling in the mesolimbic dopaminergic system (Heinz et al., 2004; Wang et al., 2002). Consumption of high-sugar or high-fat food results in DA release in the striatum in animals (Avena et al., 2009) and humans, with the amount released correlating with meal pleasantness ratings (Small et al., 2003) and energy density (Ferreira et al., 2012). In humans, consumption of palatable food is associated with increased activation in the reward-related circuitry, including the dorsal- and ventral striatum and orbitofrontal cortex (OFC; Stice et al., 2013a).

Likewise, all addictive drugs lead to DA release in the striatum and associated mesolimbic regions (Kalivas and O'Brien, 2008).

Utilizing functional magnetic resonance imaging (fMRI), previous research has observed parallels in neural responsivity to food/drug cues and intake between obesity and substance-use disorders (Tang et al., 2012). Obese versus lean humans show greater responsivity of brain regions associated with reward (e.g., striatum, amygdala, OFC) and attention (e.g., anterior cingulate cortex, ACC) to pictures of high-fat/sugar foods (versus control stimuli; Martin et al., 2010; Stice et al., 2010b) and to pictorial cues that signal impending palatable food receipt (Ng et al., 2011; Stice et al., 2008). Similarly, humans with, versus without, substance-use disorders show greater activation of reward regions (e.g., VTA, amygdala) and attention regions (e.g., ACC) to drug-related cues (Due et al., 2002; Myrick et al., 2004).

One distinct feature of addictive disorders is the transition from initially consuming drugs of abuse for their reinforcing properties to compulsive, habitual self-administration (Everitt and Robbins, 2005). Consuming drugs of abuse or highly palatable foods for hedonic effects (liking) activates the ventral striatum, whereas habitual, compulsive self-administration (wanting) appears to differentially implicate dorsal striatal regions (Everitt and Robbins, 2005; Volkow et al., 2006).

One proposed mechanism underlying the transition from "liking" to "wanting" is incentive sensitization (Robinson and Berridge, 1993). This explanation suggests that chronic consumption of addictive substances or highly palatable foods may result in sensitization of the DA system and increased salience of drug- and food-specific cues for some individuals (Berridge, 2009). In support, animal experiments indicate that firing of striatal and ventral pallidal DA neurons initially occurs in response to receipt of a novel palatable food, but that after repeated pairings of palatable food intake and cues that signal impending receipt of that food, DA neurons begin to fire in response to reward-predictive cues and no longer fire in response to food receipt (Tobler et al., 2005). In humans, midbrain and medial OFC activity in response to milkshake receipt positively correlated with subsequent *ad libitum* milkshake consumption, and BOLD response in the ventral striatum during exposure to food images positively correlated with later snack consumption (Lawrence et al., 2012; Nolan-Poupart et al., 2013). Healthy weight adolescents who were eating beyond basal metabolic needs (per objective measures) versus those who were not showed greater BOLD response during cues predicting impending palatable food receipt in regions that encode reward (striatum), salience (precuneus), and visual processing and attention (visual and anterior cingulate cortices; Burger and Stice, 2013). These latter data suggest that overeating, even if it has not yet resulted in excess weight gain, may be accompanied by elevated responsivity to food-predictive cues in reward and attentional regions. For persons susceptible to these neuroplastic changes, sensitization to relevant cues can trigger "wanting" and potentially lead to addictive-like consumption.

Reactivity in brain regions associated with reward appraisal like the striatum, amygdala, and OFC may be indicative of the incentive salience for drug and food cues, where increased salience is associated with propensities for obesity and unsuccessful abstinence in addictive disorders (Tang et al., 2012). In support, elevated striatal response to monetary reward related prospectively to substance-use onset over a 1-year follow-up (Stice et al., 2013b) and elevated responsivity of reward regions (striatum, amygdala, OFC) to palatable food images (Demos et al., 2012), palatable food commercials (Yokum et al., 2014), cues that predict palatable food image presentation (Yokum et al., 2011), and palatable food receipt (Geha et al., 2013) related prospectively to future weight gain. Further, individual differences observed in animal models provide insight to certain characteristics that may predict who will sensitize to food cues. Individuals who exhibit greater motivation to engage with cues that predict a drug or food reward, an indication of increased incentive salience (so-called sign trackers), than to elements of reward receipt like the location of reward delivery (so-called goal trackers) appear to be at greater risk for sensitization (Flagel et al., 2009). Thus, while sensitization appears to contribute to continued overconsumption of palatable foods, greater engagement with food-predictive cues may predict which individuals will sensitize.

It has also been hypothesized that a reward deficiency may predispose some individuals to develop compulsive overeating or drug-taking behavior (Blum et al., 2014). In some forms of obesity and addictive disorders, individuals may be motivated to consume highly palatable food or drugs of abuse to compensate for diminished DA receptor availability (Koob and Le Moal, 2001). Consistent with the reward deficiency theory, adults with versus without alcohol, cocaine, heroin, and methamphetamine dependence show reduced striatal D2-like receptor availability and sensitivity (Volkow et al., 2001; Wang et al., 1997) and lower D2-like receptor density may increase risk for relapse after treatment (Heinz et al., 2004). Further, low striatal D2-like receptor availability in primates relates prospectively to increased future drug self-administration (Nader et al., 2006). Likewise, obese versus lean adults show lower striatal DA D2-like receptor availability (de Weijer et al., 2011; Volkow et al., 2008), although two other studies found no significant group differences (Eisenstein et al., 2013; Haltia et al., 2007), with differences across studies possibly attributable to small sample sizes, differences in degrees of obesity, different radioligands used, or other factors. Obese versus lean adults show lower capacity of nigrostriatal neurons to synthesize DA (Wilcox et al., 2009) and less striatal responsivity to tastes of high-fat/sugar beverages (Babbs et al., 2013; Stice et al., 2008). Obese versus lean rats likewise have lower basal DA levels and D2-like receptor availability and less *ex vivo* DA release in response to electrical stimulation in the nucleus accumbens and dorsal striatum (Geiger et al., 2009; Thanos et al., 2008).

It is unknown whether hypoactivation in reward circuitry in response to acute administration may be a cause or consequence in either addictive disorders or obesity as prolonged overconsumption of drugs and rewarding food may decrease striatal DA availability. Animal experiments show that regular substance use reduces striatal D2-like receptors (Nader et al., 2006) and sensitivity of reward circuitry (Kenny et al., 2006), and humans with cocaine abuse show blunted DA release in response to stimulant drugs relative to nonaddicted comparison subjects (Volkow et al., 2005) and tolerance to the euphoric effects of cocaine (O'Brien et al., 2006). Chronic cocaine use has been associated with downregulated dopaminergic responses to both cocaine and food cues (Tomasi et al., 2015). These data imply that substance use contributes to the downregulated reward circuitry observed in the cross-sectional studies. Similarly, prospective human (Stice et al., 2010a) and experimental animal studies (Geiger et al., 2009; Johnson and Kenny, 2010; Thanos et al., 2008) indicate that overeating may contribute to reward region hyporesponsivity during food consumption. Further, studies have suggested that individuals vulnerable to substance use (Stice et al., 2013b) and obesity (Stice et al., 2011; Verbeken et al., 2012) may initially exhibit hyperresponsiveness in reward-related brain regions to rewards in general. This hyperresponsiveness may increase motivation to seek out highly rewarding, palatable foods and these individuals may appear to be hyporesponsive to reward after neuroplastic changes associated with chronic overeating behavior have occurred.

3 SHARED NEURAL SYSTEMS: IMPULSIVITY

Another domain implicated in both obesity and addictive disorders is an executive-control deficiency, often evident by impulsive behavior. Obese individuals and persons with addictions appear to favor short-term rewards of food or drug instead of long-term health benefits (e.g., weight reduction in obesity; Mole et al., 2014). In decision-making tasks, obese women not only make more impulsive decisions than healthy women (Davis et al., 2010), but also exhibit decreased activation in executive-control brain regions (e.g., middle frontal gyri, medial prefrontal cortex [PFC]) during decision-making processes (Kishinevsky et al., 2012; Stoeckel et al., 2013). Similar patterns of impulsive decision-making coupled with diminished activation in executive-control regions have also been observed in addictive disorders (MacKillop et al., 2011), and decreased inhibitory-control activation may be predictive of relapse (Paulus et al., 2005). Similarly, Kishinevsky and colleagues (2012) observed that diminished activity in executive-control structures related to future weight gain.

One proposed explanation for this potentially maladaptive decision-making process is abnormal striato-cortical connectivity similarly observed in addictive disorders (Hanlon et al., 2011; Liu et al., 2009) and obesity (Garcia-Garcia et al., 2013; Tomasi and Volkow, 2013). In this dysfunctional connection, brain structures implicated in reward appraisal (e.g., OFC, ventromedial PFC) may be capable of overwhelming inhibitory-control regions (e.g., dorsolateral PFC) to result in impulsive decision-making (Weygandt et al., 2013; Zhang et al., 2015). Additionally, recent studies observed lower brain volumes in the OFC in obese women (Shott et al., 2014) and a lack of synchronicity between the OFC and PFC in fasted obese men (Zhang et al., 2015), providing further evidence that diminished inhibitory control may lead to overeating in obesity (Zhang et al., 2015) in a manner akin to compulsive drug-taking in addictive disorders (Ma et al., 2010). For some obese individuals, food cues may activate regions associated with reward salience, and the executive-control neural system may be inefficient at suppressing the drive to seek certain foods. Though neuroimaging research should continue to examine deficits in striato-cortical connectivity in obesity, it appears likely that deficits in executive-control neural circuitry may similarly contribute to impulsive decision-making in obesity and addiction.

4 SHARED NEURAL SYSTEMS: EMOTION DYSREGULATION

In both addiction and obesity, neural systems underlying emotion-regulation processes seem to be impaired. Strong emotional states frequently precipitate drug use and overeating behavior, which may suggest that some individuals utilize addictive substances and highly palatable foods to compensate for deficient emotion-regulation processes (Singh, 2014; Sinha and Jastreboff, 2013). Notably, existing research has observed that humans typically consume foods high in fat and/or refined carbohydrates in response to emotional states like stress or negative affect, which may be particularly relevant for the development of obesity (Morris et al., 2014).

Similar to addictive disorders, negative affect appears to implicate neural systems associated with increased craving and compulsive consumption in obesity (Sinha and Jastreboff, 2013). Jastreboff and colleagues (2013) found that obese, compared to lean, individuals exhibited greater activity in striatal regions when exposed to stress and highly palatable food cues, relative to neutral-relaxing cues, and increased activation of the dorsal striatum in response to stress and food cues was related to stronger food cravings. This suggests that emotional states may activate brain regions associated with habitual behavior, like the dorsal striatum, which motivates certain obese individuals to consume highly palatable foods (Jastreboff et al., 2013). Coupled with increased activation in reward motivation structures, Tryon et al. (2013) observed that chronic stress was associated with less prefrontal activation in response to high-calorie food cues, and this pattern of activation related to greater consumption of high-calorie foods. Collectively, it appears that emotional states may be related to neural activation subserving increased craving and diminished inhibitory control in obesity, which is also observed in addictive disorders (Sinha, 2008).

5 SUMMARY OF SHARED NEURAL SYSTEMS

Overlapping neural systems appear to be implicated in both obesity and addictive disorders, including reward dysfunction, executive-control deficiencies, and emotion regulation. The existing neuroimaging data suggests that addictive-like mechanisms may contribute to obesity for some individuals. Thus, exploration of the "food addiction" construct may be clinically useful for understanding overeating behavior and informing intervention approaches for certain individuals with obesity.

6 DIFFERENCES BETWEEN OBESITY AND ADDICTIVE DISORDERS

While existing neuroimaging studies have observed similarities between obesity and substance-use disorders, important differences exist between "food addiction" and traditional addictive disorders. Notably, food is necessary for survival. However, many highly palatable foods are not in their natural state and have instead been processed with added amounts of potentially rewarding ingredients like fat and refined carbohydrates (Gearhardt et al., 2011a). Similar to the word "drug" which includes addictive (e.g., cocaine) and nonaddictive (e.g., aspirin) substances, future research is warranted to examine whether foods in a natural state (e.g., banana) are equally implicated in problematic, addictive-like eating behavior as highly processed foods (e.g., pizza).

Similarly, the current conceptualization of "food addiction" differs from traditional addictive disorders because an addictive agent in certain foods and the "dose" that may increase certain foods' addictive potential has not been investigated (Ziauddeen and Fletcher, 2013). For drugs of abuse, an addictive agent has been defined (e.g., ethanol in alcohol, nicotine in cigarettes) and high concentrations of those ingredients are linked to an increased addictive potential (Henningfield and Keenan, 1993). In contrast, although highly processed foods have been hypothesized to be most likely implicated in addictive-like eating due to high levels of fat and/or refined carbohydrates (Gearhardt et al., 2011a), it has not been examined whether specific food attributes (e.g., sugar content) may be capable of triggering an addictive-like response in certain individuals; however, there are preliminary data suggesting that sugar more effectively recruits reward and gustatory regions compared to fat (Stice et al., 2013a). Additionally, no previous research has evaluated whether a particular "dose" or quantity of the "addictive" food attribute would increase the abuse potential of an "addictive" food (Ziauddeen et al., 2012). For example, if future studies indicate that sugar may be implicated in "food addiction," a threshold may be set to describe the concentration of sugar that significantly elevates a food's addictive potential (e.g., 30% calories from sugar). Further, specific sugars may differentially elicit brain responses that may associate differentially with appetite and reward pathways and lead to addictive patterns of eating (Page et al., 2013). However, unlike drugs of abuse, certain foods may have multiple addictive agents, like fat and sugar, which may both increase the food's addictive potential, but perhaps through different mechanisms.

Another difference that may exist between "food addiction" and substance-use disorders is the presentation of withdrawal symptoms. Although limited, preclinical literature suggests that dieters may experience headaches and psychological preoccupation with food (Gearhardt et al., 2009a), but it is unlikely that individuals with "food addiction" would experience severe, life-threatening physiological withdrawal symptoms if highly processed foods were removed from their diet (Ziauddeen et al., 2012). Consequently, "food addiction" may vary from traditional addictive disorders like opioid-dependence that produce acute physiological withdrawal symptoms, such as vomiting and sweating. On the other hand, removing certain foods from the diet may be more likely to trigger psychological withdrawal symptoms like anxiety. Such psychological features are experienced by a subset of individuals with pathological gambling (or gambling disorder) and are included in a withdrawal-related inclusionary criterion for the condition (American Psychiatric Association, 2013). Although withdrawal appears to manifest differently across addictive disorders, no previous research has examined whether withdrawal symptoms involving psychological states may contribute to addictive-like eating behavior.

Although "food addiction" is less likely to be associated with life-threatening withdrawal symptoms, there may be severe risks associated with "food addiction" that are not relevant in substance-use disorders. Unlike substance-use disorders, many individuals are exposed to highly processed foods within the first year of life (Fox et al., 2004). If certain foods have addictive potential, exposure to these foods during critical developmental periods may contribute to the onset of persistent, lifelong obesity for some people (Epstein et al., 1985). Unlike drugs of abuse that are illegal, expensive, or age-restricted, highly processed foods are easily accessible and affordable in our modern food environment. In this respect, "food addiction" may be considered a greater threat than substance-use disorders because there are fewer societal restrictions on the consumption of highly processed foods. Thus, while "food addiction" may not be associated with severe physiological withdrawal, it may pose a unique risk in infancy and childhood due to early first exposure that may increase its severity.

7 DIFFERENCES BETWEEN ADDICTIVE DISORDERS

Though premature acceptance of "food addiction" has been cautioned against due to inconsistencies with other addictive disorders (Ziauddeen and Fletcher, 2013; Ziauddeen et al., 2012), it is important to note differences that exist among addictive disorders included in the DSM-5 (American Psychiatric Association, 2013). Notably, the characteristics required for a substance to be considered addictive have changed over time. As previously mentioned, alcohol and opioid dependence may trigger intense physiological withdrawal symptoms (Leshner, 1997; Skinner and Allen, 1982), whereas cocaine and nicotine dependence appear more likely to produce psychological symptoms like involving anxiety and irritability (Brower and Paredes, 1987; Weddington et al., 1990). Additionally, the inclusion of the behavioral addiction gambling disorder in the DSM-5 reflects a shift away from physiological withdrawal as a necessary component of addiction (American Psychiatric Association, 2013). Importantly, regardless of whether the symptoms are physical or psychological, the experience of withdrawal appears to increase the probability of relapse across addictive disorders (Kenford et al., 2002; Ray, 1961). Thus, a key, shared component of withdrawal within addiction may involve the increased chance of relapse associated with experiencing withdrawal symptoms. In this respect, it will be important for future research to examine whether individuals with "food addiction" experience withdrawal symptoms that may trigger addictive-like eating behavior.

Another difference that exists within addictive disorders is the experience of intoxication. Some drugs of abuse, such as heroin, may result in intoxication where the individual enters a mind-altered state and may behave recklessly or break the law (Inciardi, 1979). Other examples of heroin intoxication symptoms are unpredictable mood swings, risky behavior, and breathing problems. However, other substances like nicotine may not trigger similar intoxication symptoms and allow individuals to function while using the drug. For example, it is often legal to smoke cigarettes while at work or operating a vehicle. Although nicotine does not produce acute negative consequences due to intoxication in the same way heroin does, prolonged nicotine use often leads to long-term negative health consequences like coronary heart disease, stroke, lung cancer, and emphysema (US Surgeon General, 1982). In behavioral addictions like pathological gambling or gambling disorder, the negative consequences may not manifest in acute intoxication, but rather in long-term outcomes like loss of money and familial stress. Similarly, while certain foods are likely incapable of triggering intoxication, long-term negative consequences may be associated with chronic addictive-like eating behavior, such as obesity, heart disease, and diabetes (Bray, 2004). In summary, while addictions vary in their degree of intoxication, individuals with addictive disorders share a characteristic of continued behavioral engagement in the addictive process despite negative consequences.

If certain foods are identified as "addictive," it is unlikely that all individuals who consume these foods will develop addictive-like eating behavior. In traditional addictive disorders, a small fraction of individuals who use addictive substances or engage in addictive behaviors develop dependence (Anthony et al., 1994). Applying this logic to addictive-like eating behavior, it would follow that only a subset of individuals who are exposed to potentially "addictive" foods would later develop "food addiction." Additionally, overconsumption of "addictive" foods would likely not lead to addictive-like eating behavior for all individuals. This is similarly observed in addictive disorders, where a larger percentage of individuals overconsume drugs of abuse than develop dependence (Dawson et al., 2004; Hasin et al., 2007). Further, it is unlikely that clinically significant symptoms of addictive-like eating would occur exclusively in obesity, and symptoms have been observed across a range of body mass indexes (Gearhardt et al., 2011c). Thus, subtyping may be one fruitful avenue that emerges from examining whether an addiction perspective can be applied to problematic eating behavior for some individuals. This may increase the efficacy of treatment and prevention efforts for obesity and for disordered eating more generally.

8 FUTURE DIRECTIONS IN FOOD ADDICTION RESEARCH

One important gap in the existing literature on "food addiction" is the examination of which foods are most likely associated with addictive-like eating. An addiction framework suggests that an addictive agent in some foods would interact with individual vulnerabilities to result in "food addiction." It follows that identifying whether certain foods or food attributes (e.g., specific sugars) may be capable of triggering an addictive-like process is essential to evaluating this perspective. Further, a drug of abuse has a greater addictive potential when a high "dose" of an addictive agent is rapidly absorbed by the system (Verebey and Gold, 1988). In our modern food environment, many highly processed foods contain added fat and/or refined carbohydrates (like white flour and sugar) in quantities exceeding what is found in naturally occurring foods. Akin to addictive disorders, the way in which these ingredients are absorbed in the body may also contribute to a food's addictive potential. Foods that have been processed by adding fat and/or refined carbohydrates while simultaneously stripping nutrients that slow digestion, like fiber and water, may have an elevated addictive potential. For example, a highly processed food, like chocolate cake, with added amounts of fat and rapidly absorbed refined carbohydrates may be expected to have a greater addictive potential than an apple, which contains natural sugars, but fiber and water to slow digestion. In highly processed foods, the increased concentration of rewarding ingredients coupled with the rapid rate that refined carbohydrates are absorbed may contribute to an increased abuse potential. In support, Schulte et al. (2015) found that highly processed foods were most likely to be associated with addictive-like eating behaviors, particularly for individuals who endorsed symptoms of "food addiction." Future studies should examine whether these highly processed foods are capable of producing neuroplastic changes in the brain, akin to those associated with consumption of drugs of abuse. Demonstrating that certain foods are capable of changing neural systems in a similar manner as addictive substances would provide further support for the validity and unique explanatory power of "food addiction" for some individuals with obesity.

A potential avenue to investigate how certain foods may change reward-related circuitry may involve examining neural mechanisms across the lifespan. Interestingly, childhood-onset obesity typically persists into adulthood (Epstein et al., 1985). Similarly substance-use disorders that emerge during early adolescence also often last throughout one's life (Chen et al., 2009). One potential explanation is that similar changes in reward mechanisms may be occurring, resulting in chronic addictive responses. While hypoactive reward-responsiveness may motivate obese individuals to seek highly rewarding food in adulthood (Blum et al., 2006), recent research has suggested that obese children actually exhibit greater functional connectivity between reward-related brain regions, including the left lateral OFC, and executive-control structures, such as the left ventromedial prefrontal cortex (Black et al., 2014). Increased input from reward neural systems to regions of cognitive control may make obese children particularly responsive to food cues (Bruce et al., 2010). However, no previous studies have examined reward-related neural mechanisms in infancy that may predispose certain individuals to become obese as children. Given the existing literature, it is uncertain whether increased functional connectivity between reward and executive-control regions may represent a cause or consequence of obesity in children. This is an essential area to explore, since children are exposed to highly rewarding food at young ages. If certain foods may be associated with addictive-like eating behaviors, it is possible that marketing restrictions may help decrease potential risks that some children have to develop "food addiction" and persistent obesity.

Finally, applying an addiction perspective to problematic eating behavior for some obese individuals may increase treatment options and efficacy. Behavioral treatments for addictive disorders that focus on craving management and relapse prevention may be adapted for the treatment of addictive-like eating behaviors. Pharmacological approaches may also be relevant. For example, naltrexone and buproprion, which are used for the treatment of addictive disorders, also appear to be a successful intervention technique for obesity (Greenway et al., 2010). Additionally, neurobiological treatments, such as neurofeedback, have recently been explored as a treatment method for substance-use disorders. This technique uses real-time fMRI (rtfMRI) feedback to help individuals reduce cue-induced craving (Sokunbi et al., 2014). For example, individuals are first shown a substance-relevant cue (e.g., cigarette) and receive feedback about increased brain activity in regions associated with craving, like the ACC. Next, the patient is asked to reduce activity in this region and self-report craving for the substance. The existing literature suggests that neurofeedback may be an effective technique for reducing craving for addictive substances, including nicotine (Li et al., 2013) and opiates (Dehghani-Arani et al., 2013). Since craving appears to be implicated in both substance-use disorders and "food addiction," neurofeedback may be a useful treatment tool for obese individuals reporting addictive-like eating behavior.

9 CONCLUDING REMARKS

Obesity appears to share common neural systems with traditional addictive disorders, suggesting that an addictive-like process may contribute to problematic eating behavior for some obese individuals. If certain foods are identified as "addictive," treatment and prevention efforts that adopt an addiction framework would likely be efficacious for the subtype of obese individuals endorsing "food addiction." However, future research is needed to determine whether an addiction perspective is clinically useful to explain some forms of obesity. Important next directions may involve investigating if certain foods can produce neuroplastic changes in the brain, characterizing the potentially addictive agents in these foods, looking longitudinally at changes in reward mechanisms, and examining whether treatments developed for drug addiction translate to effective interventions for food addiction.

ACKNOWLEDGMENTS

This was supported by P20 DA027844, CASAColumbia and the National Center for Responsible Gaming.

Disclosures: The authors report that they have no financial conflicts of interest with respect to the content of this manuscript. Dr. Potenza has received financial support or compensation for the following: Dr. Potenza has consulted for and advised Somaxon, Boehringer Ingelheim, Lundbeck, Ironwood, Shire, INSYS and RiverMend Health; has received research support from the National Institutes of Health, Veteran's Administration, Mohegan Sun Casino, the National Center for Responsible Gaming, and Forest Laboratories, Ortho-McNeil, Oy-Control/Biotie, Glaxo-SmithKline, Pfizer, and Psyadon pharmaceuticals; has participated in surveys, mailings or telephone consultations related to drug addiction, impulse control disorders or other health topics; has consulted for gambling entities, law offices and the federal public defender's office in issues related to impulse control disorders; provides clinical care in the Connecticut Department of Mental Health and Addiction Services Problem Gambling Services Program; has performed grant reviews for the National Institutes of Health and other agencies; has edited or guest-edited journals or journal sections; has given academic lectures in grand rounds, CME events and other clinical or scientific venues; and has generated books or book chapters for publishers of mental health texts.

REFERENCES

American Psychiatric Association, Diagnostic and statistical manual of mental disorders: DSM-5. http://dsm.psychiatryonline.org/book.aspx?bookid=556%3E. Available from.

Anthony, J.C., Warner, L.A., Kessler, R.C., 1994. Comparative epidemiology of dependence on tobacco, alcohol, controlled substances, and inhalants: basic findings from the national comorbidity survey. Exp. Clin. Psychopharmacol. 2 (3), 244–268.

Avena, N.M., Rada, P., Hoebel, B.G., 2009. Sugar and fat bingeing have notable differences in addictive-like behavior. J. nutr. 139 (3), 623–628.

Babbs, R.K., Sun, X., Felsted, J., Chouinard-Decorte, F., Veldhuizen, M.G., Small, D.M., 2013. Decreased caudate response to milkshake is associated with higher body mass index and greater impulsivity. Physiol. Behav. 121, 103–111.

Berridge, K.C., 2009. 'Liking' and 'wanting' food rewards: brain substrates and roles in eating disorders. Physiol. Behav. 97 (5), 537–550.

Black, W.R., Lepping, R.J., Bruce, A.S., Powell, J.N., Bruce, J.M., Martin, L.E., Davis, A.M., Brooks, W.M., Savage, C.R., Simmons, W.K., 2014. Tonic hyper-connectivity of reward neurocircuitry in obese children. Obesity (Silver Spring) 22 (7), 1590–1593.

Blum, K., Chen, T.J., Meshkin, B., Downs, B.W., Gordon, C.A., Blum, S., Mengucci, J.F., Braverman, E.R., Arcuri, V., Varshavskiy, M., Deutsch, R., Martinez-Pons, M., 2006. Reward deficiency syndrome in obesity: a preliminary cross-sectional trial with a Genotrim variant. Adv. Ther. 23 (6), 1040–1051.

Blum, K., Thanos, P.K., Gold, M.S., 2014. Dopamine and glucose, obesity, and reward deficiency syndrome. Front. Psychol. 5, 919.

Bray, G.A., 2004. Medical consequences of obesity. J. Clin. Endocrinol. Metab. 89 (6), 2583–2589.

Brower, K.J., Paredes, A., 1987. Cocaine withdrawal. Arch. Gen. Psychiatry 44 (3), 297–298.

Bruce, A.S., Holsen, L.M., Chambers, R.J., Martin, L.E., Brooks, W.M., Zarcone, J.R., Butler, M.G., Savage, C.R., 2010. Obese children show hyperactivation to food pictures in brain networks linked to motivation, reward and cognitive control. Int. J. Obes. (Lond) 34 (10), 1494–1500.

Burger, K.S., Stice, E., 2013. Elevated energy intake is correlated with hyperresponsivity in attentional, gustatory, and reward brain regions while anticipating palatable food receipt. Am. J. Clin. Nutr. 97 (6), 1188–1194.

Chen, C.Y., Storr, C.L., Anthony, J.C., 2009. Early-onset drug use and risk for drug dependence problems. Addict. Behav. 34 (3), 319–322.

Davis, C., Patte, K., Curtis, C., Reid, C., 2010. Immediate pleasures and future consequences. A neuropsychological study of binge eating and obesity. Appetite 54 (1), 208–213.

Dawson, D.A., Grant, B.F., Stinson, F.S., Chou, P.S., 2004. Toward the attainment of low-risk drinking goals: a 10-year progress report. Alcohol. Clin. Exp. Res. 28 (9), 1371–1378.

Dehghani-Arani, F., Rostami, R., Nadali, H., 2013. Neurofeedback training for opiate addiction: improvement of mental health and craving. Appl. Psychophysiol. Biofeedback 38 (2), 133–141.

Demos, K.E., Heatherton, T.F., Kelley, W.M., 2012. Individual differences in nucleus accumbens activity to food and sexual images predict weight gain and sexual behavior. J. Neurosci. 32 (16), 5549–5552.

de Weijer, B.A., van de Giessen, E., van Amelsvoort, T.A., Boot, E., Braak, B., Janssen, I.M., van de Laar, A., Fliers, E., Serlie, M.J., Booij, J., 2011. Lower striatal dopamine D2/3 receptor availability in obese compared with non-obese subjects. EJNMMI Res. 1 (1), 37.

Due, D.L., Huettel, S.A., Hall, W.G., Rubin, D.C., 2002. Activation in mesolimbic and visuospatial neural circuits elicited by smoking cues: evidence from functional magnetic resonance imaging. Am. J. Psychiatry 159 (6), 954–960.

Eisenstein, S.A., Antenor-Dorsey, J.A., Gredysa, D.M., Koller, J.M., Bihun, E.C., Ranck, S.A., Arbelaez, A.M., Klein, S., Perlmutter, J.S., Moerlein, S.M., Black, K.J., Hershey, T., 2013. A comparison of D2 receptor specific binding in obese and normal-weight individuals using PET with (N-[(11)C]methyl)benperidol. Synapse 67 (11), 748–756.

Epstein, L.H., Wing, R.R., Valoski, A., 1985. Childhood obesity. Pediatr. Clin. North Am. 32 (2), 363–379.

Everitt, B.J., Robbins, T.W., 2005. Neural systems of reinforcement for drug addiction: from actions to habits to compulsion. Nat. Neurosci. 8 (11), 1481–1489.

Ferreira, J.G., Tellez, L.A., Ren, X., Yeckel, C.W., de Araujo, I.E., 2012. Regulation of fat intake in the absence of flavour signalling. J. physiolog. 590 (4), 953–972.

Flagel, S.B., Akil, H., Robinson, T.E., 2009. Individual differences in the attribution of incentive salience to reward-related cues: implications for addiction. Neuropharmacology 56 (Suppl 1), 139–148.

Fox, M.K., Pac, S., Devaney, B., Jankowski, L., 2004. Feeding infants and toddlers study: what foods are infants and toddlers eating? J. Am. Diet. Assoc. 104, 22–30.

Garcia-Garcia, I., Jurado, M.A., Garolera, M., Segura, B., Sala-Llonch, R., Marques-Iturria, I., Pueyo, R., Sender-Palacios, M.J., Vernet-Vernet, M., Narberhaus, A., Ariza, M., Junque, C., 2013. Alterations of the salience network in obesity: a resting-state fMRI study. Hum. Brain Mapp. 34 (11), 2786–2797.

Gearhardt, A.N., Corbin, W.R., Brownell, K.D., 2009a. Food addiction: an examination of the diagnostic criteria for dependence. J. Addict. Med. 3 (1), 1–7.

Gearhardt, A.N., Corbin, W.R., Brownell, K.D., 2009b. Preliminary validation of the Yale Food Addiction Scale. Appetite 52 (2), 430–436.

Gearhardt, A.N., Davis, C., Kuschner, R., Brownell, K.D., 2011a. The addiction potential of hyperpalatable foods. Curr. Drug Abuse Rev. 4 (3), 140–145.

Gearhardt, A.N., White, M.A., Potenza, M.N., 2011b. Binge eating disorder and food addiction. Curr. Drug Abuse Rev. 4 (3), 201–207.

Gearhardt, A.N., Yokum, S., Orr, P.T., Stice, E., Corbin, W.R., Brownell, K.D., 2011c. Neural correlates of food addiction. Arch. Gen. Psychiatry 68 (8), 808–816.

Geha, P.Y., Aschenbrenner, K., Felsted, J., O'Malley, S.S., Small, D.M., 2013. Altered hypo thalamic response to food in smokers. Am. J. Clin. Nutr. 97 (1), 15–22.

Geiger, B.M., Haburcak, M., Avena, N.M., Moyer, M.C., Hoebel, B.G., Pothos, E.N., 2009. Deficits of mesolimbic dopamine neurotransmission in rat dietary obesity. Neuroscience 159 (4), 1193–1199.

Greenway, F.L., Fujioka, K., Plodkowski, R.A., Mudaliar, S., Guttadauria, M., Erickson, J., Kim, D.D., Dunayevich, E., C-IS Group, 2010. Effect of naltrexone plus bupropion on weight loss in overweight and obese adults (COR-I): a multicentre, randomised, double-blind, placebo-controlled, phase 3 trial. Lancet 376 (9741), 595–605.

Haltia, L.T., Rinne, J.O., Merisaari, H., Maguire, R.P., Savontaus, E., Helin, S., Nagren, K., Kaasinen, V., 2007. Effects of intravenous glucose on dopaminergic function in the human brain in vivo. Synapse 61 (9), 748–756.

Hanlon, C.A., Wesley, M.J., Stapleton, J.R., Laurienti, P.J., Porrino, L.J., 2011. The association between frontal–striatal connectivity and sensorimotor control in cocaine users. Drug Alcohol Depend. 115 (3), 240–243.

Hasin, D.S., Stinson, F.S., Ogburn, E., Grant, B.F., 2007. Prevalence, correlates, disability, and comorbidity of DSM-IV alcohol abuse and dependence in the united states: results from the national epidemiologic survey on alcohol and related conditions. Arch. Gen. Psychiatry 64 (7), 830–842.

Heinz, A., Siessmeier, T., Wrase, J., Hermann, D., Klein, S., Grüsser-Sinopoli, S.M., Flor, H., Braus, D.F., Buchholz, H.G., Gründer, G., et al., 2004. Correlation between dopamine D2 receptors in the ventral striatum and central processing of alcohol cues and craving. Am. J. Psychiatry 161 (10), 1783–1789.

Henningfield, J.E., Keenan, R.M., 1993. Nicotine delivery kinetics and abuse liability. J. Consult. Clin. Psychol. 61 (5), 743–750.

Inciardi, J.A., 1979. Heroin use and street crime. Crime Delinq. 25 (3), 335–346.

Jastreboff, A.M., Sinha, R., Lacadie, C., Small, D.M., Sherwin, R.S., Potenza, M.N., 2013. Neural correlates of stress- and food cue-induced food craving in obesity: association with insulin levels. Diabetes Care 36 (2), 394–402.

Johnson, P.M., Kenny, P.J., 2010. Dopamine D2 receptors in addiction-like reward dysfunction and compulsive eating in obese rats. Nat. Neurosci. 13 (5), 635–641.

Kalivas, P.W., O'Brien, C., 2008. Drug addiction as a pathology of staged neuroplasticity. Neuropsychopharmacology 33 (1), 166–180.

Kenford, S.L., Smith, S.S., Wetter, D.W., Jorenby, D.E., Fiore, M.C., Baker, T.B., 2002. Predicting relapse back to smoking: contrasting affective and physical models of dependence. J. Consult. Clin. Psychol. 70 (1), 216–227.

Kenny, P.J., Chen, S.A., Kitamura, O., Markou, A., Koob, G.F., 2006. Conditioned withdrawal drives heroin consumption and decreases reward sensitivity. J. Neurosci. 26 (22), 5894–5900.

Kishinevsky, F.I., Cox, J.E., Murdaugh, D.L., Stoeckel, L.E., Cook 3rd, E.W., Weller, R.E., 2012. fMRI reactivity on a delay discounting task predicts weight gain in obese women. Appetite 58 (2), 582–592.

Koob, G.F., Le Moal, M., 2001. Drug addiction, dysregulation of reward, and allostasis. Neuropsychopharmacology 24 (2), 97–129.

Lawrence, N.S., Hinton, E.C., Parkinson, J.A., Lawrence, A.D., 2012. Nucleus accumbens response to food cues predicts subsequent snack consumption in women and increased body mass index in those with reduced self-control. Neuroimage 63 (1), 415–422.

Leshner, A.I., 1997. Addiction is a brain disease, and it matters. Science 278 (5335), 45–47.

Li, X., Hartwell, K.J., Borckardt, J., Prisciandaro, J.J., Saladin, M.E., Morgan, P.S., Johnson, K.A., Lematty, T., Brady, K.T., George, M.S., 2013. Volitional reduction of anterior cingulate cortex activity produces decreased cue craving in smoking cessation: a preliminary real-time fMRI study. Addict. Biol. 18 (4), 739–748.

Liu, J., Liang, J., Qin, W., Tian, J., Yuan, K., Bai, L., Zhang, Y., Wang, W., Wang, Y., Li, Q., Zhao, L., Lu, L., von Deneen, K.M., Liu, Y., Gold, M.S., 2009. Dysfunctional connectivity patterns in chronic heroin users: an fMRI study. Neurosci. Lett. 460 (1), 72–77.

Ma, N., Liu, Y., Li, N., Wang, C.X., Zhang, H., Jiang, X.F., Xu, H.S., Fu, X.M., Hu, X., Zhang, D.R., 2010. Addiction related alteration in resting-state brain connectivity. Neuroimage 49 (1), 738–744.

MacKillop, J., Amlung, M.T., Few, L.R., Ray, L.A., Sweet, L.H., Munafo, M.R., 2011. Delayed reward discounting and addictive behavior: a meta-analysis. Psychopharmacology (Berl) 216 (3), 305–321.

Martin, L.E., Holsen, L.M., Chambers, R.J., Bruce, A.S., Brooks, W.M., Zarcone, J.R., Butler, M.G., Savage, C.R., 2010. Neural mechanisms associated with food motivation in obese and healthy weight adults. Obesity (Silver Spring) 18 (2), 254–260.

Mole, T.B., Irvine, M.A., Worbe, Y., Collins, P., Mitchell, S.P., Bolton, S., Harrison, N.A., Robbins, T.W., Voon, V., 2014. Impulsivity in disorders of food and drug misuse. Psychol. Med. 45, 771–782.

Morris, M.J., Beilharz, J., Maniam, J., Reichelt, A., Westbrook, R.F., 2014. Why is obesity such a problem in the 21st century? The intersection of palatable food, cues and reward pathways, stress, and cognition. Neurosci. Biobehav. Rev. In press.

Myrick, H., Anton, R.F., Li, X., Henderson, S., Drobes, D., Voronin, K., George, M.S., 2004. Differential brain activity in alcoholics and social drinkers to alcohol cues: relationship to craving. Neuropsychopharmacology 29 (2), 393–402.

Nader, M.A., Morgan, D., Gage, D.H., Nader, S.H., Calhoun, T.L., Buchheimer, N., Ehrenkaufer, R., Mach, R.H., 2006. PET imaging of dopamine D2 receptors during chronic cocaine self-administration in monkeys. Nat. Neurosci. 9 (8), 1050–1056.

Ng, J., Stice, E., Yokum, S., Bohon, C., 2011. An fMRI study of obesity, food reward, and perceived caloric density. Does a low-fat label make food less appealing? Appetite 57 (1), 65–72.

Nolan-Poupart, S., Veldhuizen, M.G., Geha, P., Small, D.M., 2013. Midbrain response to milkshake correlates with ad libitum milkshake intake in the absence of hunger. Appetite 60 (1), 168–174.

O'Brien, C.P., Volkow, N., Li, T.K., 2006. What's in a word? Addiction versus dependence in DSM-V. Am. J. Psychiatry 163 (5), 764–765.

Page, K.A., Chan, O., Arora, J., Belfort-Deaguiar, R., Dzuira, J., Roehmholdt, B., Cline, G.W., Naik, S., Sinha, R., Constable, R.T., Sherwin, R.S., 2013. Effects of fructose vs glucose on regional cerebral blood flow in brain regions involved with appetite and reward pathways. JAMA 309 (1), 63–70.

Paulus, M.P., Tapert, S.F., Schuckit, M.A., 2005. Neural activation patterns of methamphetamine-dependent subjects during decision making predict relapse. Arch. Gen. Psychiatry 62 (7), 761–768.

Ray, M.B., 1961. The cycle of abstinence and relapse among heroin addicts. Soc. Probl. 9 (2), 132–140.

Robinson, T.E., Berridge, K.C., 1993. The neural basis of drug craving: an incentive-sensitization theory of addiction. Brain Res. Brain Res. Rev. 18 (3), 247–291.

Schulte, E.M., Avena, N.M., Gearhardt, A.N., 2015. Which foods may be addictive? The roles of processing, fat content, and glycemic load. PLoS One 10 (2), e0117959. http://dx.doi.org/10.1371/journal.pone.0117959.

Shott, M.E., Cornier, M.A., Mittal, V.A., Pryor, T.L., Orr, J.M., Brown, M.S., Frank, G.K., 2014. Orbitofrontal cortex volume and brain reward response in obesity. Int. J. Obes. (Lond) 39, 214–221.

Singh, M., 2014. Mood, food, and obesity. Front. Psychol. 5, 925.

Sinha, R., 2008. Chronic stress drug use, and vulnerability to addiction. Ann. N. Y. Acad. Sci. 1141, 105–130.

Sinha, R., Jastreboff, A.M., 2013. Stress as a common risk factor for obesity and addiction. Biol. Psychiatry 73 (9), 827–835.

Skinner, H.A., Allen, B.A., 1982. Alcohol dependence syndrome: measurement and validation. J. Abnorm. Psychol. 91 (3), 199–209.

Small, D.M., Jones-Gotman, M., Dagher, A., 2003. Feeding-induced dopamine release in dorsal striatum correlates with meal pleasantness ratings in healthy human volunteers. Neuroimage 19 (4), 1709–1715.

Sokunbi, M.O., Linden, D.E., Habes, I., Johnston, S., Ihssen, N., 2014. Real-time fMRI brain-computer interface: development of a "motivational feedback" subsystem for the regulation of visual cue reactivity. Front. Behav. Neurosci. 8, 392.

Stice, E., Spoor, S., Bohon, C., Veldhuizen, M.G., Small, D.M., 2008. Relation of reward from food intake and anticipated food intake to obesity: a functional magnetic resonance imaging study. J. Abnorm. Psychol. 117 (4), 924–935.

Stice, E., Yokum, S., Blum, K., Bohon, C., 2010a. Weight gain is associated with reduced striatal response to palatable food. J. Neurosci. 30 (39), 13105–13109.

Stice, E., Yokum, S., Bohon, C., Marti, N., Smolen, A., 2010b. Reward circuitry responsivity to food predicts future increases in body mass: moderating effects of DRD2 and DRD4. Neuroimage 50 (4), 1618–1625.

Stice, E., Yokum, S., Burger, K.S., Epstein, L.H., Small, D.M., 2011. Youth at risk for obesity show greater activation of striatal and somatosensory regions to food. J. Neurosci. 31 (12), 4360–4366.

Stice, E., Burger, K.S., Yokum, S., 2013a. Relative ability of fat and sugar tastes to activate reward, gustatory, and somatosensory regions. Am. J. Clin. Nutr. 98 (6), 1377–1384.

Stice, E., Yokum, S., Burger, K.S., 2013b. Elevated reward region responsivity predicts future substance use onset but not overweight/obesity onset. Biol. Psychiatry 73 (9), 869–876.

Stoeckel, L.E., Murdaugh, D.L., Cox, J.E., Cook 3rd, E.W., Weller, R.E., 2013. Greater impulsivity is associated with decreased brain activation in obese women during a delay discounting task. Brain Imaging Behav. 7 (2), 116–128.

Tang, D.W., Fellows, L.K., Small, D.M., Dagher, A., 2012. Food and drug cues activate similar brain regions: a meta-analysis of functional MRI studies. Physiol. Behav. 106 (3), 317–324.

Thanos, P.K., Michaelides, M., Piyis, Y.K., Wang, G.J., Volkow, N.D., 2008. Food restriction markedly increases dopamine D2 receptor (D2R) in a rat model of obesity as assessed with in-vivo muPET imaging ([11C] raclopride) and in-vitro ([3H] spiperone) autoradiography. Synapse 62 (1), 50–61.

Tobler, P.N., Fiorillo, C.D., Schultz, W., 2005. Adaptive coding of reward value by dopamine neurons. Science 307 (5715), 1642–1645.

Tomasi, D., Volkow, N.D., 2013. Striatocortical pathway dysfunction in addiction and obesity: differences and similarities. Crit. Rev. Biochem. Mol. Biol. 48 (1), 1–19.

Tomasi, D., Wang, G.J., Wang, R., Caparelli, E.C., Logan, J., Volkow, N.D., 2015. Overlapping patterns of brain activation to food and cocaine cues in cocaine abusers: association to striatal D2/D3 receptors. Hum. Brain Mapp. 36 (1), 120–136.

Tryon, M.S., Carter, C.S., Decant, R., Laugero, K.D., 2013. Chronic stress exposure may affect the brain's response to high calorie food cues and predispose to obesogenic eating habits. Physiol. Behav. 120, 233–242.

US Surgeon General, 1982. The Health Consequences of Smoking: Chronic Obstructive Lung Disease. US Department of Health and Human Resources, Washington, DC.

Verbeken, S., Braet, C., Lammertyn, J., Goossens, L., Moens, E., 2012. How is reward sensitivity related to bodyweight in children? Appetite 58 (2), 478–483.

Verebey, K., Gold, M.S., 1988. From coca leaves to crack: the effects of dose and routes of administration in abuse liability. Psychiatr. Ann. 18 (9), 513–520.

Volkow, N.D., Chang, L., Wang, G.J., Fowler, J.S., Ding, Y.S., Sedler, M., Logan, J., Franceschi, D., Gatley, J., Hitzemann, R., Gifford, A., Wong, C., Pappas, N., 2001. Low level of brain dopamine D2 receptors in methamphetamine abusers: association with metabolism in the orbitofrontal cortex. Am. J. Psychiatry 158 (12), 2015–2021.

Volkow, N.D., Wang, G.J., Ma, Y., Fowler, J.S., Wong, C., Ding, Y.S., Hitzemann, R., Swanson, J.M., Kalivas, P., 2005. Activation of orbital and medial prefrontal cortex by methylphenidate in cocaine-addicted subjects but not in controls: relevance to addiction. J. Neurosci. 25 (15), 3932–3939.

Volkow, N.D., Wang, G.J., Telang, F., Fowler, J.S., Logan, J., Childress, A.R., Jayne, M., Ma, Y., Wong, C., 2006. Cocaine cues and dopamine in dorsal striatum: mechanism of craving in cocaine addiction. J. Neurosci. 26 (24), 6583–6588.

Volkow, N.D., Wang, G.J., Telang, F., Fowler, J.S., Thanos, P.K., Logan, J., Alexoff, D., Ding, Y.S., Wong, C., Ma, Y., Pradhan, K., 2008. Low dopamine striatal D2 receptors are associated with prefrontal metabolism in obese subjects: possible contributing factors. Neuroimage 42 (4), 1537–1543.

Wang, G.J., Volkow, N.D., Fowler, J.S., Logan, J., Abumrad, N.N., Hitzemann, R.J., Pappas, N.S., Pascani, K., 1997. Dopamine D2 receptor availability in opiate-dependent subjects before and after naloxone-precipitated withdrawal. Neuropsychopharmacology 16 (2), 174–182.

Wang, G.J., Volkow, N.D., Fowler, J.S., 2002. The role of dopamine in motivation for food in humans: implications for obesity. Expert Opin. Ther. Targets 6 (5), 601–609.

Weddington, W.W., Brown, B.S., Haertzen, C.A., Cone, E.J., Dax, E.M., Herning, R.I., Michaelson, B.S., 1990. Changes in mood, craving, and sleep during short-term abstinence reported by male cocaine addicts. A controlled, residential study. Arch. Gen. Psychiatry 47 (9), 861–868.

Weygandt, M., Mai, K., Dommes, E., Leupelt, V., Hackmack, K., Kahnt, T., Rothemund, Y., Spranger, J., Haynes, J.D., 2013. The role of neural impulse control mechanisms for dietary success in obesity. Neuroimage 83, 669–678.

Wilcox, C.E., Braskie, M.N., Kluth, J.T., Jagust, W.J., 2009. Overeating behavior and striatal dopamine with 6-[F]-fluoro-L-m-tyrosine PET. J. Obes. 2010, 12–20.

Yokum, S., Ng, J., Stice, E., 2011. Attentional bias to food images associated with elevated weight and future weight gain: an fMRI study. Obesity (Silver Spring) 19 (9), 1775–1783.

Yokum, S., Gearhardt, A.N., Harris, J.L., Brownell, K.D., Stice, E., 2014. Individual differences in striatum activity to food commercials predict weight gain in adolescents. Obesity (Silver Spring) 22 (12), 2544–2551.

Zhang, B., Tian, D., Yu, C., Zhang, J., Tian, X., von Deneen, K.M., Zang, Y., Walter, M., Liu, Y., 2015. Altered baseline brain activities before food intake in obese men: a resting state fMRI study. Neurosci. Lett. 584C, 156–161.

Ziauddeen, H., Fletcher, P.C., 2013. Is food addiction a valid and useful concept? Obes. Rev. 14 (1), 19–28.

Ziauddeen, H., Farooqi, I.S., Fletcher, P.C., 2012. Obesity and the brain: how convincing is the addiction model? Nat. Rev. Neurosci. 13 (4), 279–286.

Index

Note: Page numbers followed by "*f*" indicate figures, and "*t*" indicate tables.

B

I

L

M

N

Volume 167: Stress Hormones and Post Traumatic Stress Disorder: Basic Studies and Clinical Perspectives, by E.R. de Kloet, M.S. Oitzl and E. Vermetten (Eds.) – 2008, ISBN 978-0-444-53140-7.
Volume 168: Models of Brain and Mind: Physical, Computational and Psychological Approaches, by R. Banerjee and B.K. Chakrabarti (Eds.) – 2008, ISBN 978-0-444-53050-9.
Volume 169: Essence of Memory, by W.S. Sossin, J.-C. Lacaille, V.F. Castellucci and S. Belleville (Eds.) – 2008, ISBN 978-0-444-53164-3.
Volume 170: Advances in Vasopressin and Oxytocin – From Genes to Behaviour to Disease, by I.D. Neumann and R. Landgraf (Eds.) – 2008, ISBN 978-0-444-53201-5.
Volume 171: Using Eye Movements as an Experimental Probe of Brain Function—A Symposium in Honor of Jean Büttner-Ennever, by Christopher Kennard and R. John Leigh (Eds.) – 2008, ISBN 978-0-444-53163-6.
Volume 172: Serotonin–Dopamine Interaction: Experimental Evidence and Therapeutic Relevance, by Giuseppe Di Giovanni, Vincenzo Di Matteo and Ennio Esposito (Eds.) – 2008, ISBN 978-0-444-53235-0.
Volume 173: Glaucoma: An Open Window to Neurodegeneration and Neuroprotection, by Carlo Nucci, Neville N. Osborne, Giacinto Bagetta and Luciano Cerulli (Eds.) – 2008, ISBN 978-0-444-53256-5.
Volume 174: Mind and Motion: The Bidirectional Link Between Thought and Action, by Markus Raab, Joseph G. Johnson and Hauke R. Heekeren (Eds.) – 2009, 978-0-444-53356-2.
Volume 175: Neurotherapy: Progress in Restorative Neuroscience and Neurology — Proceedings of the 25th International Summer School of Brain Research, held at the Royal Netherlands Academy of Arts and Sciences, Amsterdam, The Netherlands, August 25–28, 2008, by J. Verhaagen, E.M. Hol, I. Huitinga, J. Wijnholds, A.A. Bergen, G.J. Boer and D.F. Swaab (Eds.) –2009, ISBN 978-0-12-374511-8.
Volume 176: Attention, by Narayanan Srinivasan (Ed.) – 2009, ISBN 978-0-444-53426-2.
Volume 177: Coma Science: Clinical and Ethical Implications, by Steven Laureys, Nicholas D. Schiff and Adrian M. Owen (Eds.) – 2009, 978-0-444-53432-3.
Volume 178: Cultural Neuroscience: Cultural Influences On Brain Function, by Joan Y. Chiao (Ed.) – 2009, 978-0-444-53361-6.
Volume 179: Genetic models of schizophrenia, by Akira Sawa (Ed.) – 2009, 978-0-444-53430-9.
Volume 180: Nanoneuroscience and Nanoneuropharmacology, by Hari Shanker Sharma (Ed.) – 2009, 978-0-444-53431-6.
Volume 181: Neuroendocrinology: The Normal Neuroendocrine System, by Luciano Martini, George P. Chrousos, Fernand Labrie, Karel Pacak and Donald W. Pfaff (Eds.) – 2010, 978-0-444-53617-4.
Volume 182: Neuroendocrinology: Pathological Situations and Diseases, by Luciano Martini, George P. Chrousos, Fernand Labrie, Karel Pacak and Donald W. Pfaff (Eds.) – 2010, 978-0-444-53616-7.
Volume 183: Recent Advances in Parkinson's Disease: Basic Research, by Anders Björklund and M. Angela Cenci (Eds.) – 2010, 978-0-444-53614-3.
Volume 184: Recent Advances in Parkinson's Disease: Translational and Clinical Research, by Anders Björklund and M. Angela Cenci (Eds.) – 2010, 978-0-444-53750-8.
Volume 185: Human Sleep and Cognition Part I: Basic Research, by Gerard A. Kerkhof and Hans P.A. Van Dongen (Eds.) – 2010, 978-0-444-53702-7.
Volume 186: Sex Differences in the Human Brain, their Underpinnings and Implications, by Ivanka Savic (Ed.) – 2010, 978-0-444-53630-3.
Volume 187: Breathe, Walk and Chew: The Neural Challenge: Part I, by Jean-Pierre Gossard, Réjean Dubuc and Arlette Kolta (Eds.) – 2010, 978-0-444-53613-6.
Volume 188: Breathe, Walk and Chew; The Neural Challenge: Part II, by Jean-Pierre Gossard, Réjean Dubuc and Arlette Kolta (Eds.) – 2011, 978-0-444-53825-3.
Volume 189: Gene Expression to Neurobiology and Behaviour: Human Brain Development and Developmental Disorders, by Oliver Braddick, Janette Atkinson and Giorgio M. Innocenti (Eds.) – 2011, 978-0-444-53884-0.

Volume 190: Human Sleep and Cognition Part II: Clinical and Applied Research, by Hans P.A. Van Dongen and Gerard A. Kerkhof (Eds.) – 2011, 978-0-444-53817-8.
Volume 191: Enhancing Performance for Action and perception: Multisensory Integration, Neuroplasticity and Neuroprosthetics: Part I, by Andrea M. Green, C. Elaine Chapman, John F. Kalaska and Franco Lepore (Eds.) – 2011, 978-0-444-53752-2.
Volume 192: Enhancing Performance for Action and Perception: Multisensory Integration, Neuroplasticity and Neuroprosthetics: Part II, by Andrea M. Green, C. Elaine Chapman, John F. Kalaska and Franco Lepore (Eds.) – 2011, 978-0-444-53355-5.
Volume 193: Slow Brain Oscillations of Sleep, Resting State and Vigilance, by Eus J.W. Van Someren, Ysbrand D. Van Der Werf, Pieter R. Roelfsema, Huibert D. Mansvelder and Fernando H. Lopes da Silva (Eds.) – 2011, 978-0-444-53839-0.
Volume 194: Brain Machine Interfaces: Implications For Science, Clinical Practice And Society, by Jens Schouenborg, Martin Garwicz and Nils Danielsen (Eds.) – 2011, 978-0-444-53815-4.
Volume 195: Evolution of the Primate Brain: From Neuron to Behavior, by Michel A. Hofman and Dean Falk (Eds.) – 2012, 978-0-444-53860-4.
Volume 196: Optogenetics: Tools for Controlling and Monitoring Neuronal Activity, by Thomas Knöpfel and Edward S. Boyden (Eds.) – 2012, 978-0-444-59426-6.
Volume 197: Down Syndrome: From Understanding the Neurobiology to Therapy, by Mara Dierssen and Rafael De La Torre (Eds.) – 2012, 978-0-444-54299-1.
Volume 198: Orexin/Hypocretin System, by Anantha Shekhar (Ed.) – 2012, 978-0-444-59489-1.
Volume 199: The Neurobiology of Circadian Timing, by Andries Kalsbeek, Martha Merrow, Till Roenneberg and Russell G. Foster (Eds.) – 2012, 978-0-444-59427-3.
Volume 200: Functional Neural Transplantation III: Primary and stem cell therapies for brain repair, Part I, by Stephen B. Dunnett and Anders Björklund (Eds.) – 2012, 978-0-444-59575-1.
Volume 201: Functional Neural Transplantation III: Primary and stem cell therapies for brain repair, Part II, by Stephen B. Dunnett and Anders Björklund (Eds.) – 2012, 978-0-444-59544-7.
Volume 202: Decision Making: Neural and Behavioural Approaches, by V.S. Chandrasekhar Pammi and Narayanan Srinivasan (Eds.) – 2013, 978-0-444-62604-2.
Volume 203: The Fine Arts, Neurology, and Neuroscience: Neuro-Historical Dimensions, by Stanley Finger, Dahlia W. Zaidel, François Boller and Julien Bogousslavsky (Eds.) – 2013, 978-0-444-62730-8.
Volume 204: The Fine Arts, Neurology, and Neuroscience: New Discoveries and Changing Landscapes, by Stanley Finger, Dahlia W. Zaidel, François Boller and Julien Bogousslavsky (Eds.) – 2013, 978-0-444-63287-6.
Volume 205: Literature, Neurology, and Neuroscience: Historical and Literary Connections, by Anne Stiles, Stanley Finger and François Boller (Eds.) – 2013, 978-0-444-63273-9.
Volume 206: Literature, Neurology, and Neuroscience: Neurological and Psychiatric Disorders, by Stanley Finger, François Boller and Anne Stiles (Eds.) – 2013, 978-0-444-63364-4.
Volume 207: Changing Brains: Applying Brain Plasticity to Advance and Recover Human Ability, by Michael M. Merzenich, Mor Nahum and Thomas M. Van Vleet (Eds.) – 2013, 978-0-444-63327-9.
Volume 208: Odor Memory and Perception, by Edi Barkai and Donald A. Wilson (Eds.) – 2014, 978-0-444-63350-7.
Volume 209: The Central Nervous System Control of Respiration, by Gert Holstege, Caroline M. Beers and Hari H. Subramanian (Eds.) – 2014, 978-0-444-63274-6.
Volume 210: Cerebellar Learning, Narender Ramnani (Ed.) – 2014, 978-0-444-63356-9.
Volume 211: Dopamine, by Marco Diana, Gaetano Di Chiara and Pierfranco Spano (Eds.) – 2014, 978-0-444-63425-2.
Volume 212: Breathing, Emotion and Evolution, by Gert Holstege, Caroline M. Beers and Hari H. Subramanian (Eds.) – 2014, 978-0-444-63488-7.
Volume 213: Genetics of Epilepsy, by Ortrud K. Steinlein (Ed.) – 2014, 978-0-444-63326-2.
Volume 214: Brain Extracellular Matrix in Health and Disease, by Asla Pitkänen, Alexander Dityatev and Bernhard Wehrle-Haller (Eds.) – 2014, 978-0-444-63486-3.

Volume 215: The History of the Gamma Knife, by Jeremy C. Ganz (Ed.) – 2014, 978-0-444-63520-4.
Volume 216: Music, Neurology, and Neuroscience: Historical Connections and Perspectives, by François Boller, Eckart Altenmüller, and Stanley Finger (Eds.) – 2015, 978-0-444-63399-6.
Volume 217: Music, Neurology, and Neuroscience: Evolution, the Musical Brain, Medical Conditions, and Therapies, by Eckart Altenmüller, Stanley Finger, and François Boller (Eds.) – 2015, 978-0-444-63551-8.
Volume 218: Sensorimotor Rehabilitation: At the Crossroads of Basic and Clinical Sciences, by Numa Dancause, Sylvie Nadeau, and Serge Rossignol (Eds.) – 2015, 978-0-444-63565-5.
Volume 219: The Connected Hippocampus, by Shane O'Mara and Marian Tsanov (Eds.) – 2015, 978-0-444-63549-5.
Volume 220: New Trends in Basic and Clinical Research of Glaucoma: A Neurodegenerative Disease of the Visual System, by Giacinto Bagetta and Carlo Nucci (Eds.) – 2015, 978-0-444-63566-2.
Volume 221: New Trends in Basic and Clinical Research of Glaucoma: A Neurodegenerative Disease of the Visual System, by Giacinto Bagetta and Carlo Nucci (Eds.) – 2015, 978-0-12-804608-1.
Volume 222: Computational Neurostimulation, by Sven Bestmann (Ed.) – 2015, 978-0-444-63546-4.

Edwards Brothers Malloy
Ann Arbor MI. USA
August 16, 2016